11th Brazilian Congress on Metrology (Metrologia 2021)

Journal of Physics: Conference Series Volume 2606

Online
18 – 21 October 2021

ISBN: 978-1-7138-8198-8
ISSN: 1742-6588

Printed from e-media with permission by:

Curran Associates, Inc.
57 Morehouse Lane
Red Hook, NY 12571

Some format issues inherent in the e-media version may also appear in this print version.

Institute of Physics
Dirac House, Temple Back
Bristol BS1 6BE UK

Phone: 44 1 17 929 7481
Fax: 44 1 17 920 0979

techtracking@iop.org

Additional copies of this publication are available from:

Curran Associates, Inc.
57 Morehouse Lane
Red Hook, NY 12571 USA
Phone: 845-758-0400
Fax: 845-758-2633
Email: curran@proceedings.com
Web: www.proceedings.com

TABLE OF CONTENTS

11th Brazilian Congress on Metrology (Metrologia 2021)

PREFACE

The Metrology 2021 congress was held from October 18th to 21st, 2021. It was a remote event, with all activities done in virtual rooms. The general president of Metrology 2021 was Americo Bernardes, president of the Brazilian Society of Metrology and professor at the Federal University of Ouro Preto.

Metrology 2021 Magna session was the lecture Dr. Claire M. Saundry (NIST) gave on the first day, October 18th. Dr. Saundry spoke about the main activities of the Interamerican Metrology System (or "Sistema Interamericano de Metrologia" – SIM) regarding capacity building and knowledge transfer.

Following the Magna session, on the morning of October 18th, the work began with the plenary lecture given by Dr. Daniel Varela Magalhães from the Mechanical Engineering Department at USP. The speaker gave a general outline of the new edition of the International Vocabulary of Metrology (VIM).

Further after, on the afternoon of October 18th, two virtual rooms were held in parallel with the sessions related to CBM (general metrology) and CBMRI (ionizing radiations). CBMRI sessions also featured four thematic discussions on specific topics of interest.

The second day of the congress, October 19th, started with the lecture of Dr. Thomas Hartung from the Johns Hopkins Bloomberg School of Public Health. He spoke about assessing the health effects of drugs and chemicals, which is considered a scientific revolution and a critical application of Metrology in the Health Area. Continuing in the Health area, Dr. Cameron Miller from NIST presented his lecture on documentary standards and the development of measurement methods for UV disinfection technologies. The plenary lectures were closed by Dr. Renato Garcia from UFSC, who spoke about quality programs in medical equipment used in primary health care, specifically about post-market safety and performance trials.

Later, on the afternoon of October 19th, two virtual rooms were held in parallel sessions related to CBM (general metrology) and CBMO (optical metrology).

The third day of the congress started with an exciting round table about female researchers' presence in ionizing radiations. A plenary lecture by Stephane Solve from the "Bureau International de Poids et Mesures" (BIPM) Radiometry-Photometry group followed. The main

Journal of Physics: Conference Series **2606** (2023) 011001 doi:10.1088/1742-6596/2606/1/011001

subject was an extension of the BIPM on-site primary voltage standards comparisons to AC voltages, an important topic in Electrical Metrology. Further on, the morning works ended with a round table about Metrology in Health, featuring Renata Souza from FIOCRUZ and Roberto Macoto Ichinose from the Biomedical Engineering of COPPE/UFRJ.

The afternoon of the third day started with two virtual rooms that held parallel sessions related to SEMETRO (electrical metrology) and to CIMMEC (mechanical metrology). Later on, there was another round table about Metrology in Health, specifically about metrological traceability of laboratory tests, with lectures by Dr. Marcelo Medeiros from INMETRO, Debora Michele Morone D'Aiuto from FIOCRUZ, Rafael Monsores from the company CONTROLLAB and MD Luisane Maria Falci Vieira, manager of the laboratory Álvaro Apoio.

The event's last day, October 21st, started with a plenary lecture by Dr. Jim Huggett from National Measurement Laboratory (NML), talking about the role of measurement science in supporting a molecular diagnosis of SARS-COV-2, a very current and hot topic. A vivid discussion on Metrology Education followed the plenary lecture.

The afternoon of the last day started with two virtual rooms that held parallel sessions related to REMEQ (chemical metrology) and CBM (general metrology). Later, round tables were held in the two parallel virtual rooms, discussing hot topics such as metrological traceability, quality control on COVID-19 diagnostics, alternative methods applied to toxicology, and a web app developed for the performance evaluation of laboratories.

Metrologia 2021 disclosed about 170 articles. After the second round of revision, the editorial committee selected the best 35 to be published by IOP JPCS.

Metrology 2021 had sponsors from the industrial sector (Fluke Calibration and Sartorius do Brasil (Gold sponsors) and institutional support (ARBB Consultoria em Metrologia e Qualidade, RMMG, Canal Metrologia, REDE, SBLUZ and GM Metrologia). The event's next edition will be face-to-face in 2023, from 28th to 30th November, in Itaipava, a gorgeous area belonging to the "Imperial City" of Petrópolis, Rio de Janeiro state. You are all welcome to join us!

Rio de Janeiro, RJ, Brasil, August 28th, 2023.

Editorial committee

Carlos Roberto Hall Barbosa

Journal of Physics: Conference Series **2606** (2023) 011001 doi:10.1088/1742-6596/2606/1/011001

Daniel Ramos Louzada

Rodrigo P.B Costa-Felix

Organizing committee

Américo Tristão Bernardes (UFOP and SBM) – Presidente

Alexandre Mendes (IFRJ and SBM) – Organizing committee

Aurea Valadares – Coordinator of the V Workshop of the Brazilian Network of Chemical Metrology (REMEQ-I)

Daniel Louzada (PUC – RJ) – Coordinator of the V International Congress of Mechanical Metrology (CIMMEC)

Gelson Rocha – Coordinator of the XIII International Congress of Electric Metrology (SEMETRO)

Iakyra Couceiro – Coordinator of the III Brazilian Congress of Optical Metrology (CBMO)

José Guilherme Pereira Peixoto (IRD) – Coordinator of the Brazilian Congress of Ionizing Radiation Metrology (CBMRI)

Jose Mauro Granjeiro (Inmetro) – Organizing committee

Rodrigo P.B. Costa-Felix (Inmetro and SBM) – Coordinator of the 10th Brazilian Congress of Metrology (CBM)

Thais Medeiros Rodrigues (SBM) – Executive secretary

Vanderlea de Souza (Inmetro) – Organizing committee

Journal of Physics: Conference Series **2606** (2023) 011002 doi:10.1088/1742-6596/2606/1/011002

Peer Review Statement

All papers published in this volume have been reviewed through processes administered by the Editors. Reviews were conducted by expert referees to the professional and scientific standards expected of a proceedings journal published by IOP Publishing.

- **Type of peer review:** Single Anonymous
- **Conference submission management system:** Morressier
- **Number of submissions received:** 174
- **Number of submissions sent for review:** 91
- **Number of submissions accepted:** 43
- **Acceptance Rate (Submissions Accepted / Submissions Received × 100):** 24.7
- **Average number of reviews per paper:** 3
- **Total number of reviewers involved:** 90
- **Contact person for queries:**
  Name: Rodrigo Costa-Felix
  Email: rpfelix@inmetro.gov.br
  Affiliation: Inmetro

Portable EDXRF spectrometer: determination of linearity, LOD, LOQ, and working range for the analysis of base and precious metals in liquid matrices

A Krummenauer[1], L H Lapazin[1], M P Cenci[1] and H M Veit[1]

[1] Programa de Pós-Graduação em Engenharia de Minas, Metalúrgica e de Materiais (PPGE3M), Universidade Federal do Rio Grande do Sul (UFRGS), Av. Bento Gonçalves, 9500, Porto Alegre, RS, Brazil.
E-mail: alex_krum@yahoo.com.br

Abstract. Utilizing a portable energy dispersive X-ray fluorescence (EDXRF) spectrometer and certified reference materials, this study aimed to determine the linearity, limit of detection (LOD), limit of quantification (LOQ), and working range of the method for quantitative analysis of base and precious metals in liquid matrices. As the EDXRF method lacks standardization and is highly susceptible to matrix effects, these calculations play a crucial role in the analytical validation process. The obtained R^2 values in this study exceeded 0.99, and the residuals exhibited a random distribution pattern. The LOD and LOQ values were determined to be exceptionally low, approaching the resolution limit of the spectrometer. Consequently, our calculations confirmed the analytical validation of the portable EDXRF method for analyzing elements in liquid matrices. The utilization of portable EDXRF equipment in quantitative measurements of elements in liquid matrices represents an innovative approach, unveiling new areas of application for this type of instrument.

1. Introduction

Atomic spectrometric techniques have a wide range of applications and are essential for many research fields, including environmental assessment, soil characterization, mineral exploration, archaeology, material sciences, and quality control of various types of matrices [1-6]. Usually, the determination of the atomic composition mostly relies on lab-based methods, such as mass or optical emission spectrometry by inductively coupled plasma (ICP-MS and ICP-OES) and atomic absorption spectroscopy (AAS) [2, 3, 6-8]. Although being very effective, these methods require considerable sample preparation, large quantities of chemical inputs, and are time-consuming, uneconomical, and indirect [2]. Therefore, there is a strong demand from all stakeholders in the aforementioned fields for routine applications of methods that are faster, more cost-effective, practical, and environmentally friendly [9].

Energy dispersive X-ray fluorescence (EDXRF) is a multi-elemental quantitative analytical technique of atomic spectroscopy, which detects the number of element-characteristic X-rays by a unit of time, and rapidly provides results [4, 10, 11]. Precisely, the bombardment of the samples with X-rays results in the ionization of the elements, and the corresponding emission of fluorescent energy (particular for each element) is detected and converted to elemental composition [9]. The technique is recommended for the determination of elemental composition in solid and liquid samples, being a cheap, non-destructive, simple sample preparation, practical, direct, wide range, and eco-friendly method [2, 9].

In addition to the aforementioned advantages of EDXRF, portable EDXRF devices provide further simplification of the analysis process and a significant reduction in expenses, by the portability advantages [2-4]. In the last two decades, portable EDXRF devices have been improved through new accessories, peripherals, and hardware and software enhancements, providing researchers with an effective and practical tool for a wide range of analytes, facilitating routine analyses [2-4]. However, some characteristics and limitations of portable devices are still a matter of research in the metrology field. Mainly, compared with other techniques, portable EDXRF devices have worse detection limits than ICP-OES or AAS [2, 3].

In this study, a portable EDXRF device was used to investigate the linearity, the limit of detection (LOD), the limit of quantification (LOQ), and the working range for quantitative analyses in liquid matrices. Liquid matrices are particularly important for many research fields (such as materials characterization) because they promote sample homogeneity. These calculations contribute to achieve higher metrological reliability of the results and to obtain the validation of the analytical method in liquid matrices. The portable EDXRF method is not standardized. Thus, these calculations are essential prerequisites to metrological validation and open new application fields for this type of instrument.

2. Materials and methods

2.1. Equipment and Certified Reference Materials (CRMs)

The analysis was conducted using a portable energy-dispersive X-ray fluorescence (EDXRF) spectrometer, specifically the Thermo Fisher Scientific brand, model NITON XL3t GOLDD+, in conjunction with its mobile test stand accessory. Figure 1 presents a visual depiction of the utilized equipment.

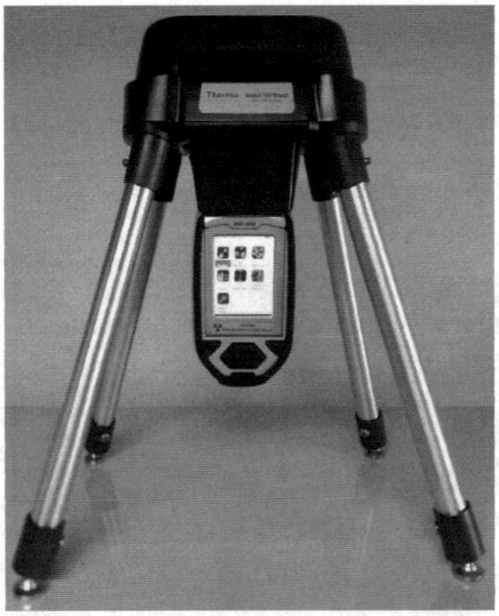

Figure 1. Thermo Fisher Scientific NITON model, accompanied by a mobile test stand.

The monoelementar certified reference materials (CRMs) utilized in the analyses are presented in table 1. These CRMs are specifically designed for ICP-MS and OES analysis and were employed within their specified expiration dates. These analytical techniques, based on inductively coupled plasma (ICP), offer significantly lower measurement uncertainties compared to EDXRF and serve as validated reference methods for spectrometric analysis of elements in liquid matrices [12]. Table 2 provides the concentration values (*C*), densities (*d*), measurement uncertainties (*U*), and coverage factors (*k*) for each utilized CRM. Since the EDXRF instrument measures concentrations in $\%_{m/m}$, equation (1) is required to convert mg/L to $\%_{m/m}$ for the preparation of the calibration curves.

$$C(\%m/m) = \frac{C(mg/L)}{10000 \times d} \tag{1}$$

Table 1. List of the CRMs and their traceability.

Element	Lot Number	Certification Issue Date	Lot Expiration Date	Producer
Ag	F20C0718H	Aug. 2020	Aug. 2022	Specsol
Au	F20C0375H	Aug. 2020	Aug. 2022	Specsol
Cu	N2-CU672882	Dec. 2018	Dec. 2022	Inorganic Ventures
Fe	N2-FE672883	Nov.2018	Nov.2022	Inorganic Ventures
Ni	N2-NI665352	Mar. 2018	Mar. 2022	Inorganic Ventures
Pb	N2-PB666052	Mar. 2018	Mar. 2022	Inorganic Ventures
Sn	F19L0531H	Aug. 2020	Aug. 2022	Specsol

Table 2. Concentrations, densities, measurement uncertainties and coverage factors of the CRMs.

Element	C (mg/L)	d (g/mL)	U (mg/L)	k
Ag	998	1.0236	± 4	2
Au	1000	1.0257	± 4	2
Cu	1001	1.016	± 5	2
Fe	1000	1.010	± 4	2
Ni	1000	1.011	± 5	2
Pb	998	1.002	± 4	2
Sn	998	1.0308	± 4	2
Zn	1000	1.011	± 5	2

2.2. Equipment configuration and procedures for determining linearity, LOD, LOQ, and working range

The CRMs, listed in table 1, were independently diluted with ultrapure water type I (>18 MΩ.cm) to obtain, at least, five concentration levels for each calibration curve, highlighting that one concentration level was positioned close to the predicted LOD of each element [4, 13, 14]. The dilution of the certified reference materials (CRMs) was conducted following the identical procedures employed for the preparation of calibration curves for the ICP-OES equipment. The process involved the use of ultrapure water (Type I) and previously decontaminated Class A glassware, exclusively designated for calibration curve preparation in the ICP-OES system. Additionally, a calibrated micropipette with disposable tips and an analytical balance with a resolution of 0.1 mg were employed. Prior to the dilution process, the calibration of the glassware, micropipette, and balance was verified using reference masses and volumes to ensure their accuracy and precision. To prevent the possibility of any leakage, the sample holder of the X-ray fluorescence (XRF) equipment was configured with a layered arrangement of three polypropylene films (4 µm each). This design effectively ensured the containment of the sample during analysis and maintained the integrity of the experimental setup. For each analysis, a volume of 4 mL of liquid sample was carefully positioned within the sample holder.

In the equipment NITON XL3t GOLDD+, the mode "mining Cu/Zn" was selected from the main menu because this mode enables the configuration of calibration curves and analysis of all elements of interest. The selected elements for this study include base metals (Cu, Fe, Ni, Pb, Sn, and Zn) and precious metals (Au and Ag). For Au, Cu, Fe, Ni, Pb, Sn, and Zn, only the main filter was used (as it is the only one that measures these elements), with an analysis of 40 seconds. Determination of Ag and Sn needs two filters: main (with 30 seconds) and high (with 30 seconds). All measurements were taken by using the following procedure quoted in the equipment user's guide: perform a system check once every working day and warm up for ten minutes after start-up [4, 5, 15].

The concentration levels of the eight CRMs (Ag, Au, Cu, Fe, Ni, Pb, Sn, and Zn) were measured five times each, always shaking the liquid before a new measurement, with exception of the lowest analyte concentration levels (closer to the predicted LOD), which were measured ten times (to determine LOD and LOQ). Through these analyses, the calculations of linearity, LOD, LOQ, and working range were conducted based on the orientations described by DOQ-CGCRE-008 [13], Magnusson and Örnemark [14], and Krummenauer [4, 5]. The determination of LOD and LOQ was conducted based on ten replicate measurements of test samples featuring low analyte concentrations [4, 5, 13, 14, 16], under repeatability conditions. It is important to note that these LOD and LOQ values are exclusively validated for liquid matrices. Since the samples involved low analyte concentrations, the LOD was estimated using Equation 2 [5, 14].

$$LOD = 3.66 \cdot \frac{s_0}{\sqrt{n}} \tag{2}$$

Where s_0 represents the estimated standard deviation obtained from n individual measurements conducted at or near zero concentration, while n denotes the total number of repeated observations.

The determination of LOQ involves the use of a factor, denoted as k_Q, which is assigned a value of 10 based on the recommendation by the International Union of Pure and Applied Chemistry (IUPAC) [5, 14]. This value is incorporated into the calculation, as expressed in equation 3.

$$LOQ = k_Q \cdot \frac{s_0}{\sqrt{n}} \tag{3}$$

3. Results and discussion

Initially, the calibration curves were constructed using the certified reference materials (CRMs) listed in table 1, wherein the corresponding concentrations are provided in tables 2 and 3 specifically for the element Cu, converted to %m/m. Subsequently, the focus of this study shifts to presenting the results for the element Cu, while the results for the remaining elements are presented in a summarized format due to the adoption of the same methodology as employed for Cu. Consequently, Table 3 displays the concentration levels (true values) employed to prepare the calibration curve for the element Cu, alongside the corresponding measured values obtained using the portable EDXRF spectrometer, model NITON XL3t GOLDD+. The values provided in table 3 were specifically utilized to assess the linearity of the element Cu. In accordance with the literature, linearity is determined by means of a simple linear regression analysis, as represented by Equation 4 [13, 14, 17].

$$Y = b \cdot X + a \tag{4}$$

Where Y represents the measured response in terms of the instrument signal, X denotes the concentration as the true value, a represents the linear coefficient, which signifies the intercept value of the linear equation on the Y-axis, and b represents the angular coefficient, also known as the analytical sensitivity.

Before constructing the linear regression, it is necessary to verify the presence of outliers for each concentration level and the data homoscedasticity (homogeneity of the variance residues) [13, 14, 18, 19]. The Grubbs' test can be used to the analysis of the outliers, and the Levene's test to verify the homoscedasticity [19-21]. These tests are widely available in statistical software packages and are based on hypothesis testing methodologies. In the case of the Grubbs' test, the H0 hypothesis assumes that data come from a normal distribution, and for the H1 hypothesis the values that are above or below the confidence level are outliers. H0 hypothesis is accepted if the p-value is higher than 0.05 (significance level of 5 %), and it is concluded that the data are from a normal probability distribution [18, 20, 21]. In the case of the Levene's test, the H0 hypothesis establishes that the calculated variances of each level are similar, and the H1 hypothesis establishes that at least one variance is different from the others. H0 hypothesis is accepted if the p-value is higher than 0.05, and the data are considered homoscedastic [19-21]. The Grubbs' and Levene's tests were conducted on the calibration curves of the eight elements using Minitab 17 software, and the findings are presented in table 4.

Table 3. Concentration levels of the calibration curve of CRM Cu.

Level	Measured values by EDXRF	True value CRM Cu
1	0.000 $\%_{m/m}$	0.000 $\%_{m/m}$
2	0.006 $\%_{m/m}$	0.005 $\%_{m/m}$
3	0.020 $\%_{m/m}$	0.015 $\%_{m/m}$
4	0.028 $\%_{m/m}$	0.020 $\%_{m/m}$
5	0.036 $\%_{m/m}$	0.025 $\%_{m/m}$
6	0.055 $\%_{m/m}$	0.037 $\%_{m/m}$
7	0.111 $\%_{m/m}$	0.074 $\%_{m/m}$
8	0.148 $\%_{m/m}$	0.099 $\%_{m/m}$

Table 4. Results for the Grubbs' and Levene's tests for the calibration curves.

Element	Grubbs' test		Levene's test	
	p-value	Result for outliers	*p*-value	Result for homoscedasticity
Ag	0.557	No outliers	0.368	Yes
Au	0.079	No outliers	0.130	Yes
Cu	0.884	No outliers	0.492	Yes
Fe	1.000	No outliers	0.711	Yes
Ni	0.975	No outliers	0.322	Yes
Pb	0.194	No outliers	0.185	Yes
Sn	0.744	No outliers	0.719	Yes
Zn	1.000	No outliers	0.431	Yes

Table 4 demonstrates the absence of outliers and homoscedasticity, indicating that all data points conform to linearity assumptions. However, it is important to note that the calculation of the calibration curve for the EDXRF spectrometer involves a unique aspect. The linearity of the analyzed elements is considered acceptable only when the instrument signal is plotted on the *X*-axis and the concentration of the certified reference material (CRM) is plotted on the *Y*-axis. This specific approach is explicitly recommended by the manufacturer and emphasized in the user's guide of the spectrometer. Moving forward, the linear regression model will be customized to align with the variable *Y* representing the true value, and the variable *X* representing the measured values obtained through the EDXRF equipment. This adjustment will be illustrated in subsequent instances, such as in figure 2.

Figure 2 displays the Cu calibration curve, which is plotted using the values from table 3. The graph exhibits the linear equation and the R^2 coefficient. Typically, an R^2 value exceeding 0.98 signifies strong linearity in spectrometric measurements [22], whereas a value exceeding 0.99 is desirable for analytical methods [17]. Notably, all linear regression models performed in this study yielded R^2 values greater than 0.99 for each element. A graphic of residuals is shown in figure 3 (residuals are the differences between the observed *Y* value and the calculated *Y* value predicted by a straight line, for each *X* value).

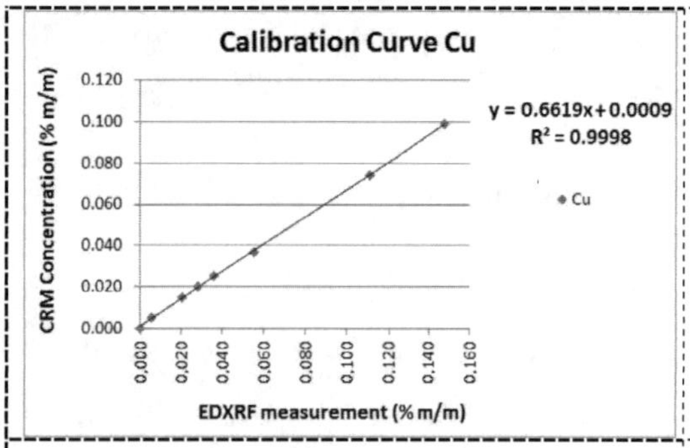

Figure 2. Calibration curve of Cu.

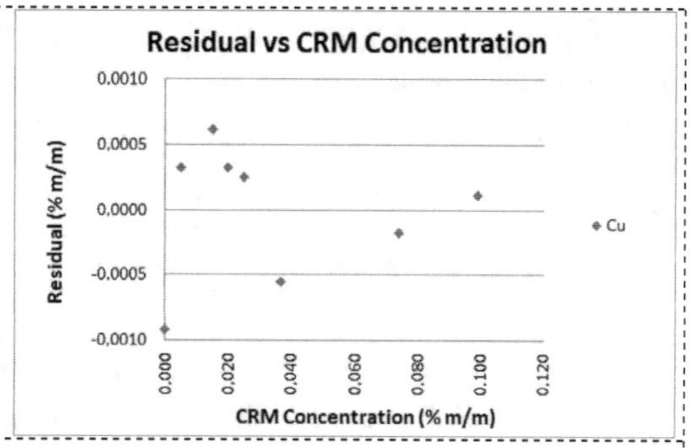

Figure 3. Residuals of the Cu calibration curve.

In the case of a presence of tendency in the graphic of residuals, there may be indications that the linear model is inadequate [4, 13, 14, 20, 21]. For the confirmation that the residuals present a random distribution, the Anderson-Darling's test (AD) is applied. The AD test considers that a random distribution is confirmed if the resultant p-value is lower than 0.05 (hypothesis H0 of a normal distribution is rejected) [20, 21].

The linearity of the regression curve must also be analyzed. In the test F (Fischer), also known as F-Snedecor test in the variance analysis of the regression (ANOVA), the linearity is significant if the value F_{calc} is higher than F_{tab} in the calculations of the statistical software [13, 14, 18-21]. Another test that was performed is the Variance Inflation Factor test (VIF). It measures the effect of multicollinearity among the predictors. Multicollinearity is a term used to describe when two or more predictors in the regression are highly correlated. The rules of thumb to analyze VIF are: if VIF = 1, there is no multicollinearity; if 1 < VIF < 5, there is small multicollinearity; if VIF $\geq$ 5, there is medium multicollinearity; and, if VIF $\geq$ 10, there is large multicollinearity [20, 21].

Finally, the Durbin-Watson test (DW) was employed to assess the presence of autocorrelation in the residuals of the linear regression model. The DW value was calculated using statistical software and compared to the tabulated lower limit (DL) and upper limit (DU) values. When DW exceeds DU, it indicates the absence of autocorrelation in the residuals. If the DW value falls within the range DL $\leq$ DW $\leq$ DU, the test results are inconclusive. However, if DW is less than DL, it signifies the presence of autocorrelation in the residuals [20, 21].

Table 5 summarizes the tests AD, F, VIF, and DW for the calibration curves of the elements Ag, Au, Cu, Fe, Ni, Pb, Sn e Zn, using the Minitab 17 software.

For each element, the analysis verifies the random distribution of the residuals. The results presented in table 5 demonstrate that the Anderson-Darling (AD) test confirms the normal distribution of the data. Furthermore, the F test confirms the statistical significance of the linearity observed in the regression models. The Variance Inflation Factor (VIF) test reveals the absence of multicollinearity, and the Durbin-Watson (DW) test indicates that the residuals are not autocorrelated. Consequently, the linear regression models provide a satisfactory fit for the calibration curves of all elements. Table 6 presents a summary of the values obtained from the interception with the *Y*-axis (*a*), the slope (*b*), and R^2 of each calibration curve.

Table 5. Resultant values for the tests AD, F, VIF, and DW

Element	AD test		*F* test			VIF test		DW test			
	p-value	Normal distribution	F_{calc}	F_{tab}	Significative	VIF	Multicollinearity	DW	DU	DL	Autocorrelation
Ag	0.725	Yes	15109	6	Yes	1	No	2.918	0.763	1.332	No
Au	0.742	Yes	67991	6	Yes	1	No	2.211	0.763	1.332	No
Cu	0.136	Yes	27445	6	Yes	1	No	1.402	0.763	1.332	No
Fe	0.552	Yes	763	8	Yes	1	No	2.994	0.610	1.400	No
Ni	0.859	Yes	914	10	Yes	1	No	2.989	0.610	1.400	No
Pb	0.053	Yes	58074	6	Yes	1	No	1.552	0.824	1.320	No
Sn	0.436	Yes	10479	7	Yes	1	No	1.502	0.700	1.356	No
Zn	0.268	Yes	32420	7	Yes	1	No	2.795	0.700	1.356	No

Table 6. Parameters of the calibration curves.

Element	Intercept (a)	Slope (b)	R^2
Ag	-0.0006	2.2347	0.9996
Au	0.0000	0.8523	0.9999
Cu	0.0009	0.6619	0.9998
Fe	0.0159	0.6546	0.9948
Ni	0.0176	0.6631	0.9967
Pb	0.0003	0.6690	0.9999
Sn	-0.0003	1.5913	0.9995
Zn	0.0004	0.7122	0.9998

Upon analyzing the values of the coefficients "*a*" and "*b*" for each element, as presented in table 6, a substantial variation among these values becomes apparent ("*a*" ranging from -0.0006 to 0.0176, and "*b*" ranging from 0.6546 to 2.2347). Consequently, the construction of calibration curves using certified reference materials (CRMs) assumes significant importance in ensuring reliable measurements by EDXRF spectrometers. This is attributed to the utilization of default coefficients $a = 0.000$ and $b = 1.000$ in the standard configuration of the equipment for calculating the outcome variable. Utilizing these default coefficients may lead to measurement errors of up to 50 %, as observed, for instance, in the cases of Cu, Pb, and Ag.

Furthermore, a comparison of the "*a*" values (intercept with the Y-axis) enables the prediction of differences in the lower limits of the working ranges. Specifically, it is expected that the elements Fe and Ni will exhibit significantly higher lower limits of the working ranges due to their comparatively elevated "*a*" values (0.00159 and 0.0176, respectively), in comparison to the other elements.

Table 7 shows the LOD and LOQ values obtained for each element, in a liquid matrix, through the procedures described in section 2.2. The resulting values can be considered low as the resolution of the portable EDXRF equipment is 0.001 $\%_{m/m}$.

Table 7. LODs and LOQs values for a liquid matrix.

Element	LOD (% $_{m/m}$)	LOQ (% $_{m/m}$)	Lower End (% $_{m/m}$)	Upper End (% $_{m/m}$)
Ag	0.001	0.002	0.002	0.097
Au	0.001	0.002	0.002	0.097
Cu	0.001	0.002	0.002	0.099
Fe	0.001	0.004	0.020	0.099
Ni	0.001	0.003	0.025	0.099
Pb	0.001	0.002	0.002	0.100
Sn	0.001	0.002	0.002	0.097
Zn	0.001	0.002	0.002	0.099

Lastly, the determination of the working range is presented in table 7. The working range represents the span between the lowest and highest concentrations of the analyte that can be reliably measured using the analytical procedure, while ensuring an acceptable level of precision, linearity, accuracy, and uncertainty [4, 13, 14, 17]. In this study, a reliable level of linearity (R^2 higher than 0.98 and graphical analysis of the residuals) was achieved. The lower end of the working range is the LOQ value for the elements Ag, Au, Cu, Pb, Sn, and Zn. For these elements, the LOQ value is situated into the calibration curve. However, in the case of elements Fe and Ni, the lower limit of the working range surpasses the LOQ value due to the inability of the zero analyte concentration point (0;0) to properly fit into the calibration curve. This inadequate fit resulted in unfavorable values for R^2, residuals, and uncertainty, necessitating the inclusion of an initial point with a higher-than-zero analyte concentration. Consequently, for Fe and Ni, the lower limit of the working range is defined as the initial point of the calibration curve (ensuring it is higher than the LOQ). In all instances, the concentrations measured by the equipment must be adjusted using the calibration curve parameters listed in table 6. Similarly, the upper limit of the curve corresponds to the highest concentration point, also adjusted using the calibration curve parameters listed in table 6, reflecting the maximum concentration level of each CRM analyzed [4, 13, 14, 17].

4. Conclusion

The calibration curves exhibited excellent linearity in liquid matrices, as evidenced by R^2 values exceeding 0.99 for all analyzed base and precious metals. This signifies that over 99 % of the measurements conform to the linear equation. The randomness observed in the residual values further validates the linearity of the calibration curves, supported by the results obtained from various statistical tests such as Grubbs' tests, Levene's tests, Anderson-Darling's tests, F-Snedecor's tests, VIF tests, and Durbin-Watson's tests. It is important to emphasize that in the absence of specific calibration curves tailored for liquid matrices, measurement errors can be as high as 50 %. Therefore, the quantitative nature of measurements obtained using the portable EDXRF spectrometer can only be established if they align with a successful application of a linear regression model (calibration curve). The determination of LOD and LOQ yielded values that were considerably low, approaching the equipment's resolution (LOQs ranged between 0.002 and 0.004 %m/m or between 20 and 40 mg/L, while the equipment's resolution stood at 0.001 %$_{m/m}$ or 10 mg/L). However, it is important to note that the LOD and LOQ values are influenced by several factors, including the number of measurements, equipment and sample holder configuration, instrumental errors, sample quality, variations in analysis procedures, and the nature of the liquid matrix. Therefore, the estimated LOD and LOQ values for each element should be regarded as indicative. Furthermore, when determining the lower limit of the working range, it is crucial to apply the coefficients from the calibration curves, including for the LOQ values. Finally, it should be noted that the determined values for linearity, LOD, LOQ, and working range are specific to EDXRF method and the procedures outlined in this study. These calculations play a crucial role in validating the effectiveness of this method when applied to liquid matrices. The utilization of a portable EDXRF spectrometer for quantitatively measuring elements in liquid matrices represents an innovative approach that opens up new avenues for application in this instrumental domain.

Acknowledgments
The authors would like to express their gratitude to CAPES for the support provided through the doctoral program. The authors would also like to acknowledge the technical and instrumental assistance provided by UFRGS, PPGE3M, and LACOR. Additionally, the authors extend their appreciation to FINEP for the resources obtained through the SIBRATEC - Technological Services Provision Networks, which facilitated the acquisition of the portable EDXRF spectrometer by LACOR.

References
[1] Singh V, Padalia D and Devlal K 2017 *J. Nucl. Physics, Mater. Sci. Radiat. Appl.* **4** 383-91
[2] Zhou S, Weindorf D C, Cheng Q, Yang B, Yuan Z and Chakraborty S 2020 *Spectrochim. Acta Part* B *At. Spectrosc.* **174** 105999
[3] Korf N, Løvik A N, Figi R, Schreiner C, Kuntz C,Mählitz P M, Rösslein M, Wäger P and Rotter V S 2019 *Waste Management* **92** 124-136
[4] Krummenauer A 2017 *Desenvolvimento e Validação de Metodologia Analítica para Análise de Aços por Espectrometria de Fluorescência de Raios X por Dispersão de Energia (EDXRF),* Master Thesis (Porto Alegre: UFRGS)
[5] Krummenauer A, Dias P R and Veit H M 2021 *J. Phys.: Conf. Ser.* **1826** 012008
[6] Cenci M P, Dal Berto F C, Camargo P S S and Veit H M 2021 *Waste Management* **120** 136-145
[7] Kasper A C, Berselli G B T, Freitas B D, Tenório J A S, Bernardes A M and Veit H M 2011 *Waste Management* **31** 2536-2545
[8] Yamane L, Moraes V, Espinosa D and Tenório J 2011 *Waste Management* **31** 2553-2558
[9] Herreros-Chavez L, Cervera M L and Morales-Rubio A 2019 *Food Chem.* **278** 373-79
[10] Beckhoff B, Kanngiefer B, Langhoff N, Wedell R and Wolff H (Eds) 2006 *Handbook of practical X-ray fluorescence analysis* (Berlin: Springer)
[11] POTTS P J and WEST M 2008 *Portable X-ray Fluorescence Spectrometry: capabilities for in situ analysis* (Cambridge: Royal Society of Chemistry)
[12] Floor G H, Queralt I, Hidalgo M and Marguí E 2015 *Spectrochimica Acta Part B* **111** 30–37
[13] INMETRO 2020 *DOQ-CGCRE-008 Orientação sobre Validação de Métodos Analíticos* (Rio de Janeiro: INMETRO)
[14] Magnusson B and Örnemark U (Eds.) 2014 *Eurachem Guide: the fitness for purpose of analytical methods – a laboratory guide to method validation and related topics* (Available from http://www.eurachem.org)
[15] Thermo Fisher Scientific 2010 *Niton XL3 Analyzer User's Guide Revision C*
[16] Uhrovčík J 2013 *Talanta* **119** 178-180
[17] Albano F M and Raya-Rodriguez M T 2009 *Validação e garantia da qualidade de ensaios laboratoriais: guia prático* (Porto Alegre: Rede Metrológica RS)
[18] Cardoso M H W M, Gouvêa A V, Da Nóbrega A W and ABRANTES S de M P 2010 *Ciênc. Tecnol. Aliment.* **30 supl. 1** 63-72
[19] Souza, S V C and Junqueira R G 2005 *Analytica Chimica Acta* **552** 25-35
[20] Montgomery D C 1997 *Design and Analysis of Experiments* (New York: John Wiley & Sons)
[21] Montgomery D C 2009 *Introduction to Statistical Quality* (New York: John Wiley & Sons)
[22] Drglin T 2003 *Accreditation and Quality Assurance* **8** 130-33

New measuring system for locating firearm projectiles in the human body based on eddy currents and GMI sensors

V Seixas, C Hall Barbosa and E Costa Monteiro

Pontifical Catholic University of Rio de Janeiro, Postgraduate Programme in Metrology, Rua Marquês de São Vicente, 225, Rio de Janeiro, Brazil

E-mail: hall@puc-rio.br

Abstract. Noninvasive magnetic mapping for locating non-ferromagnetic metallic foreign bodies, such as firearm projectiles, has a promising application for guiding accurate surgical removals. The strategy consists of inducing eddy currents in the metallic object by applying an external alternating magnetic field. A secondary magnetic field generated by the induced currents in the foreign body can be detected by a high-sensitivity transducer. Previously proposed systems presented constraints to the measurement capability of the transducer imposed by the primary magnetic field interference on their sensing elements. Overcoming these limitations, this work proposes a new configuration of the system that decouples the excitation and measuring stages, arranging their elements symmetrically to generate a spatial distribution of the field lines in such a way that adequate magnetic flux density levels are produced in the region of the foreign body, with values nearly not-detectable at the sensor location. Thereby, the secondary field generated by the induced eddy currents can be amplified, allowing the detection of smaller fragments located at greater depth distances from the skin. The measuring system concept, the spatial distribution of the primary and secondary magnetic fields, and the new approach's performance to localize foreign bodies are presented.

1. Introduction

The occurrence of foreign bodies accidentally inserted into the human body is common in clinical practice, and these, in most cases, require a surgical procedure for removal. Before removal, the conventional approaches employed for their location involve X-ray imaging methods, including radioscopy performed during the surgical procedure [1,2]. These strategies, however, do not precisely identify the object position and expose both the patient and staff to radiation [1]. Employing a Superconducting Quantum Interference Device (SQUID), the most sensitive magnetometer available, a new technique for noninvasive and accurate location of foreign body in patients was developed [1,3]. This method, based on measuring the magnetic flux density generated by the ferromagnetic object, was applied successfully in six procedures for surgical removal of hypodermic and sewing needles, which, as a result, were performed quickly and accurately [1]. However, this superconducting technology presents high manufacturing and operating costs, working at cryogenic temperatures, preventing, therefore, their compliance with the Biometrological Principles [4], which require aspects such as low-cost, portability, noninvasiveness, innocuousness, high accuracy, and precision.

In a pursuit to develop biomedical techniques complying with these biometrological principles, recent studies have been carried out to develop low-cost, high-sensitivity magnetic transducers for safe and noninvasive measurements in humans, operating at room temperature, based on the Giant Magnetoimpedance (GMI) and Giant Magnetoresistance (GMR) effects to assist, among others, in performing foreign bodies location [5-11].

Firearm projectiles are metallic foreign bodies also frequent in medical practice. However, unlike hypodermic needles, these bullets are mostly made of lead, a diamagnetic material, having no remanent magnetic field. Since lead is a good conductor of electricity, the induction of Eddy Currents, using an alternating magnetic field (primary) for excitation, will generate an alternating magnetic field (secondary) that a magnetometer with adequate sensitivity and resolution can detect [12].

Based on the model of the magnetic flux density produced by eddy currents induced in a spherical object, a transducer composed of a long solenoid and two GMI sensors located at the longitudinal axis limits, in a first-order gradiometric configuration, was developed and tested [8,9]. The system proved to be effective in estimating the radius of the projectile and its depth relative to the patient's skin. However, due to its design, a limiting factor to the measuring capability was the coupling of the excitation and measurement stages. The excitation magnetic field (primary) was limited to the maximum value of the linear operating range of the GMI sensor, which is ±1 μT. Consequently, the secondary field generated by the induction of Eddy Currents is also limited, reducing the detection capability of the instrument depending on the volume and depth of the metallic foreign body.

This study aims at improving the configuration of the magnetic measuring system designed to locate non-ferromagnetic metallic foreign bodies, overcoming limitations to the measuring capability of the transducer imposed by the primary magnetic field interference on the sensing elements operation. The proposed configuration decouples the excitation and measuring stages of the system, arranging their elements symmetrically to generate a spatial distribution of the field characterized by adequate magnetic flux density levels in the region of the foreign body and nearly not-perceptible levels at the sensor location. Thereby, the secondary field generated by the induced eddy currents can be amplified, allowing the detection of smaller projectile fragments located at greater depth distances from the skin. The performance of the proposed measuring system is determined considering safety limits to the primary magnetic field levels according to the International Commission on Non-Ionizing Radiation Protection (ICNIRP) recommendations.

2. Materials and Methods

2.1. Instrument Concept

The instrument designed to locate non-magnetic foreign bodies, such as firearm projectiles, is based on inducing eddy currents on the target and measuring the secondary magnetic field generated by such currents. As shown in Fig. 1, it is composed of two stages: excitation and measurement. The excitation of the foreign body is provided by two parallel solenoids winded in opposite directions, thus generating a quasi-circular magnetic flux through both solenoids and the target, traversing it almost horizontally. The measurements are based on a GMI sensor placed between the solenoids, in two possible orientations: horizontal and vertical. The whole system is positioned as close as possible to the foreign body.

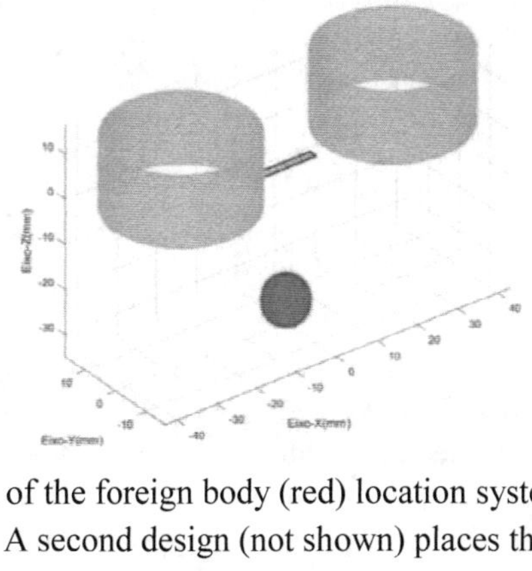

Figure 1. Schematic diagram of the foreign body (red) location system, with the GMI sensor (blue) in a horizontal orientation. A second design (not shown) places the GMR sensor vertically.

2.2. GMI Sensor

The design herein presented is based on a GMI sensor manufactured by Aichi Micro Intelligent Corporation, model MI-CB-1DJ-M-B. It has small dimensions (1.2 mm x 13.5 mm) and operates at room temperature, measuring the magnetic flux density parallel to its length.

The frequency range is between 0.1 Hz and 10 kHz, with a sensitivity of 5 V/μT, saturation field of ±1 μT and noise figure of 10 pT/√Hz at 1 Hz, with an estimated resolution of 1 nT.

2.3. Excitation System

Two identical solenoids are separated by a horizontal gap of 20 mm, in which the GMI sensor is to be positioned. Each solenoid has a radius $a_s = 1.65$ cm and length $h_s = 1.65$ cm, so as to yield a portable device. The wires have a diameter of 0.4547 mm, allowing $N = 36$ turns in each solenoid and being able to carry currents of 0.457 A amplitude [13].

Both solenoids are powered by sinusoidal currents with 8 kHz frequency, same intensity, and opposite directions. This design generates a quasi-horizontal magnetic field with large intensity in the region of the foreign body, and a very low magnetic field in the region where the GMI sensor is located. In fact, for the case where the GMI sensor is positioned vertically, the primary fields generated by both solenoids cancel, allowing the increase in the excitation without saturating the sensor. A similar scenario occurs for the horizontal orientation, but in this case, there is a non-zero resultant magnetic field that must be kept under ±1 μT. The defined excitation parameters also consider the reference levels for limiting general public exposure to time-varying magnetic fields established by ICNIRP guidelines [14].

In this work, the primary magnetic field generated by the solenoids is simulated by a numerical integration of the Biot-Savart Law applied to a current dipole over the path of the wires [15], which are discretized with a resolution of 1 mm. As the frequency is low, the sinusoidal magnetic field can be calculated by considering just the amplitude of the current, without the need to consider dynamic effects.

Fig. 2 shows an example of the magnetic flux density B lines over the region of the solenoids, foreign body and GMI sensor, considering that the center of the foreign body is at a depth $h = 50$ mm below the base of the solenoids. In Fig. 2a, the length of the vectors is proportional to the magnetic flux density magnitude (considering only the components B_{0x} and B_{0z}). In Fig. 2b, the vectors are normalized, so as to facilitate their visualization in the regions away from the solenoids.

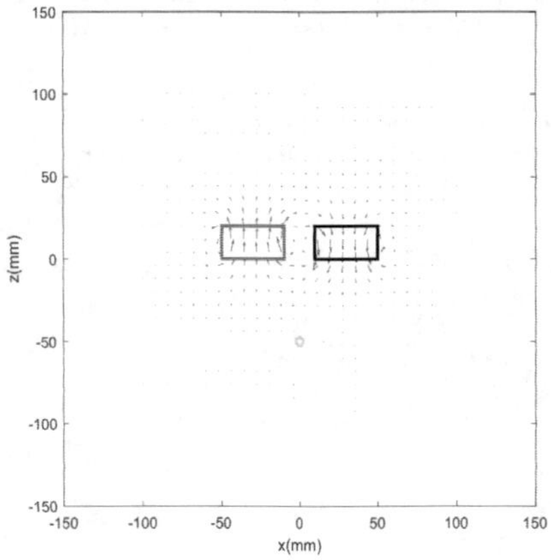

(a)

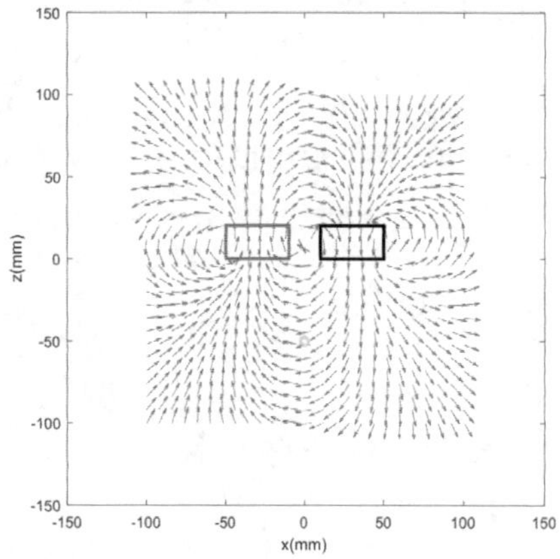

(b)

Figure 2. (a) Magnetic flux density vectors and (b) normalized magnetic flux density vectors. The rectangles show the profiles of the solenoids, with the different colors indicating different directions of the current. The green circle shows an example of a 2.5 mm radius foreign body positioned 50 cm below the solenoids.

Considering the horizontal (*x-y*) plane at the depth z = -50 mm, where the center of the foreign body is located, Fig. 3 shows the x component of the amplitude of the magnetic flux density (B_{0x}) generated by the pair of solenoids with N = 36 turns and a current with 0.457 A of amplitude and 8 kHz of frequency. This map shows the primary magnetic field that induces eddy currents in the foreign body, as explained in the next section. As expected, the magnetic flux density is not uniform, but the different values can be taken into account when calculating the secondary magnetic field generated by the eddy currents.

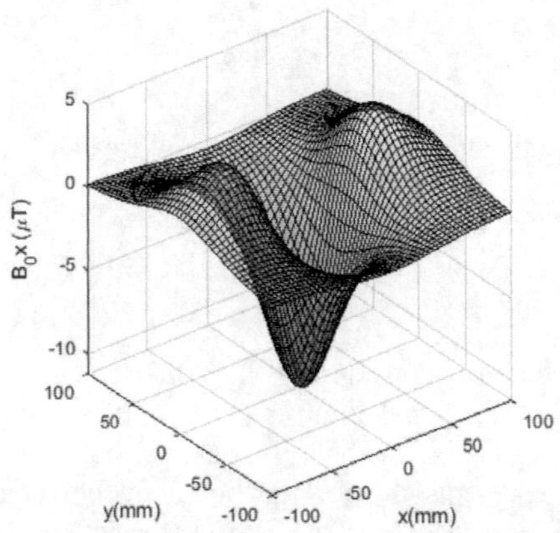

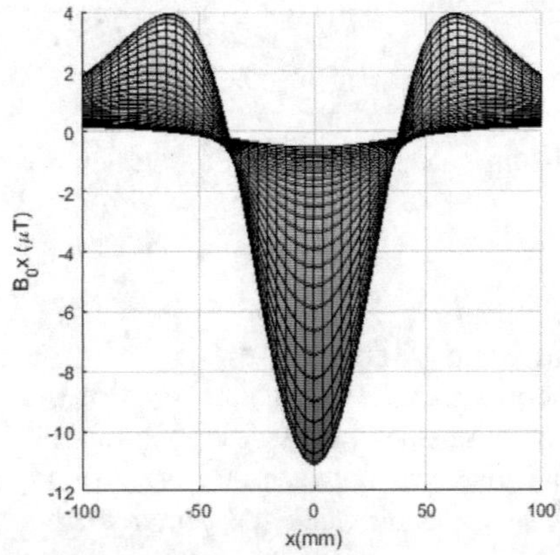

Figure 3. Primary magnetic flux density B_{0x} over a plane *x-y* at the depth z = -50 mm.

2.4. Eddy Currents

The primary magnetic field generated by the excitation system induces eddy currents in the foreign body, herein modeled as a sphere with radius a. These eddy currents, in turn, generate a secondary magnetic field that can be detected by the GMI sensor. This secondary magnetic field has been theoretically deduced in other works [8,9,12], considering a vertical excitation field B_{0z} and the detection of the vertical component B_z of the secondary field.

In the present work, the primary magnetic field is approximated as horizontal, in the x direction, while the secondary magnetic field components of interest are B_x or B_z, depending on the orientation of the GMI sensor.

Thus, consider a conducting sphere of radius a, resistivity ρ and permeability μ, subject to a uniform alternating magnetic field with angular frequency $\omega = 2\pi f$ directed along the x axis, with magnetic flux density given by

$$\vec{B}_0 = B_{0x} \cdot e^{j\omega t} \cdot \hat{x} \tag{1}$$

This alternating magnetic flux density induces circular eddy currents in the *y-z* plane, also alternating with the same frequency f, which in turn generate a secondary alternating magnetic field whose amplitude is similar to a magnetic dipole oriented in the x direction. Taking as the origin the center of the sphere, the amplitude of the three components of the secondary magnetic flux density are given by

$$B_x(x,y,z) = \frac{B_{0x}}{2} \cdot |V(a,f)| \cdot (x^2+y^2+z^2)^{-5/2} \cdot [2x^2 - (y^2+z^2)], \tag{2}$$

$$B_y(x,y,z) = \frac{3B_{0x}}{2} \cdot |V(a,f)| \cdot (x^2+y^2+z^2)^{-5/2} \cdot (y^2+z^2)^{-1/2} \cdot (x \cdot y \cdot z) \text{ and} \tag{3}$$

$$B_z(x,y,z) = \frac{3B_{0x}}{2} \cdot |V(a,f)| \cdot (x^2+y^2+z^2)^{-5/2} \cdot (y^2+z^2)^{-1/2} \cdot (x \cdot z^2), \tag{4}$$

where $V(a,f)$ is a complex number that imposes a phase shift φ between the primary and secondary magnetic fields and is given by

$$V(a,f) = \frac{(2\mu_r+1)-(2\mu_r+v(a,f)^2+1)\cdot\frac{\tanh(v(a,f))}{v(a,f)}}{(\mu_r-1)-(\mu_r-v(a,f)^2-1)\cdot\frac{\tanh(v(a,f))}{v(a,f)}} = |V(a,f)| \cdot e^{j\varphi}, \tag{5}$$

$\mu_r = \mu/\mu_0$ is the relative permeability and

$$v(a,f) = \frac{(1+j)\cdot a}{\delta}. \tag{6}$$

Finally, δ is the skin depth, given by

$$\delta = \sqrt{\frac{\rho}{\mu\cdot\pi f}}. \tag{7}$$

Thus, the secondary magnetic flux density (generated by the eddy currents) can be written as

$$\vec{\boldsymbol{B}} = \left(B_x \cdot \hat{x} + \boldsymbol{B}_y \cdot \hat{\boldsymbol{y}} + B_z \cdot \hat{\boldsymbol{z}}\right) \cdot e^{j\omega t+\varphi}. \tag{8}$$

3. Results and Discussion

This section presents the main results of the simulated scenarios, considering the GMI sensor in the two orientations previously described (aligned in the x and z directions), and typical radii (0.5 mm to 5 mm) and distances between the foreign body and the sensor. This distance is measured between the base of the solenoids and the center of the sphere that represents the foreign body. In the clinical practice, the surface of the patient's skin is the maximum limit of proximity between the sensor and the foreign body, corresponding to the depth of the object to be surgically removed.

The following sections consider a 2D scanning of the instrument (solenoids+GMI sensor) in the x-y plane at a known distance h from the center of the foreign body. As shown in Fig. 2, the primary magnetic flux density amplitude B_{0x} varies with both x and y, so these non-uniform values are used as inputs to equations (2)-(4). Also, it has been considered a detection threshold of 1 nT for the GMI sensor used, so the detection is deemed viable if the secondary magnetic flux density (generated by the eddy currents) is larger than 1 nT.

3.1. *GMI Sensor Aligned in the X Direction*

Considering the GMI sensor in the x direction and aligned with the base of the solenoids (as shown in Fig. 1), the x component of the secondary magnetic flux density, B_x, has been calculated on the x-y plane over a 200 mm x 200 mm area, for various sphere depths and radii. Fig. 4 shows one example of such a magnetic field map, for z = -50 mm and a = 5 mm. The secondary magnetic field is symmetric over both directions, x and y, with a prominent peak over the position of the foreign body. Also, the maximum amplitude is larger than 2 nT, thus exceeding the detection threshold of 1 nT.

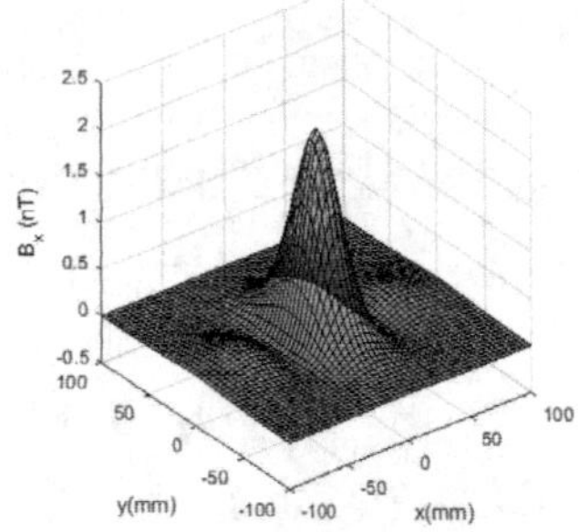

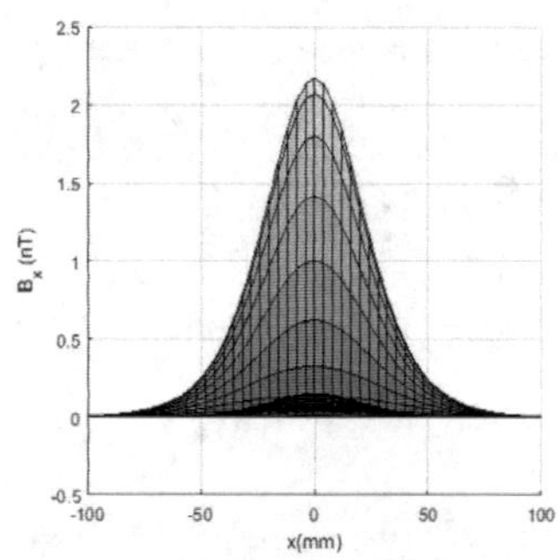

Figure 4. Secondary magnetic flux density B_{0x} over a plane x-y, generated by eddy currents induced in a sphere at the depth z = -50 mm and radius a = 5 mm.

By fixing the radius and systematically increasing the depth until the maximum amplitude is below the detection threshold of 1 nT, it is possible to estimate the maximum depth in which the system detects that foreign body with a specific dimension. Fig. 5 shows the results of this estimative for sphere radii varying from 0.5 mm to 5 mm. As expected, the smaller the object, the closer the sensor needs to be.

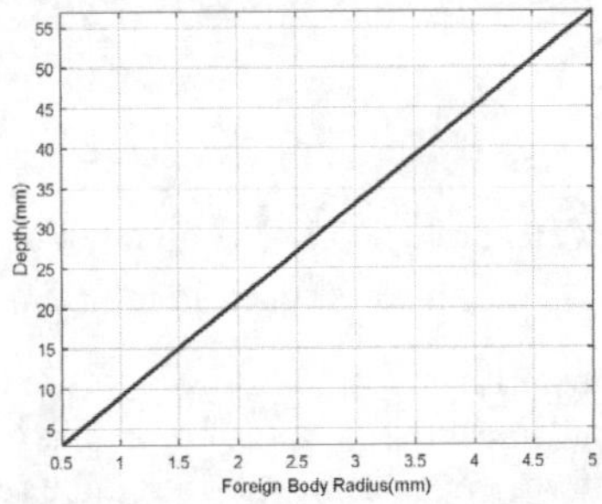

Figure 5. Maximum detectable depth as a function of the foreign body radius, for the GMI sensor aligned in the *x* direction.

3.2. GMI Sensor Aligned in the Z Direction

Now considering the GMI sensor aligned in the *z* direction, thus vertical, with its base aligned with the base of the solenoids, the *z* component of the secondary magnetic flux density, B_z, has been calculated on the same x-y plane over a 200 mm x 200 mm area, for various sphere depths and radii. Fig. 6 shows one example of these magnetic field maps, obtained for z = -50 mm and a = 5 mm. In this case, the secondary magnetic field presents a dipolar configuration, with symmetrical peaks aligned with the primary magnetic field B_{0x}, and a zero magnitude over the position of the foreign body. The maximum amplitude is larger than 1.2 nT, thus exceeding the detection threshold of 1 nT.

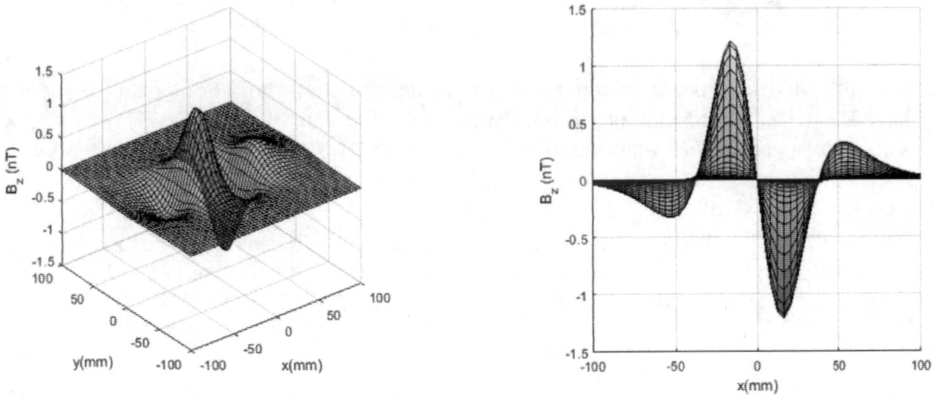

Figure 6. Secondary magnetic flux density B_{0z} over a plane *x-y*, generated by eddy currents induced in a sphere at the depth z = -50 mm and radius a = 5 mm.

By employing the same procedure described in the previous subsection, it was estimated the maximum distance in which the GMI sensor aligned in the z direction can detect foreign bodies as a function of their dimensions. Fig. 7 shows the results of this estimative for radii varying from 0.5 mm to 5 mm. As expected, and also observed in the previous subsection, for the smaller the sphere, the closer the sensor needs to be.

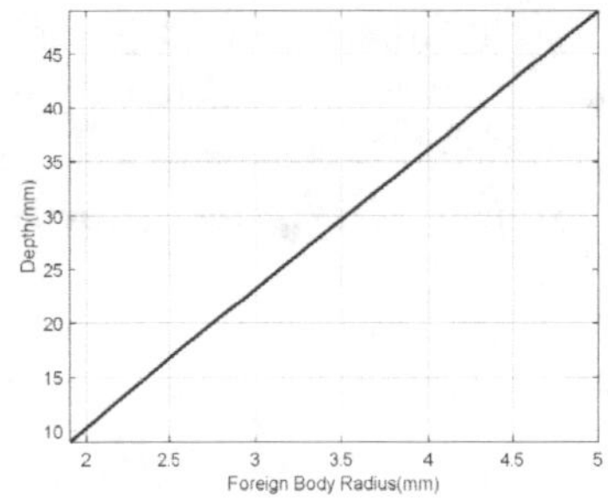

Figure 7: Maximum detectable depth as a function of the foreign body radius, for the GMI sensor aligned in the *z* direction.

Both the proposed configurations, either with the GMI sensor aligned in the x-direction or the z-direction, presented a very similar linear behavior of the maximum detectable depth curve as a function of the foreign body diameter. However, the performance reached by the setup with the sensor horizontally positioned reached higher levels, detecting, for instance, a 2 mm radius sphere at a distance twice superior as that required when using a vertically positioned sensor.

4. Conclusions

This work proposed a novel configuration of a magnetic measuring system designed to locate non-ferromagnetic metallic foreign bodies. The arrangement decouples the excitation and measuring stages, positioning their elements symmetrically. This approach aimed at overcoming limitations to the measuring capability of the former transducer imposed by the primary magnetic field interference on the sensing elements operation. The proposed design considered the GMI sensor in two possible orientations: horizontal and vertical.

Simulated results of the primary magnetic field generated by the solenoid pair arrangement have shown an adequate spatial distribution of the magnetic flux density levels in the region of the foreign body and at the sensor location. Therefore, solving the limitations of the previous approach, the new strategy allowed amplifying the secondary magnetic field generated by the eddy currents induced in the foreign body.

Estimates of the maximum depth in which the system detects the foreign body as a function of its dimension have shown a linear behavior for both approaches for sensor positioning, horizontally and vertically, with better performance than the previous design [8,9]. Despite the similarity between the results obtained by the two sensor orientation setups, the best performance was accomplished by the configuration in which the GMI sensor was horizontally oriented.

The study results point toward the promising feature of the proposed measuring system arrangement, which provides a step forward towards the noninvasive and accurate location of non-ferromagnetic metallic foreign bodies in patients.

Acknowledgments

The authors thank for the financial support provided by the Brazilian funding agencies CNPq, CAPES, FINEP and FAPERJ. This study was financed in part by the Coordenação de Aperfeiçoamento de Pessoal de Nível Superior - Brasil (CAPES) - Finance Code 001.

References

[1] Costa Monteiro E, Barbosa C, Lima E, Costa Ribeiro P and Boechat P 2000 Locating steel needles in the human body using a SQUID magnetometer *Phys Med. Bio.*, v. 45, n. 8, pp. 2389

[2] Aras M H, Miloglu O, Barutcugil C, Kantarci M, Ozcan E and Harorli A 2010 Comparison of the sensitivity for detecting foreign bodies among conventional plain radiography, computed tomography and ultrasonography *Dentomaxillofacial Radiology* 39 n 2 pp 72-78

[3] Barbosa C, Costa Monteiro E, Lima E A, Santos S, Cavalcanti E and Costa Ribeiro P 2001 Improvement of a technique for localization of steel needles in humans using a SQUID magnetometer *IEEE Trans. App Sup*, v. 11, n. 1, p. 677-680

[4] Costa Monteiro E and Leon L 2015 Metrological Reliability of Medical Devices *Journal of Physics: Conference Series*, v. 588, pp. 012032

[5] Pompéia F, Gusmão L A P, Barbosa C, Costa Monteiro E, Gonçalves L and Machado F 2008 Ring shaped magnetic field transducer based on the GMI effect *Meas. Sci. Technol.*, v. 19, pp. 025801

[6] Silva E C, Gusmão L A P, Barbosa C, Costa Monteiro E and Machado F 2011 High sensitivity giant magneto-impedance (GMI) magnetic transducer: Magnitude versus phase sensing *Meas. Sci. Technol.*, v. 29, pp. 035106

[7] Silva E, Gusmão L, Barbosa C and Costa Monteiro E 2014 An enhanced electronic topology aimed at improving the phase sensitivity of GMI sensors *Meas. Sci. Technol.*, v. 25, pp. 115010

[8] Fortaleza L G S, Barbosa C, Costa Monteiro E, Silva E and Gusmão L 2015 Detecting non-magnetic metallic foreign bodies by GMR sensors through the use of eddy currents *Proc. XI International Congress on Electrical Metrology (SEMETRO)*, pp.1-4

[9] Fortaleza L G S, Costa Monteiro E, Barbosa C, Silva E C, Gusmão L A P 2018 Biomedical comparison of magneto-meters for non-ferromagnetic metallic foreign body detection *Journal of Physics: Conf. Series*, pp.012013

[10] Benavides L S, Silva E C and Costa Monteiro E 2018 Pressure transducer based on the phase characteristics of GMI effect for measuring the arterial pulse wave *Journal of Physics: Conf. Series*, v. 1065, pp. 072039

[11] Oliveira B, Louzada D, Costa Monteiro E and Barbosa C 2021 Automatic system for locating magnetic foreign bodies using GMI magnetometer *Journal of Physics: Conf. Series*, v. 1826, pp. 012085

[12] Barbosa C 2004 Localization of firearm projectiles in the human body using a superconducting quantum interference device magnetometer: A theoretical study *Review of Scientific Instruments*, vol. 75, no 6, p. 2098–2106

[13] ASTM B258-14 Standard *Specification for Standard Nominal Diameters and Cross-sectional Areas of AWG Sizes of Solid Round Wires Used as Electrical Conductors*, West Conshohocken: ASTM International.

[14] International Commission on Non-Ionizing Radiation Protection 2010 Guidelines for limiting exposure to time-varying electric and magnetic fields (1 Hz – 100 kHz) *Health Physics* 99(6):818-836

[15] Palatnik de Sousa I, Barbosa C and Costa Monteiro E 2018 Safe exposure distances for transcranial magnetic stimulation based on computer simulations. *PeerJ*, v.6, p.e5034

Journal of Physics: Conference Series **2606** (2023) 012003 doi:10.1088/1742-6596/2606/1/012003

Economic distortion as an indicator to measure the impact of legal metrology activities

Bruno Amado Rodrigues Filho, Alexandre Nixon Raulino Soratto, Léssio Lourenço Nunes, Carlos Otávio de Almeida Afonso

Inmetro, Av. Nossa Senhora das Graças 50, Prédio 20, Duque de Caxias, RJ, Brasil

bafilho@inmetro.gov.br

Abstract. The economic distortion represents the monetary unit involved in measurement deviations in trade. The greater are the deviations of measuring instruments used in the market, the greater are the money under uncertainty in the economy. That market asymmetry can be reduced by legal metrology processes, decreasing losses to both seller and buyers, ensuring trust. The Brazilian National Metrology Institute, Inmetro, in 2021, introduced the economic distortion as one of the indicator to evaluate the performance of legal metrology services. The preliminary results were obtained for the fuel market, showing that the economic distortion in 2019 and 2020, represented R$ 87.3 million for the final demand.

1. Introduction

Metrology is part of our everyday life in several aspects, for example: in health for blood pressure measurement and weight; environmental control as gases and particle emissions; commerce, measuring mass, volume, lengthiness. The instruments whose measurements may affect either the society or the economy are usually regulated by legal metrology, which enforces requirements, as maximum permissible errors (MPE) in order to control those impacts. Especially in the commerce, MPE are set to ensure trust and decrease losses as the parties have no means to check the reliability of measurements, considering international standards as references. In legal metrology, the International Organization of Legal Metrology is the organization responsible for issuing recommendations, documents and reports. Brazil is one of the countries that adopt the OIML recommendations as a reference for legal metrology regulation.

Measuring instruments in the commerce, the so called trade metrology, may have a great impact in the economy, since a great amount of the countries production is measured by instruments. For example, £622 billion worth of goods whose value was based on measurement were sold in the United Kingdom in 2002 [1]. Consequently, small deviations may cause great economic impacts due to the volume of measurements considering a country's economy. It is estimate that an increase of 1% of deviation would generate an impact of 5% to the GDP [2]. This monetary unit associate to deviations in measurements is called economic distortion (ED), and can be used as an indicator to measure the impact of instruments used in trade metrology [2–4].

Another important aspect worth mentioning regards the necessity to demonstrate the impacts of legal metrology activities to justify the allocation of resources in legal metrology processes to control measuring instruments in the market, ensuring trust and decreasing losses due to inaccurate

Journal of Physics: Conference Series **2606** (2023) 012003 doi:10.1088/1742-6596/2606/1/012003

measurements [4]. In that regard, Inmetro, the National Institute of Metroloyy (NIM) in Brazil, set, in 2021, the development of the ED as one of the indicator to measure the performance of legal metrology activities [5]. And, once the indicator is fully implemented, the NMI expects to evaluate the performance of legal metrology activities, as well as how improvements within its processes reflect on the economy, and consequently the society.

Finally, this study aims to present the preliminary results of the economic distortion as an indicator showing the impact of legal metrology activities in Brazil. The fuel sector is used as an example once it has an important impact to the Brazilian economy, as well as the availability of the information necessary to computation of the ED.

2. Methodology

The economic distortion represents the amount of monetary unit (money) involved in measurement deviation from instruments used in trade metrology. Equation 1 shows the proposed formulation for the ED [6]:

$$ED = VxIMI.e$$

Where V represents the amount of monetary unit traded, IMI the ratio between rejected measuring instruments over total, and e the mean error of the ration between measurement error over goods traded for rejected measuring instruments, hereinafter called error for simplification. As we are computing the ED as a macroeconomics indicator, we used mean values for both V and e. In summary, ED formulation evaluates the mean value of money associate to deviations of inaccurate measuring instruments.

It is worth noting that the present study compute only the errors outside the legal's MPE, in order to compute e. Once Inmetro is using the ED, there is a need that the indicator shall be influenced according to the legal metrology activities conducted by the NMI. As the measuring instruments working according to the MPE cannot be withdrawn from the market, the money involved in error within MPE was not considered.

Brazilian regulation for fuel dispensers establishes MPE as - 0.5% and + 0.3%. The MPE test measures the error for a 20 l delivery, using a 20 l standard test measure for both maximum and minimum flowrate of tested dispensers. Figure 1 summarizes MPE according to Brazil regulation.

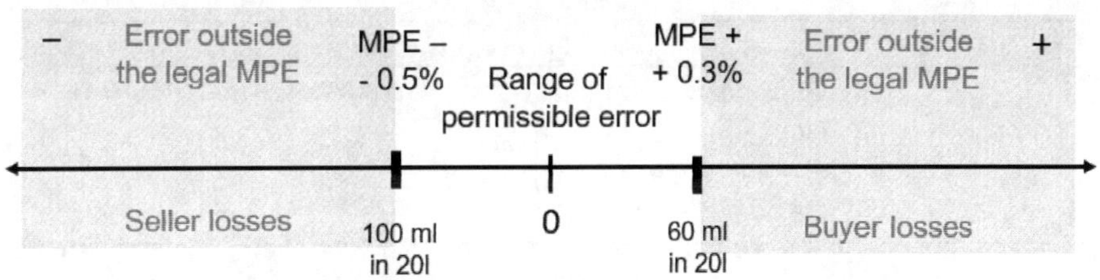

Figure 1. MPE according to Brazilian regulation for fuel dispensers.

Finally, we compute the ED in two different steps, where the first represents the ED considering only the errors that exceed the positive MPE, representing losses to the buyer and the second step considering the errors that exceed the negative MPE, losses to seller. This approach allowed us to identify the losses to each part involved in trade. The total ED represent by the sum of both parts.

It is also important to highlight that the ED regards random deviation of measuring instruments. Electronic tampered devices were not contemplated in the present study [7] since the results from those devices are not identified by verification tests.

The information regarding the fuel market is from the Brazilian National Agency for Petroleum, Natural Gas and Biofuels (ANP) [8]. V was obtained from mean values of traded gasoline (liters), ethanol and diesel, multiplied by the mean price of each fuel, respectively, for each year (R$). Both IMI

and e and were obtained from Inmetro's information system used for legal metrology tests (SGI), that has information about verifications of tested dispensers, as well as the number of instruments rejected. We comprise the information for the years of 2019 and 2020 allowing comparisons between two different periods. The ED in this study considers only the final demand trade, not the aggregate value of interconnected sector along the economy. The current year was not considered as information regarding tests and volume of fuel traded is still not consolidated.

3. Results

Tables 1 and 2 give the information about the fuel market in Brazil, comprising the mean price for gasoline, ethanol and diesel, as well as traded values in 2019 and 2020, respectively. Total traded value V is also given as a multiplication of traded volume by mean price, for each fuel.

Table 1. Fuel market information and V, in 2019.

Fuel	Volume (l)	Mean price (R$/l)	V (Billion R$)
Gasoline	66,025,515,626	4.419	291.7
Ethanol	22,544,050,095	3.168	71.4
Diesel	57,298,400,000	3.655	209.4

Table 2. Fuel market information and V, in 2020.

Fuel	Volume (l)	Mean price (R$/l)	V (Billion R$)
Gasoline	61,813,594,528	4.301	265.8
Ethanol	19,257,902,020	3.178	61.2
Diesel	57,472,100,000	3.474	199.6

Tables 3 and 4 show the values for IMI and e, according to the information available from Inmetro's database for instruments used in trade for 2019 and 2020. Both parameters are given for positive and negative measuring errors, so it is possible to compute either losses to buyers and sellers.

Table 3. IMI and e parameters for both negative and positive errors, in 2019.

Negative error		Positive error	
IMI	e (ml/l)x0.001	IMI	e (ml/l)x0.001
0.003	0.006	0.019	0.004
0.003	0.005	0.012	0.006
0.004	0.007	0.017	0.004

Table 4. IMI and e parameters for both negative and positive errors, in 2020.

Negative error		Positive error	
IMI	e (ml/l)x0.001	IMI	e (ml/l)x0.001
0.004	0.004	0.010	0.006
0.004	0.005	0.006	0.010
0.008	0.004	0.008	0.003

The ED is given as the sum of absolute values of buyer losses and seller losses, representing the total amount of monetary unit in measuring deviations. Table 5 shows the values for seller losses, buyer losses and the economic distortion, for 2019 and 2020.

Table 5. Seller and buyer losses, and the ED, in 2019 and 2020.

Year	Seller losses	Buyer losses	ED (Million R$)
2019	R$ 11.8	R$ 39.8	R$ 51.6
2020	R$ 11.6	R$ 24.0	R$ 35.7

Table 5 shows that buyer losses are bigger when compared to sellers losses in 2019 and 2020. It is also possible to notice a decrease in the economic distortion from in 2020, comparing to 2019. Tables1 and 2 illustrate that this decrease is due to the decrease of traded value of fuel in 2020 when comparing to 2019. We hypothesize that the decrease of traded volume V reflects the economy period due to the pandemic that initiated in 2020.

According to table 5, a total of R$ 87.3 million is the amount of monetary unit under uncertainty from random errors in the fuel market in Brazil, in 2019 and 2020. Since the legal metrology processes such as verification tests, surveillance and after repair verification, inaccurate measuring instruments causing market asymmetry were withdrawn from the market, ensuring trust, for both seller and buyer parties.

4. Conclusion

The economic distortion represents the amount of monetary unit, money, under uncertainty in trade. Consequently, the bigger are the economic distortion, the bigger are the losses to the economy, and consequently to the society. Legal metrology processes is a tool used to prevent losses due to verification tests, surveillance, after repair tests and others. Improvement in the legal metrological control decreases deviations in the market consequently decreasing economic distortion, improving the economy.

Since the economy distortion can be associated to legal metrology processes, Inmetro, in 2021, set it as one of the indicator to measure the benefits of legal metrology to the economy and society. The presented formulation computes the economic distortion as a function of traded value of goods, measuring errors and the ratio of inaccurate devices to the total.

The results show that in 2019, an economic distortion of R$ 51.6 million. A significant decrease to R$ 35.7 million in 2020 was observed. Even though the measuring error is a component of the economic distortion, the observed decrease can be also associate to traded values differences between 2019 and 2020. It represents that economic aspects as mean price of products and amount traded can also have an important influence over the economic distortion.

Finally, it is worth noting the importance of data availability for studies involving metrology macroeconomic indicator, as the economic distortion. Inmetro has been consolidating a database regarding measuring instruments used in trade since 2011, which allowed us to collect information for the present study.

Future steps regard the study of the economic distortion applied to different sector of the economy, comprising key products to the Brazilian economy.

References

[1] Sanders R. Why do we regulate measuring instruments used fortrade. OIML Bulletin 2011; LII: 13–15.

[2] Birch J. Benefit of legal metrology for the economy and society – A study for the International Committee of Legal Metrology.

[3] Rodrigues Filho BA, Gonçalves RF. Legal metrology, the economy and society: A systematic literature review. Measurement 2015; 69: 155–163.

[4] Stiefel W. Management Assistance for Weights and Measures Progress, Measuring Inaccuracy's Economic Distortion. 1973.

[5] INMETRO. Planejamento Estratégico do INMETRO 2021-2023, https://www.gov.br/pt-br/noticias/financas-impostos-e-gestao-publica/2021/03/conheca-o-plano-estrategico-do-instituto-para-202120132023 (2021).

[6] Dias RF, Caldas LAC, Costa RRF. A distorção econômica evitada pelas ações do Inmetro: o caso do Rio Grande do Sul. Rio de Janeiro, 2010.

[7] Leitão FO, Vasconcellos MT, Brandão PCR. Contramedidas de hardware y softwa-re sobre el fraude de alta tecnología al Surtidor de combustible bajo el alcance de Metro-logía Legal. In: 9th International Symposium. In: 9th International Symposium "Metrologia 2014" Proceedings. Havana/Cuba, 2014.

[8] ANP. Síntese Mensal de Comercialização de Combustíveis, https://www.gov.br/anp/pt-br/centrais-de-conteudo/publicacoes/sinteses/sintese-de-comercializacao-de-combustiveis.

Technical Feasibility Study for Development of Reference Material for Nitrate in Meat Product

M C de Oliveira[1], M D de Almeida[2], A C P Osorio[2], R C de Sena[2], V Souza[3]

[1]Professional Master's Course in Metrology and Quality, Inmetro, Duque de Caxias, RJ, Brazil 25250-020; [2]Inorganic Analysis Laboratory, Chemical Metrology Division, Inmetro, Duque de Caxias, RJ, Brazil 25250-020; [3]Capacity Centre, Inmetro, Duque de Caxias, RJ, Brazil 25250-020

E-mail: marianacardoso41@hotmail.com

Abstract. Nitrate measurement is a fundamental part of quality control in the meat products industries. Ensuring the technical validity of the results of these measurements can be achieved through the adoption of appropriate metrological tools such as reference materials. The technical feasibility for producing a nitrate reference material in a meat product was evaluated by preparing a batch of material on a small scale. The adequacy of the preparation procedure was evaluated by means of pertinent physicochemical and microbiological measurements. A batch of 175 units containing approximately 5 g of lyophilized cooked ham in each package was produced. The batch was submitted to preliminary studies to assess homogeneity and stability under transport conditions, in accordance with ISO 17034 and the applicable ISO guides.

1. Introduction

Nitrite and nitrate salts are widely used in the meat products industries as additives due to the various technological advantages [1]. However, indiscriminate ingestion of these salts can carry serious risks to human health [2]. Due to the possible nitrate and nitrite adverse health effects, several countries establish legal limits for the use of these salts in food, including meats and their derivatives. In Brazil, the legislation for the addition of food additives in meats and industrialized meat products was revised in 2019 and passed to be officially regulated by Resolution of the Collegiate Director (RDC) n° 272 of the National Sanitary Vigilance Agency (ANVISA) [3].

The measurement of nitrite and nitrate is a mandatory part of quality control of meat products and the results obtained must be reliable and technically valid. The importance of guaranteeing the quality of measurements of nitrite and nitrate in food had already been highlighted within the worldwide scope of metrology by the Consultative Committee for Amount of Substance: Metrology in Chemistry and Biology (CCQM) [4]. For this, it is necessary to adopt appropriate metrological tools such as reference materials (RMs) and certified reference materials (CRMs). Despite worldwide concern, RMs for nitrite or nitrate are still difficult to obtain, mainly because of the difficulty in preserving the values of the RM properties of interest [5].

The aim of this work was to evaluate the technical feasibility for the development of a reference material (RM) for nitrate in lyophilized cooked ham. This study was conducted in accordance with the requirements of the ISO 17034 [6] and following ISO guidelines applicable to the production process and certification of reference materials.

2. Experimental

2.1. Material processing for feasibility study

Twenty samples of commercially available low-fat cooked ham, each one containing 200 g, were grinded and homogenized together in a mini cutter (Dal Pont Otimização Industrial, Brazil), totalizing in 4,000 g of starting material. The batch was prepared using fresh cooked ham, without any pre-treatment or fortification. The material obtained was lyophilized for 72 hours (model LH 0400/2L, Terroni Equipamentos Científicos, Brazil). The lyophilized cooked ham was manually homogenized and filled in aluminized packages. A total of 175 units were produced, containing approximately 5 g of lyophilized cooked ham in each package. The irradiation was performed using a 7 kGy radiation dose at the Technology Center of the Brazilian Army (CTEx, Guaratiba, Rio de Janeiro).

2.2. Measurements to control batch processing

Physical-chemical and microbiological measurements were performed in order to assess the influence of the techniques adopted for batch preparation on the properties of the material obtained.

Measurements of moisture content [7] and water activity [8] in the lyophilized cooked ham were performed to evaluate the irradiation efficiency. The counts of lactic acid bacteria and mesophilic aerobic and facultative anaerobic bacteria [9] were performed in irradiated and non-irradiated units to evaluate the efficiency of the irradiation step.

2.3. Preliminary study to assess homogeneity

The preliminary assessment of homogeneity was performed by analyzing nitrate content in 9 batch units. A stratified random sampling strategy was used to select the units.

The measurement of the nitrate content in each unit was performed in duplicate by IC-UV. Moisture content was also measured by coulometric Karl Fischer (KF) titration. Nitrate IC-UV measurements were corrected considering the respective dilution factor used in the preparation of extracts for chromatographic analysis, as well as the water content of each selected unit. The results were available by ANOVA single factor.

2.4. Testing of the effect of irradiation

For this study, five units from the batch that were not irradiated were selected. The measurement of the nitrate mass fraction was performed in duplicate by IC-UV, under repeatability condition. Moisture content was also measured by coulometric Karl Fischer (KF) titration.

The results of the analysis of the non-irradiated units were compared with the results of the analysis of the irradiated units in the preliminary assessment of homogeneity, considering the associated combined standard uncertainty values.

2.5. Preliminary study to assess stability under transport conditions

Stability of the batch under transport conditions was assessed for a period of 15 days using an isochronous design [10]. The values of temperatures evaluated were (21 ± 1) °C and (50 ± 2) °C. Two subsets of 10 units were selected for each of the temperatures studied. A stratified random sampling strategy was used to select the units. Three other units were kept at the reference temperature (3 ± 6 °C).

The measurement of the nitrate content in each selected unit was performed in duplicate by IC-UV under repeatability conditions. Moisture content was also measured by coulometric Karl Fischer (KF) titration. Thus, the assessment of stability was then assigned on a dry mass basis.

3. Results

3.1. Moisture content and water activity measurements

After the freeze-drying step was completed, the moisture content of the lyophilized cooked ham was then measured. The average value obtained was (3.8 ± 0.1) %, which was considered satisfactory.

The measurement of the water activity of the lyophilized cooked ham was also performed and the mean value obtained was 0.129, which is low enough to make the material unsuitable for the growth of microorganisms [11].

3.2. Microbiological measurements

Microbiological measurements were performed to evaluate the efficiency of the irradiation step in reducing the microbiological load of the material produced. The dose of 7 kGy was efficient in reducing the microbiological load of the lyophilized ham, as shown in Table 1.

Table 1. Microbiological measurements of lyophilized material before and after irradiation.

Sample	Replica	Lactic acid bacteria (UFC g^{-1})	Aerobic and facultative anaerobic mesophilic bacteria (UFC g^{-1})
Non-irradiated units	1	3.1×10^5	4.8×10^4
	2	4.0×10^5	3.5×10^4
Irradiated units	1	<1.0 × 10	<1.0 × 10
	2	<1.0 × 10	<1.0 × 10
	3	<1.0 × 10	<1.0 × 10

3.3. Preliminary study to assess homogeneity

The measurement results obtained were used to assess the homogeneity between the units using Analysis of Variance (ANOVA) single factor as the statistical tool for calculating the standard uncertainty associated with heterogeneity (u_{hom}) with equation (1), where: MS_{among} represents mean square between groups (*i.e.* mean square between the results of units); MS_{whitin} represents mean square within groups (*i.e.* mean square for replicate results of each analyzed unit); and *n* represents the number of observations per unit.

$$u_{hom} = \sqrt{\frac{MS_{among} - MS_{within}}{n}} \quad (1)$$

Moisture content was under control and would not compromise the other properties of the material produced. It was verified that the batch of material produced presented an acceptable degree of heterogeneity of 1.1 %, considering the target measurement uncertainty of 10 % for the intended use.

3.4. Testing of the effect of irradiation

A comparison of values of nitrate mass fraction with its uncertainty was performed through equation (2), where: x_{IU} is the mean measured value for irradiated units; x_{INU} is the mean measured value for non-irradiated units; u_{IU} is the standard uncertainty of measurement results for irradiated units; u_{INU} is the standard uncertainty of measurement results for non-irradiated units; and *k* is coverage factor. This equation compares the difference between the values and their uncertainty [12].

$$|x_{IU} - x_{INU}| \leq k\sqrt{(u_{IU})^2 + (u_{INU})^2} \quad (2)$$

The difference between the measurement results was not significant, at a confidence level of 95 % ($k = 2$), as shown in Table 2. Thus, the irradiation with a dose of 7 kGy did not change the value of the nitrate mass fraction.

Table 2. Results obtained to evaluate the effect of irradiation on the nitrate mass fraction.

	Irradiated units		Non-irradiated units
$w \pm u_c$ (mg kg^{-1})	1,197 ± 29		1,140 ± 21
$\lvert x_{IU} - x_{INU} \rvert$ (mg kg^{-1})		57	
$k\sqrt{(u_{IU})^2 + (u_{INU})^2}$ (mg kg^{-1})		71	

3.5. Preliminary study to assess stability under transport conditions

The presence of a trend in the measurement results of the nitrate mass fraction over the study time was verified by means of linear regression analysis. The t-test was applied to verify the significance of the slope of the regression line and the t-statistic (t_{b_1}) was calculated using the equation (3), where b_1 is the angular coefficient obtained from linear regression and $s(b_1)$ is the standard error of the slope.

$$t_{b_1} = b_1 / s(b_1) \tag{3}$$

A regression analysis of the plotted data was also performed, indicating slope (b_1) close to zero at the temperature of (21 ± 1) °C. The standard uncertainty associated with stability under transport conditions (u_{trn}) was calculated by equation (4), considering the value of the standard deviation associated with the slope of the straight line obtained in the linear regression ($s(b_1)$) and the duration of the study (t = 15 days).

$$u_{trn} = s(b_1) \times t \tag{4}$$

4. Conclusion

The preparation of the batch of material for feasibility study using the proposed techniques was adequate. The freeze-drying allowed to obtain a material with low moisture content and the irradiation promoted the satisfactory sterilization. The batch of material was sufficiently homogeneous for the intended use and the value of the contribution of heterogeneity was 1.1 %. Thus, long term stability and characterization studies can be performed to assess the feasibility of producing RM.

Acknowledgments

The authors warmly acknowledge the financial support from PRONAMETRO (Inmetro, Rio de Janeiro, Brazil) and the collaboration of the following institutions: SENAI Institute of Food and Beverage Technology (SENAI-SC, Chapecó, Brazil) and Technology Center of the Brazilian Army (CTEx, Guaratiba, Rio de Janeiro).

References

[1] Binkerd E F, Kolari O E 1975 *Food and cosmetics toxicology* **13** 655–61
[2] Merino L, Ornemark U and Toldrá F 2017 *Advances in Food and Nutrition Research* **81** 65–107
[3] Brasil. Agência Nacional de Vigilância Sanitária. Resolução - RDC nº 272, de 14 de março de 2019. Estabelece os aditivos alimentares autorizados para uso em carnes e produtos cárneos. *Diário Oficial da União*, Brasília, DF, 18 de março de 2019, Seção 1, p 194
[4] Máriássy M *et al*. 2008 *Metrologia* **45** 08015–08015.
[5] Siu DC, Henshall A 1998 *J. Chromatogr. A.* **804** 157–60
[6] ISO 17034:2017 *General requirements for the competence of reference material producers*
[7] ISO 1442:1997 *Meat and meat products — Determination of moisture content (Reference method)*
[8] ISO 18787:2017 *Foodstuffs — Determination of water activity*
[9] AOAC 990.12: 1994 *Aerobic Plate Count in Foods*

[10] ISO Guide 35: *Reference materials - General and statistical principles for certification*
[11] Stewart G F and Rockland L B 1981 *Water Activity: Influences on Food Quality* (Boston: Springer) Chapter 2 pp 1-61
[12] Linsinger, T P J 2010 Application Note 1 - Comparison of a measurement result with the certified value European Commission - Joint Research Centre Institute for Reference Materials and Measurements (IRMM)

A method to extend the calibration beyond the defined periodicity

S A Miqueleti[1] and R P Costa-Félix[1]

[1]Laboratory of Ultrasound (Labus), Directory of Scientific and Industrial Metrology (Dimci), National Institute of Metrology, Quality and Technology (Inmetro), Av. Nossa Sra. das Graças 50, Duque de Caxias, RJ, Brazil, ZIP 25250-02.

e-mail: samiqueleti@inmetro.gov.br; rpfelix@inmetro.gov.br

Abstract. The assignment of calibration intervals shall be a formal process based on the results of previous calibrations. With the coronavirus pandemic, one means to avoid a non-conformity instrument and ensure an updated calibration program is to extend the calibration interval of laboratories equipment. The ILAC-G24 has published guidelines for that purpose. What can be done if an extension on the calibration validity is further than previously defined? This article presents a way to increase instrument uncertainty and extend the calibration period based on ISO Guide 35, applying a trend analysis and stability criteria of the reference material properties to the case of a measurement system.

1. Introduction

According to the ISO/IEC-17025[1], a calibration laboratory must be able to produce reliable measurement results traceable to the International System of Units (SI). Measuring equipment shall be calibrated when the measurement accuracy or uncertainty affects the validity of the reported results, or the calibration is required to establish the metrological traceability. The laboratory shall establish a calibration program, which shall be reviewed and adjusted as necessary to maintain confidence in calibration status.

The calibration frequency needs to be determined considering, but not just restricted to:

- Save costs by calibrating less frequently but ensuring reliability
- Satisfy the requirements of the customer
- Compliance with the requirements of standards

The International Laboratory Accreditation Cooperation (ILAC) published a document to guidance on how to determine calibration intervals (ILAC-G24) [2]. In this document, the so-called "engineering intuition" which fixes the initial calibration intervals and a system that maintains fixed intervals without review, are not considered as being sufficiently reliable and are therefore not recommended. The decision about the initial calibration interval does not depend solely on you as the user. These could be based on some criteria:

- Manufacturer recommendation
- Frequency of use – the more it is used, the shorter the calibration interval
- Experience of the user with the same type of instrument
- More critical instruments to produce reliable results needs shorter calibration interval

It is recommended to use the ILAC-G24 guide to establish the calibration program. This article provides a method for assigning a new calibration and uncertainty value if the given timeframe needs to be extrapolated.

2. Methodology

The methodology proposed here follows the recommendations of clause 8.7 of ISO GUIDE 35 [3] for assigning a reference value and uncertainty using stability studies to a reference material (RM).

The value to be assigned to the measurand and their uncertainty is one process of combining the results from the trend and stability of previous calibrations. The mathematical model adopted makes it possible to estimate changes for extended periods at the cost of depreciation of measurement uncertainty.

2.1. Testing for statistically significant change

This analysis consists of weighted least squares regression ($Y = ax + b$) of the experimental results over time, and the study of their behavior, where Y is the calibrated value and x is the time that occurs the calibration. The use of statistical software can do the fitting (e.g., OriginLab®) that computes the standard error of parameters $u(a)$ and $u(b)$, respectively. The relationship between the standard error and the standard deviation is such that, for given sample size, the standard error equals the standard deviation divided by the square root of the sample size.

Assuming the validity of the basic linear regression model applies to the temporal series data (R-square about 1), the usual test for a statistically significant gradient is a t-test for slope (a) significantly different from zero. It is conducted by calculating the t statistic in the formula

$$t_a = \frac{|a|}{u(a)} \quad (1)$$

And comparing this with the two-tailed critical value of Student's t for n-2 degrees of freedom at the 95 % level of confidence. If the calculated test statistic t_a exceeds the critical value, the slope is considered to be significantly different from zero at the 95 % level of confidence and indicates a temporal trend.

2.2. Confidence interval for the regression line

The fitted values (a) and (b) may be used to provide an estimate of the calibration value $\hat{Y}$ for a particular time value $\hat{x}$ in the future, using

$$\hat{Y} = a\hat{x} + b \quad (2)$$

Two sources of uncertainty are joined to determine the combined uncertainty, stability $u(a)$ and weighted uncertainty $u(w)$

$$u(c) = \sqrt{u(a)^2 + u(w)^2} \quad (3)$$

The weighted uncertainty is assessed as [4]:

$$u(w) = \left[\sum_{i=1}^{t} \frac{1}{u(x_i)^2}\right]^{-\frac{1}{2}} \quad (4)$$

In which $u(x_i)$ is the combined uncertainty for the measurand at the year i for a total of n measurements.

To obtain a final confidence interval for the predicted value $\hat{Y}$ is used a two-sided confidence interval at the 95 % level of confidence by ISO GUIDE 35 formula (B.21)

$$\hat{y} \pm t_{95,n-2} \cdot u(c) \cdot \sqrt{\frac{1}{n} + \frac{(\hat{x}-\bar{x})^2}{\sum_{i=1}^{n}(x_i-\bar{x})^2}} \quad (5)$$

Where $t_{95,n-2}$ is the two-tailed critical value for Student's t with $n-2$ degrees of freedom, $u(c)$ the combined uncertainty, $\hat{x}$ the extended year, $\bar{x}$ the mean of years of observation. Confidence intervals for

Journal of Physics: Conference Series **2606** (2023) 012005 doi:10.1088/1742-6596/2606/1/012005

different confidence levels may be similarly obtained by using the appropriate critical value for Student's *t*.

3. Measuring system

The measurand used as an example is frequency. One system used to determine the central frequency of a piezoelectric crystal, figure 1, consists of acquiring the response with an oscilloscope, transferring the data, performing an FFT analysis, and indicating the peak frequency in the software. Calibration of the system is performed with the arrangement below:

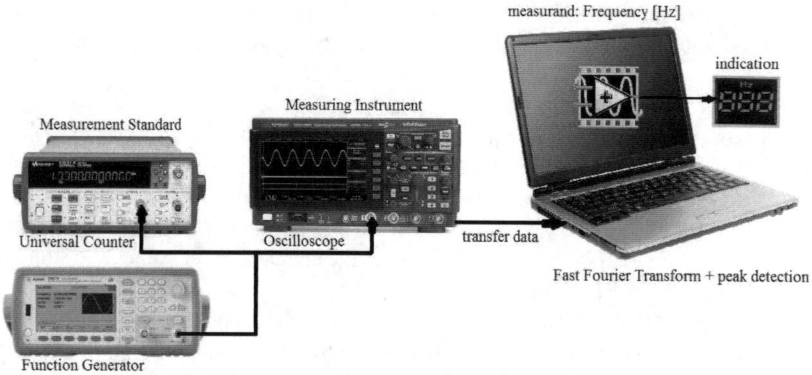

Figure 1. Measurement system to frequency calibration.

The standard instrument is a universal counter Agilent 53132A (Agilent, Malaysia) calibrated at a National Metrology Laboratory (NMI). Measurements were performed using a measurement system consisting of an oscilloscope model DSO-X 2014A (Keysight, Malaysia) and a function generator model 33250A (Agilent, Malaysia). The data generated for each acquisition were saved on a computer, where the frequency was assessed using software developed in LabView™ (National Instruments, USA).

4. Results

The calibration performs in a wide frequency range, and the uncertainty is evaluated according to the (JCGM, 2008) [5]. The results are expressed through a calibration curve fitted with a weighted least squares regression ($Y = ax + b$), see figure 2. The error bar is too small to be represented.

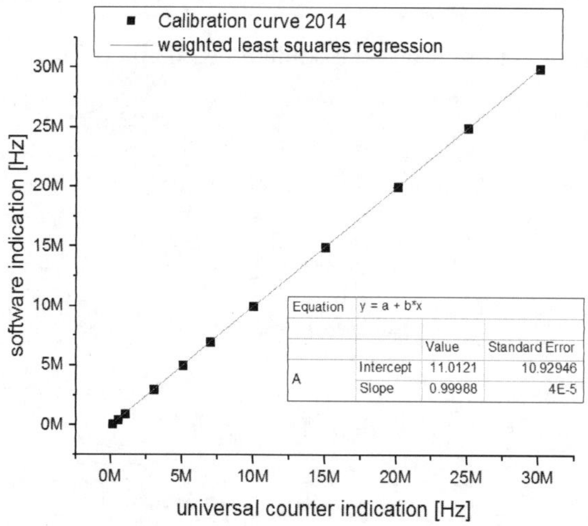

Figure 2. Calibration curve.

The calibrations curves over the years are shown in history table 1

Table 1. Curve of calibration history

Year	Xi	slope (a)	intercept (b)	u(a)	u(b)
2012	0	1.00045	9.2	0.00004	6.3
2014	2	0.99988	11.0	0.00004	10.9
2016	4	0.99998	-1.3	0.00002	3.9
2018	6	0.99999	-0.2	0.00001	2.8
2022	10	0.99967	-9.43	0.00024	11.2

The method described was applied to the value of slope (a) as a variable to be evaluated, and the process is repeated to intercept (b). The new values predicted in red in Table 1 for the year 2022 have a 95 % confidence level.

5. Discussion

The calibration curve presented in figure 1 is used to correct de indication of the software performing an automatic measurement of the central frequency of a piezoelectric crystal. According to the established calibration program, the universal counter needed recalibration for the year 2020. Due to the coronavirus pandemic, it was not possible to carry out this calibration. Then, the equipment stability analysis was carried out to assign a new valid curve until 2022 with an acceptable level of confidence for the service. The methodology described was applied to the historical data to appoint a new calibration curve.

In the curve calibration example, table 1 year-2022, the new value assigned to the slope has six times the uncertainty expected in a recalibration. This corresponding to a 0.024 % of relative uncertainty, but in this case, remains below the limit established in 0.05 % intended for this type of calibration.

6. Conclusion

The risk of extending a certificate of calibration without analysis is that the measuring instrument can exceed the specified requirements and don't achieve the measurement accuracy or uncertainty to provide a valid result. It can generate a cost of necessary correction in emitted certificates.

If it is intended to extend a calibration cycle based on an audit, you have a failure in your Quality Management System. There are no specific guidelines for determining calibration intervals that are required to follow ISO/IEC 17025. Still, it is better to have published documents like (ISO GUIDE 35) or guides (ILAC G24) as a reference to evaluate calibration intervals or extrapolating calibrations.

A mathematical model was presented to assign a calibration value and uncertainty for extrapolating the period defined in the calibration plan of an instrument or measurement system. The model disclosed the main limitation that it is only applicable when the calibration variation occurs linearly.

References

[1] ISO/IEC 17025:2017 -General requirements for the competence of testing and calibration laboratories

[2] ILAC-G24:2007 / OIML D 10:2007 (E) - Guidelines for the determination of calibration intervals of measuring instruments

[3] ISO Guide 35:2017 - Reference materials - Guidance for characterization and assessment of homogeneity and stability

[4] Douglas RJ, Steele AG. Pair-difference chi-squared statistics for key comparisons. Metrologia 2005;43:89.

[5] JCGM, 2008, Evaluation of Measurement Data – Guide of the Expression of Uncertainty in Measurements JCGM 100:2008.

Acknowledgments

The authors thank the Brazilian National Council for Scientific and Technological Development (grants 312.501/2017-0 [RCF], 430.376/2018-9 [RCF], and 307.562/2020-4 [RCF]) and the Research Support Foundation of the State of Rio de Janeiro (grant E-26/201.563/2014 [RCF] and 260.671/2021 [RCF]).

Journal of Physics: Conference Series **2606** (2023) 012006 doi:10.1088/1742-6596/2606/1/012006

Risk analysis in conformity assessment of positive displacement flowmeter

I T Chacon[1], J T G S B Oliveira[1], R O Ozanan[2] and R P B Costa-Félix[3]

[1] Fluid Measurement Sector – Directory of Legal Metrology - National Institute of Metrology, Quality and Technology, Av. Nossa Senhora das Graças, 50 – Xerém – Duque de Caxias – RJ – Brazil – 25250-020

[2] Electrical Quantities Measurement Sector – Diretory of Legal Metrology - National Institute of Metrology, Quality and Technology, Av. Nossa Senhora das Graças, 50 – Xerém – Duque de Caxias – RJ – Brazil – 25250-020

[3] Laboratory of Ultrasound - Directory of Scientific Metrology and Technology - National Institute of Metrology, Quality and Technology, Av. Nossa Senhora das Graças, 50 – Xerém – Duque de Caxias – RJ – Brazil - 25250-020

itchacon@inmetro.gov.br

Abstract. Volume measurement of manufactured liquids and liquid raw materials carried out by road tankers (RT) is usual and has economic relevance, justifying the need to ensure metrological reliability. It requires some actions, including choosing suitable standards. In this scenario and considering the practice for type evaluation of measuring instruments, the article proposes a case study, analyzing performance data from a positive displacement meter, a measurement principle widely used as a standard in RT verifications. Current criteria of legal metrology were used: for the 0.3 accuracy class, the maximum permissible error (MPE) is 0.2%. The meter was tested with three different fluids (with different viscosities), and measurement errors were contrasted with the MPE specification, being considered compliant in all situations. On the other hand, risk analysis tools for conformity assessment were incorporated (JCGM 106) based on the different consumer risks, the formulated decision rule, and the measurement uncertainty. Even though, when including the risk analysis, the flowmeter was considered non-compliant, showing the need for a better basis decision using measurement uncertainty in legal metrology actions.

1. Introduction

Positive displacement flowmeters have application in the quantification of fluids for industrial purposes in several unit processes. However, in this study, this meter will be addressed in its function as a volume totalizer, working as a reference standard for defining the volume of compartments (tanks) in road tankers (RT).

The transportation of raw materials or manufactured liquid fuels occurs, in Brazil, predominantly through the RT. Due to several isolated or combined factors, such as the extensive road network available, the economic and operational unfeasibility of handling through pipelines, the logistical ease of operation, and the use of the road tanker for purposes measurement in many cases.

Inmetro Ordinance n. 208/2016 [1] establishes the conditions to be met by tanks mounted on-road vehicles intended to transport and measure volumes of bulk liquids. It applies to the RT involved in the activities established in Conmetro Ordinance n. 8/2016 [2], in this specific case, in commercial activity. In the current mode in use, the RT tank performs the function of materialized volume measurement, having its capacity established by a reference device located inside. The RT is submitted to verifications bodies of the Brazilian Network of Legal Metrology and Quality – Inmetro (RBMLQ-I). Once verified and sealed, the RT becomes the reference for liquid volume quantification.

In addition to the technical and legal aspects involved in the verifications, from a metrological point of view, the focus is on the definition of tank volumes. Such application uses materialized volume measures to reference comparison, measurement systems, or flowmeters as volume totalizers. Flowmeters commonly used are the positive displacement type. Despite the indication on the types of references to be used and on the expression of a maximum permissible error (MPE) in the checks, equivalent to 0.25%, there are not many elements about the specification of the meters.

In this context, the motivation to better assess the performance of the meters used as a standard emerged, not limited to the simple analysis of calibration certificates or the result of type evaluations. In this regard, it should be clarified that the type evaluation of positive displacement meters for use as standards in RT checks is not explicitly mandatory either. However, some instruments available in the Brazilian market have already been evaluated for this or other purposes. In this case, the flowmeters have their specifications met, contained in Inmetro Ordinance n. 291/2021 [3].

Thus, this article presents a case study that incorporates a risk analysis procedure in the conformity assessment of positive displacement meters from the perspective of the last regulatory act aforementioned. A prominent aspect of the approach was using measurement uncertainty and data science in legal metrology (type evaluation). The theoretical basis used was documented N. 106 of the Joint Committee for Guides in Metrology (JCGM) – *Evaluation of measurement data – The role of measurement uncertainty in conformity assessment* [4].

Pendril [5] warns of the risk of incorrect acceptance in the inspection of a non-conforming object, resulting in costs for the customer associated with the product out of tolerance. A similar analogy can be made in the approval of a non-conforming flowmeter, particularizing for the present study. The author also analyzes how well the recent main guides and other publications on the role of measurement uncertainty in conformity assessment deal with steps considered essential for conformity assessment: definition of requirements to be met, the description of methods, associated uncertainties, decision criteria, and analysis of results.

The subject has deserved the attention of several authors. Thus, Desimoni and Brunetti [6] highlight the need to create or adopt unified guidelines for decision-making involving risk analysis and conformity assessment. Tanaka and Tanaka [7] show an approach to calculating the risks involved in the in-service verification/inspection of automatic check weighers. The study of the latest authors made it possible to reduce weighing, giving one more application to the theme.

1.1. Displacement positive flowmeter and type evaluation

The mechanical operating principle capable of providing the flow and volume totalized by the positive displacement type meters can be understood by the explanation presented by Lipták [8].

> *"Positive displacement (PD) meters trap a known fixed volume of fluid and transfer it from the inlet to the outlet side of the meter. The number of "packages" of fluid is a measure of the flow. Designs include rotary piston, oval gear, sliding vane, and reciprocating piston. Liquid PD meters offer good accuracy and flow range and are particularly suited to high viscosity products."*
> (Flow Measurement, Béla G. Lipták - 1993)

In the evaluation procedure, functional tests are carried out to evidence the performance in terms of technical and metrological aspects. Flow conditions are also part of the specification criteria. An accuracy class must be indicated for each instrument to identify which MPE must be met. Flowmeter compliance is achieved by meeting all requirements simultaneously and declaring certain process conditions and fluid rheology, such as temperature, pressure, and viscosity ranges for which the meter is considered approved.

1.2. Risk analysis in the type evaluation context

Legal metrology activities, by definition, include the conformity assessment of regulated products and activities. The International Vocabulary of Legal Metrology Terms (Ordinance Inmetro 150/2016) [9] can be used to substantiate this statement. The evaluation of the measurement instrument type is one of the legal metrology activities. It includes a conformity assessment action when checking compliance with the requirements established in the regulatory acts. The conditions for State intervention through a process of regulation of a measuring instrument or measurement method are presented by Ozanan et al [10]. For flowmeters of liquids other than water, the regulatory action in use in Brazil is Ordinance Inmetro n. 291/2021 [3]. In this scenario, risk analysis is relevant to better support the regulator's decision-making, optimizing the allocation of resources, mitigating economic issues and adding quality to regulated instruments. Thus, it is necessary to have well-established decision rules, and a clear definition (or option) of the risk assumed.

The book *Lucro Social: Ferramenta de Transparência e Gestão No Setor Público* [11] points out the relevance of conformity assessment aligned with legal metrology issues. Thus, according to JCGM 106 [4], for greater assurance of confidence in a conformity assessment, a preliminary analysis of the risks involved must be carried out. Therefore, several arguments that substantiate the study should be considered and answer the following question: what should be regarded to provide the most significant credibility necessary to the measurement result?

The risk analysis presented by the authors [11] considers two types of risks, with wording adapted as follows:

I. Consumer (specific) risk - in the particular case, probability of the measuring instrument submitted to the model evaluation being PASSED during the tests, being FAILED,

II. Producer (specific) risk - the probability of the flowmeter being FAILED during the tests, being PASSED.

In both cases, there are risks of losses associated with the parties: losses for the producer in the case of unnecessary rework, compliance costs, project resizing, among others, and in the case of the consumer due to setbacks caused by measuring outside specifications.

Considering p_c as the probability of compliance of the flowmeter, the risk to the consumer ($\overline{p_c}$), corresponds to the approval of a flowmeter that failed and can be understood by the following relationship:

$$\overline{p_c} = 1 - p_c \tag{1}$$

Knowing the normalized t-student statistic (z), the specified limits (T) and the measurement uncertainty (u_c), it is possible to know the minimum value of the measurement error (*VM*) to ensure that at least one measurement with the MUT at 100 ($\overline{p_c}$ = 0.01) or at 1000 ($\overline{p_c}$ = 0.001), for example, will be done out of specification. Following JCGM 106 [4]:

$$VM = \frac{T}{1 - u_c \cdot z} \tag{2}$$

Another relevant piece of information in the analysis is the concept of acceptance band, to incorporate the measurement uncertainty data in the decision rule. Figure 1 presents possibilities for decision-making based on the use of a SAB or large acceptance band (LAB). The opposite is also true for rejection. Figure 1 shows "T_U" and "T_L".

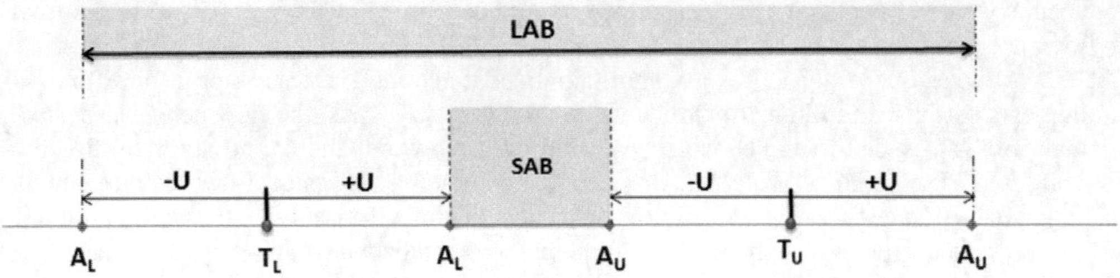

Figure 1. Acceptance Bands. SAB = stringent acceptance band, U = expanded uncertainty, *T* tolerance, and *A* acceptance limit, with the superscript "L" low and "U" Upper. *RV* corresponds to the reference value

Despite the exposition above, it is common practice in legal metrology actions to use only the error data (E) as a criterion for analysis in decision making. In the case of positive displacement meters, this pattern is followed: the decision rule complying with the MPE, according to the accuracy class. That is, we have a situation with the lower (MPE_-) and upper (MPE_+) limits explicit and defined. The following condition is mandatory for the consideration of conformity:

$$MPE_- \leq E \leq MPE_+ \tag{3}$$

However, not knowing the measurement uncertainty can lead to false compliance or false non-compliance, thus contributing to increasing the risk for the producer or the consumer. After the publication of JCGM 106 [4] and considering the need to assist the use of measurement uncertainty in legal metrology activities, OIML edited Guide G19 [12], *The role of measurement uncertainty in conformity assessment decisions in legal metrology*, being this reference base yet another motivation for the development of this work. The document brings shared risk, which is defined as an agreement between the interested parties. The test result is that none of them will have an advantage or disadvantage concerning measurement uncertainty.

The measurement result, the decision rule, and the specified requirement must be known for risk analysis. The specified condition allows the definition of the range (or limits) of tolerance. In this article, the values of MPE_- and MPE_+ establish this interval. However, special attention must be paid to the concepts applied here since the use of parameters does not mean taking *MPE* and tolerance as synonyms, following the consistency of the definition contained in the International Vocabulary of Metrology [13] and related documents [14][15].

2. Methods

This study focused on evaluating a positive displacement flow meter, tested with three different oils (different kinematic viscosities), according to Table 1. The kinematic viscosity was determined from the fluid temperature.

Table 1. Testing Oils.

Identification	Kinematic Viscosity $[mm^2 \cdot s^{-1}]$	Fluid Temperature [°C]
1	10 - 12	26.9 – 27.9
2	60 - 62	21.9 – 23.9
3	199 - 205	27.4 – 28.6

For confidentiality reasons, data from the measuring instrument are not identified, called MUT (meter under test) throughout the text. The regulation defines MPE equal to 0.2% for 0.3 accuracy class. The accuracy class is declared by the manufacturer and confirmed by the metrological performance of the flowmeter.

The flowmeter was evaluated as a volume totalizer function. The testing procedure followed the requirements established in Inmetro Ordinance n. 291/2021 [3]. Thus, the flowmeter was tested at six nominal flow values, distributed between Q_{min} (minimum flowrate) and Q_{max} (maximum flowrate), for each oil used. The totalized volumes were compared with a reference flowmeter to obtain type evaluation data. The tests performed with the MUT resulted in a dataset with additional information to the ones presented. However, only the data of interest to the intended study were transcribed.

Thus, the first analysis was based on model evaluation requirements, which presents one of the criteria for meeting the *MPE* in the metrological performance test.

Afterward, the approach was made according to JCGM 106 [4]. However, as usual in legal metrology activities, the measurement uncertainty was not evaluated. Thus, a typical value was used, identified in later experimental runs with other meters of the same type [16].

3. Results

3.1. MPE as analysis criteria

Table 2 presents the results of the MUT evaluation with each of the three oils.

Table 2. MUT error subjected to performance testing with different fluids.

Test Point	Error (%)		
	Oil 1	Oil 2	Oil 3
1	0.000	-0.040	-0.020
2	-0.050	-0.050	-0.050
3	-0.071	-0.050	-0.030
4	-0.040	-0.020	0.020
5	0.010	0.050	0.020
6	0.071	0.050	0.050

Figure 2 presents the results of the table above compared to the *MPE* limits.

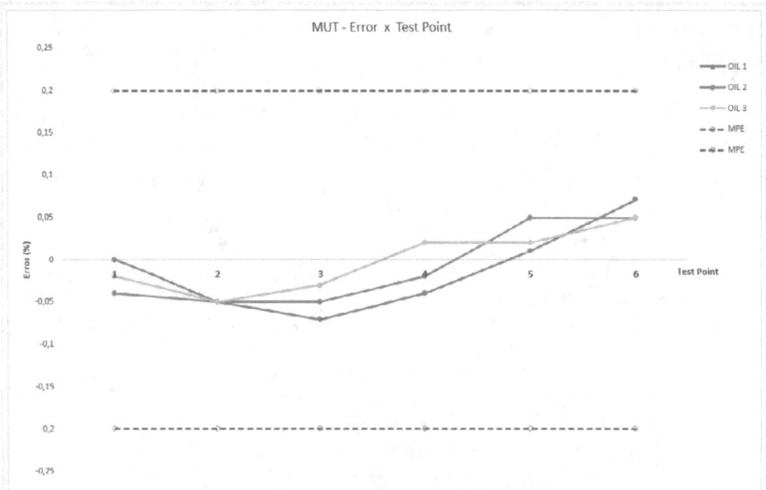

Figure 2. Analysis of the performance of MUT by the criterion of MPE limits.

3.2. Use of JCGM 106

Considering that the measurement uncertainty is not usually evaluated in legal metrology operations, as a way of experimenting with the evaluation form proposed in JCGM 106 [4], a typical uncertainty for this type of meter was used, corresponding to $u_c = 0.07\%$. This data was obtained from subsequent tests, with other meters of the same type, also evaluated based on the same normative references.

From the risks chosen for the consumer ($\overline{p_c}$) and the corresponding z values, from equation (2), having $T = 0.2\%$, $u_c = 0.07\%$, measured error values (*VM*) were established. That means that *VM* is the minimum measured error, ensuring that less than five measurements in 100 ($\overline{p_c} = 0.05 = 5\%$), 1 in 100 ($\overline{p_c} = 0.01 = 1\%$) or 1 in 1000 ($\overline{p_c} = 0.001 = 0.1\%$) performed with the MUT will be out of specification.

Table 3. Maximum measurement error considering consumer risk

$\overline{p_c}$	p	z *(double tailed)*	VM (%)
0.05	0.95	2.0	0.23
0.01	0.99	2.6	0.24
0.001	0.999	3.3	0.26

Another analysis was possible based on the model proposed by Figure 1 and replicated in Figures 3 and 4, considering U_{95} and U_{99}, for *p* equal to 0.95 and 0.99, respectively.

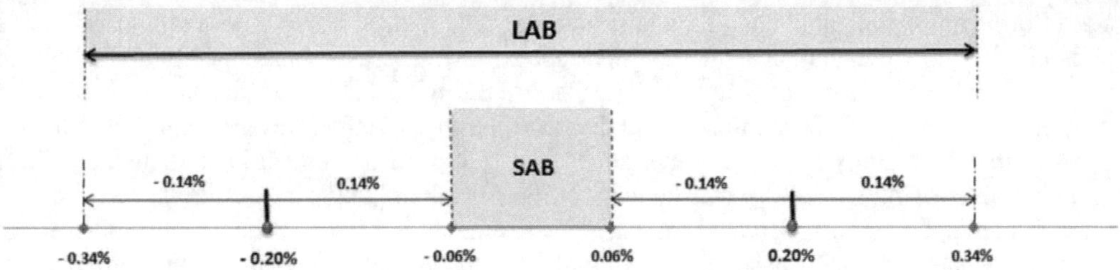

Figure 3. Acceptance band definition for MUT evaluation considering U_{95}.

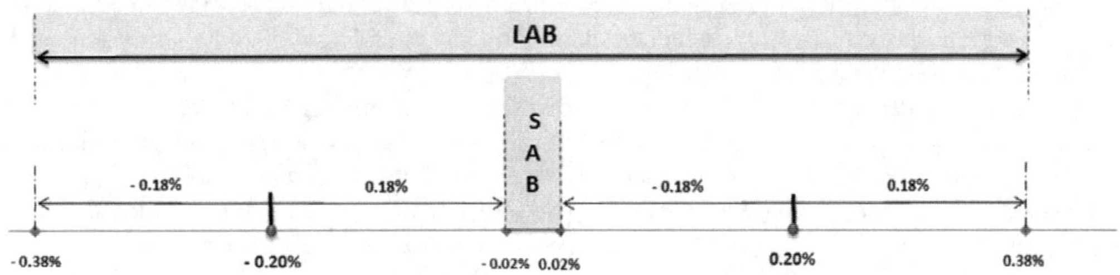

Figure 4. Acceptance band definition for MUT evaluation considering U_{99}.

As a decision rule, for the MUT to be considered compliant, the error presented at each test point, for each fluid used, must be within the range created by SAB for a stringent acceptance or by LAB for a limited approval. The proposed work considers the probabilities of 95% and 99%, obtaining values for expanded uncertainty [17].

4. Discussion

Analyzing the data based on the rule established by equation (3), the error values presented were compared with the established *MPE* (0.2%). Given these criteria, the MUT complies with the type evaluation in all three oils. Figure 2 shows the result of this analysis for the MUT, and it is possible to observe that the error values are within the *MPE* limits.

Observing Figures 3 and 4 and considering as a decision rule all measured error values are within the SAB, the data in Table 2 indicate that:

- assuming U_{95}: the MUT would be approved only for fluids 2 and 3;
- considering U_{99}: the meter would not be approved, as it presents non-conforming values in all three oils used in the test.

Returning to the discussion about the RT, it is worth remembering that one of the motivating facts for the elaboration of this article was the evaluation of the adequacy of the positive displacement meter as a standard for volume definition. As seen, the standard, in this case, the MUT, could not be considered as conforming to the specifications. Thus, the pertinence of its use must be evaluated, bearing in mind that in the definition of RT volumes, the uncertainties involved are also not assessed.

About information provided in Table 3, there is an idea for defining the risk to be admitted for the consumer ($\overline{p_c}$). According to the application, impacts of use, definition among those involved, etc., the risk to be assumed must be declared. The numerical information, as made available, helps in this decision-making.

5. Conclusion

According to JCGM 106 [4], "a measurement performed as part of a conformity assessment is designed to obtain information sufficient to enable a decision to be made with an acceptable level of risk". Given this statement, it is possible to understand the importance of using measurement uncertainty in defining a decision rule in legal metrology activities because of the need for knowledge made possible by measurement. As sufficient information, it is understood the measurement result, being the indication and its associated uncertainty. As already mentioned, the lack of measurement uncertainty contributes substantially to increasing the risk for the producer or the consumer [18].

Although the application of regulated measuring instruments is known, it is difficult to assess the consequences of incorrect decisions (risks). Thus, it is proposed that the mathematical modeling to estimate the measurement uncertainty considers some components associated with this risk. Though this article does not address this specific issue, the opportunity for future work is registered.

Even though this proposal was dedicated to studying a single case, as an initial approach to the subject, it was possible to demonstrate the relevance of measurement uncertainty in the decision-making process by legal metrology. Regarding this aspect, more in-depth research is recommended, and it is suggested to concentrate efforts on type evaluation actions, given having more data and resources for analysis than in the verifications.

If legal metrology deals with measurements directly related to commercial relations, health, safety, and the environment, decision-making must be secure. So, there is always an associated risk. From the regulator's point of view, a lack of adequate information about the measurement (measurement uncertainty) can bring negative consequences far beyond financial losses. This field of study is relevant and promising, even enabling the interface between legal, industrial, and scientific metrology.

References

[1] Instituto Nacional de Metrologia, Qualidade e Tecnologia. Ordinance Inmetro n. 208/2016: Technical metrological regulation applied to road tankers. URL: http://www.inmetro.gov.br/legislacao/rtac/pdf/RTAC002405.pdf.

[2] Conselho Nacional de Metrologia, Qualidade e Tecnologia. Ordinance Conmetro n. 8/2016: Guidelines for execution of legal metrology activities in Brazil. URL: http://www.inmetro.gov.br/legislacao/resc/pdf/RESC000258.pdf.

[3] Instituto Nacional de Metrologia, Qualidade e Tecnologia. Ordinance Inmetro n. 291/2021: Technical metrological regulation applied to measurement systems for petroleum, derivates and other fluids. URL: http://www.inmetro.gov.br/legislacao/rtac/pdf/RTAC000826.pdf

[4] Joint Committee for Guides in Metrology. Evaluation of measurement data – The role of measurement uncertainty in conformity assessment. JCGM 106. October 2012. https://www.bipm.org/en/publications/guides

[5] L R Pendrill: Using measurement uncertainty in decision making and conformity assessment – Metrologia 51, S206-S2018 – July 2014 – DOI 10.1088/0026-1394/51/4/S206

[6] Desimoni E and Brunetti B: Uncertainty of measurement and conformity assessment: a review. Anal Bioanal Chem (2011) 400:1729–1741. DOI 10.1007/s00216-011-4776-y

[7] Tanaka Y and Tanaka H: Development of stepwise tolerances for efficient verification of automatic checkweigher - Precision Engineering, Volume 72, November 2021, Pages 568-575

[8] Lipták B G. Flow Measurement – 1a edição. Pennsylvania: CRC Press, 1993. 222 p.

[9] International Organisation of Legal Metrology. International vocabulary of terms in legal metrology terms (VIML). OIML V1, Edition 2013. URL: https://www.oiml.org/en/files/pdf_v/v001-ef13.pdf

[10] Ozanan R O; Ballerini S; Alves H A; Costa-Félix R P B: Social Profit of Legal Metrology in the Brazilian Electricity Sector - J. Phys.: 2021 Conf. Ser. 1826 012028 - DOI:10.1088/1742-6596/1826/1/012028

[11] Ozanan R O; Ballerini S; Costa-Félix R P B. Lucro Social: ferramenta de transparência e gestão no setor público. 2021, 134 p. ISBN 979-8-51-051755-2. ASIN B095XNTXRL

[12] International Organization of Legal Metrology: Guide 19 - The role of measurement uncertainty in conformity assessment decisions in legal metrology, 2017. 72 p. OIML G 19: 2017 E. URL: https://www.oiml.org/en/files/pdf_g/g019-e17.pdf

[13] Joint Committee for Guides in Metrology. International vocabulary of metrology – Basic and general concepts and associated terms (VIM). JCGM 200. 3rd Edition. 2012. URL: https://www.bipm.org/en/publications/guides

[14] Santos, T.Q., Alvarenga, A.V., Oliveira, D.P., Costa-Felix, R.P.B. Metrological Validation of a Measurement Procedure for the Characterization of a Biological Ultrasound Tissue-Mimicking Material. Ultrasound in Medicine and Biology, 2017, 43(1), pp. 323–331.

[15] Baesso, R.M., Oliveira, P.A., Morais, G.C., Alvarenga, A.V., Costa-Felix, R.P.B. Using ultrasonic velocity for monitoring and analysing biodiesel production. Fuel, 2018, 226, pp. 389–399.

[16] Alvarenga, A.V., Silva, C.E.R., Costa-Félix, R.P.B. Monte Carlo uncertainty assessment of ultrasonic beam parameters from immersion transducers used to non-destructive testing. Ultrasonics, 2016, 69, pp. 144–151.

[17] Alvarenga, A.V., Costa-Félix, R.P.B. Uncertainty assessment of effective radiating area and beam non-uniformity ratio of ultrasound transducers determined according to IEC 61689:2007. Metrologia, 2009, 46(3), pp. 367–374

[18] Maggi, L.E., Silva, C.E.R., Alvarenga, A.V., Costa-Felix, R.P.B. Ultrasonic calibration and certification of V1 and V2 type reference standard blocks for use in Non-Destructive Testing. Journal of Physics: Conference Series, 2011, 279(1), 012029

Acknowledgments
The authors thank the Brazilian National Council for Scientific and Technological Development (grants 312.501/2017-0 [RCF], 430.376/2018-9 [RCF], and 307.562/2020-4 [RCF]) and the Research Support Foundation of the State of Rio de Janeiro (grant E-26/201.563/2014 [RCF] and 260.671/2021 [RCF]).

Assessment of glycerol in biodiesel using an ultrasonic measurement method

P A Oliveira[1], R M Baesso[1], A V Alvarenga[1] and R P B Costa-Felix[1]

[1]Laboratory of Ultrasound (Labus), National Institute of Metrology, Quality and Technology (Inmetro), Av. Nossa Sra. das Graças 50, Duque de Caxias, RJ, Brazil, ZIP 25250-020

passuncaooliveira@gmail.com; rpfelix@inmetro.gov.br

Abstract. Biodiesel is a renewable fuel, biodegradable and alternative to diesel oil consisting chemically of fatty acid alkyl esters and obtained by the transesterification reaction. After transesterification, undesirable components may remain in biodiesel. The presence of these components negatively influences the properties and performance of biodiesel. This research aims to evaluate biodiesel quality, using the quantitative ultrasound technique to determine the presence of the contaminant glycerol in biodiesel samples. The blends used in this study consisted of biodiesel and glycerol varying from 0 % to 0.5 % of glycerol by mass. The measurement uncertainty was evaluated according to the Guide to the Expression of Uncertainty in Measurement. The speed of sound in biodiesel/glycerol blends was measured by the pulse-echo technique (@ 5 MHz) and was considered equivalent for the different concentrations studied. The maximum measurement uncertainty obtained was 2.2 $m \cdot s^{-1}$ (0.16 %).

1. Introduction

Biodiesel is a renewable fuel, biodegradable, and an alternative to diesel oil, consisting of fatty acid alkyl esters. The transesterification process consists of the reaction of vegetable oils or animal fats with short-chain alcohol, typically methanol or ethanol, in the presence of an alkaline catalyst to produce FAME and by-product glycerol [1].

The transesterification forms diglycerides (DG) and monodiglycerides (MG) and as reaction intermediates after the reaction. In addition, unreacted triglycerides (TAG), glycerol residues, free fatty acids, residual alcohol, catalyst residues, among other undesirable components, may remain. The presence of those residual components negatively influences biodiesel properties and performance [2]. The presence of glycerol in biodiesel can result in poor engine performance, deterioration of tanks, possible increase in acrolein emissions, among other inconveniences.

Therefore, for biodiesel to be commercialized, the quality control is paramount. Quality and stability are guaranteed through a reference for physicochemical properties, predefined in quality control standards [3][4][5]. According to National Agency for Petroleum, Natural Gas and Biofuels (ANP) [6], the maximum limit of total glycerol is 0.25% (m/m).

Ultrasound has shown the potential in the characterization of pure liquids [7][8][9][10] for being a simple, non-destructive, and low-cost alternative technique. Some works have proven the ability of ultrasound to identify the presence of contaminants in fuels, such as ethanol [7], diesel, kerosene, gasoline [8][9], and biodiesel [10].

However, in the quality control of fuels by ultrasound, there is a gap to be explored. In addition, most studies do not present an assessment of measurement uncertainty according to Guide to the Expression of Uncertainty in Measurement [11]. The measurement uncertainty is a quantitative indication of the quality of the measurement result and ensures that results are traceable, reliable, comparable, and accurate [10][12].

Thus, this work aims to evaluate biodiesel quality, using the quantitative ultrasound technique to determine the presence of the contaminant glycerol in a sample of biodiesel.

2. Materials and methods

2.1. Water and glycerol solution

A preliminary test was carried out on water/glycerol blends at different concentrations to assess the impact on speed of sound. The speed of sound was determined using the frequency of 5 MHz, in water/glycerol blends at concentrations of 0.002 %, 0.01 %, 0.02 % and 0.03 % by mass. Distilled and deionized water was used due to its ultrasonic properties being well defined in the literature. For sample preparation, glycerol (99.5 % purity, Isofar) was used.

2.2. Biodiesel and glycerol blends

The study was carried out using biodiesel obtained from the transesterification reaction of soybean oil. Methanol was used at an alcohol/oil molar ratio of 6:1 and 1.5 % m/m potassium hydroxide (KOH). The reaction was carried out at 70 °C, under mechanical agitation of 500 rpm.

After the transesterification reaction, an acid solution was added to stop the reaction and neutralize the biodiesel, and then the process of separation of the organic and aqueous phases started to occur. Subsequently, the purification was carried out through successive washings with distilled water to eliminate unwanted impurities.

Mixtures of biodiesel and glycerol were prepared with concentrations ranging from 0.1 % to 0.5 % by mass. Measurements were performed at frequency of 5 MHz.

2.3. Measurement system

An arbitrary function generator model 33250A (Agilent Technologies, CA, USA) was used to excite the ultrasonic transducer, and an oscilloscope model DSO-X 3012A (Agilent Technologies, CA, USA) to digitize the ultrasonic signal. Software developed in LabVIEW ™ (National Instruments Corporation, Austin, TX, USA) was used to automate the measurements. Measurements were performed at room temperature (21 °C – 22 °C). A data acquisition unit model 34970A (Agilent Technologies, CA) coupled to a type K thermocouple was used to monitor the sample temperature. The measurement system used is described in detail in [13][14][15], in which the pulse-echo technique is used. Measurements were performed under repeatability conditions, being five repetitions for each concentration in the specific frequency. Each repetition comprises seven readings.

2.4. Measurement uncertainty

Type A uncertainty was calculated in determining the experimental speed of sound. Type B uncertainty also was considered. All measurement uncertainties were calculated according to the Guide to the Expression of Uncertainty in Measurement (GUM) [11].

2.5. Statistical analysis

Statistical test was performed to assess the equivalence of results. The normalized error E_n [16] was used to evaluate the results obtained statistically. The normalized error tests the compatibility of the results. Its evaluation criterion is given by: $|E_n| \leq 1.0$, a situation in which the sample's speeds of sound are metrologically equivalent [16].

3. Results and discussions

3.1. Speed of sound of water/glycerol solution

Propagation rate was determined in water/glycerol samples at concentrations of 0.002 %, 0.01 %, 0.02 %, and 0.03 % by mass. The tests were carried out at a constant temperature of 21.6 °C. However, there was a variation of up to 0.4 °C inherent to the measurement system between the different measurements.

Figure 1 shows the speed of sound repetitions as a function of the concentration at frequency of 5 MHz.

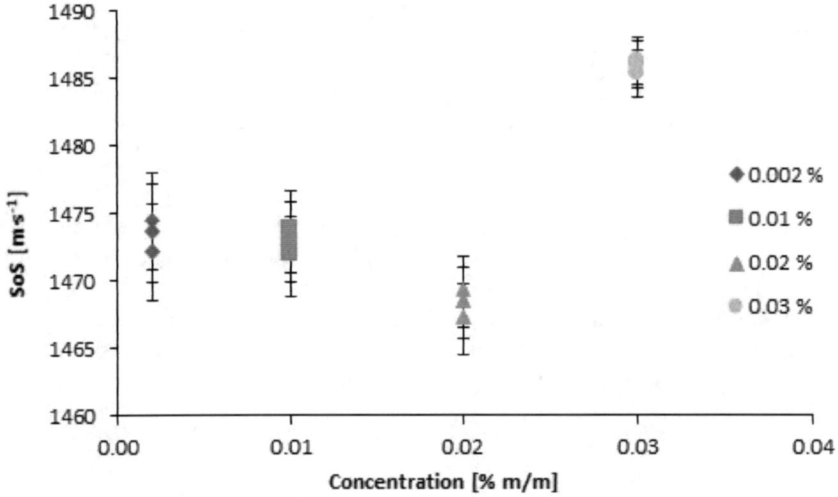

Figure 1. Speed of sound in water/glycerol solutions at different concentrations and their respective uncertainties.

Analyzing Figure 1, it is noted that samples with concentrations of 0.002%, 0.01%, and 0.02% did not present significant variations in the propagation speed. The speed of sound presented equivalent results, being equal to 1473.4 $m \cdot s^{-1} \pm 3.6\ m \cdot s^{-1}$ for sample 0.002%, 1472.8 $m \cdot s^{-1} \pm 3.0\ m \cdot s^{-1}$ for sample 0.01% and 1468.3 $m \cdot s^{-1} \pm 2.7\ m \cdot s^{-1}$ for sample 0.02%. However, it was possible to differentiate the solution with the highest concentration of glycerol (0.03 %), with an average speed of sound equal to 1485.8 $m \cdot s^{-1} \pm 1.8\ m \cdot s^{-1}$. The relative uncertainties correspond to about 0.21%.

Glycerol has three hydrophilic alcoholic groups (-OH). These groups that are considered to be responsible for unlimited solubility of glycerol in water [17]. In the mixture OH-groups form H-bonds with water molecules, making the system miscible and simple to be characterized.

The characterization by ultrasound was able to identify the increase in glycerol concentration in glycerol/water samples in the sample above 0.02% using the speed of sound parameter.

3.2. Speed of sound of biodiesel and glycerol blends

Speed of sound was determined at 5 MHz for biodiesel/glycerol blends. Figure 2 and Table 1 show the variation of the speed of sound as a function of concentration for the studied blends.

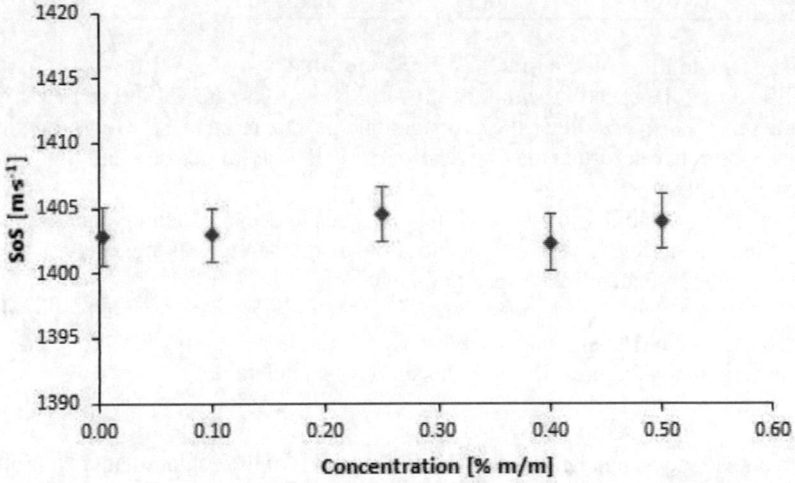

Figure 2. Speed of sound in biodiesel and biodiesel/glycerol blends at different concentrations and their respective uncertainties.

Table 1. Average speed of sound for different concentrations of biodiesel/glycerol blends and their respective uncertainties.

Concentration of glycerol in biodiesel [%]	Speed of sound [m·s^{-1}]	Uncertainty [m·s^{-1}]	Uncertainty [%]
0	1402.8	2.2	0.16
0.10	1403.0	2.0	0.14
0.25	1404.5	2.1	0.15
0.40	1402.4	2.2	0.16
0.50	1404.0	2.1	0.15

As seen in Figure 2 and Table 1, for the concentrations studied, the speed of sound is statistically equivalent. The maximum measurement uncertainty obtained was 2.2 m·s^{-1} (0.16 %). The normalized error (Table 2) was less than 1 for all pairs of samples evaluated.

Table 2. Normalized error for pairs of biodiesel/glycerol blends at different concentrations.

Pairs of blends at different concentrations [%]	E_n	Pairs of blends at different concentrations [%]	E_n
0 – 0.10	0.06	0.10 – 0.40	0.21
0 – 0.25	0.56	0.10 – 0.50	0.34
0 – 0.40	0.14	0.25 – 0.40	0.71
0 – 0.50	0.38	0.25 – 0.50	0.18

0.10 – 0.25	0.53	0.40 – 0.50	0.54

Speed of sound ranged from 1402.4 m·s^{-1} ± 2.2 m·s^{-1} to 1404.5 m·s^{-1} ± 2.1 m·s^{-1}. From the literature, it is well established that the speed of sound of glycerol is approximately 1900 m·s^{-1} [10]. Thus, with glycerol to biodiesel, it is expected that the propagation speed will increase. However, this is not the case. The equivalence between results may be due to low solubility of mixture and low variation in the concentration of glycerol.

Glycerol presents low solubility in biodiesel, hence, tends to decant when inserted into system. This low solubility allied to the low glycerol concentration in the samples is not enough to modify the sample's viscosity, and consequently, the speed of sound.

Baesso et. al. (2019) [10] used a broadband method in the measurement of pure biodiesel (B100) and a mixture of biodiesel with a contaminant (0.1 % of glycerol). The authors also concluded that it was not possible to differentiate biodiesels through the speed of sound values.

4. Conclusions

The quantitative ultrasound technique to determine the presence of the contaminant glycerol in biodiesel samples was studied. The biodiesel/glycerol blends investigated in this research showed a speed of sound equivalent over the concentration range of 0.1 % to 0.5 %. The experimental results indicate that the concentrations are not high enough to be differentiated employing the speed of sound as the parameter.

Therefore, the method presented here can be optimized to increase the sensitivity in the characterization of samples by the speed of sound. Measurement uncertainty is a tool that can be used in measurement optimization. Furthermore, additional studies at other concentrations to determine the detection limit of the current measurement system are relevant.

5. References

[1] Ma F and Hanna M A 1999 Biodiesel production: a review Bioresour. Technol **70** 1-15.

[2] Knothe G and Razon L F 2017 Biodiesel fuels *Prog. Energy Combust.* **58** 36-59. DOI:10.1016/j.pecs.2016.08.001

[3] ASTM American Society for Testing and Materials 2019 ASTM D6751.

[4] ANP 2014 Resolução Nº45.

[5] European Standards 2019 EN 14214.

[6] ANP 2008 Resolução Nº 7 de 19.3.2008

[7] Figueiredo M K-K, Costa-Félix R P B, Maggi L E, Alvarenga A V and Romeiro G A 2012 Biofuel ethanol adulteration detection using an ultrasonic measurement method *Fuel* **91** 209-12. DOI:10.1016/j.fuel.2011.08.003

[8] Leghrib R, Ouacha E, Zouida A, Faiz B and Amghar A 2020 Monitoring automobile fuel adulteration using ultrasound technique for environmental issues *Measurement* **150** 107004. DOI:10.1016/j.measurement.2019.107004

[9] Gupta A K and Sharma R K 2010 A New Method for Estimation of Automobile Fuel Adulteration *Air Pollution.* ed V Villanyi *IntechOpen*, Chapter **16.** DOI: 10.5772/10054

[10] Baêsso R M, Costa-Félix R P B, Miloro P and Zeqiri B 2019 Ultrasonic parameter measurement as a means of assessing the quality of biodiesel production *Fuel* **241** 155-63. DOI: 10.1016/j.fuel.2018.12.032

[11] JCGM 100:2008 Guide to the Expression of Uncertainty in Measurement 2008.

[12] Figueiredo M K-K, Silva C E R, Alvarenga, A V and Costa-Félix R P B 2018 Relating speed of sound and echo amplitude with biodiesel manufacture *Chem. Eng. Res. Des.* **136** 825-33. DOI:10.1016/j.cherd.2018.06.038

[13] Oliveira P A, Silva R M B, Morais G C, Alvarenga, A V and Costa-Félix R P B 2016 Speed of sound as a function of temperature for ultrasonic propagation in soybean oil *J. Phys. Conf. Ser.* **733**, 012040. DOI: 10.1088/1742-6596/733/1/012040

[14] Oliveira P A, Silva R M B, Alvarenga A V and Costa-Félix R P B 2021 Evaluation of measurement uncertainty of speed of sound in soybean oil *J. Phys. Conf. Ser.* **1826**, 012013. DOI: 10.1088/1742-6596/1826/1/012013

[15] Baesso, R.M., Oliveira, P.A., Morais, G.C., Alvarenga, A.V., Costa-Felix, R.P.B. Using ultrasonic velocity for monitoring and analysing biodiesel production. Fuel, 2018, 226, pp. 389–399

[16] Steele A G and Douglas R J 2006 Extending En for measurement science *Metrologia* **43**(4):S235–43.

[17] Egorov G I, Makarov D M and Kolker A M 2013 Volume properties of liquid mixture of water + glycerol over the temperature range from 278.15 to 348.15 K at atmospheric pressure *Thermochimica Acta* **570** 16-26. DOI: 10.1016/j.tca.2013.07.012

Acknowledgments

The authors thank the Brazilian National Council for Scientific and Technological Development (grants 312.501/2017-0 [RCF], 430.376/2018-9 [RCF], and 307.562/2020-4 [RCF]), the Research Support Foundation of the State of Rio de Janeiro (grant E-26/201.563/2014 [RCF] and 260.671/2021 [RCF]), and Pronametro/Inmetro.

Journal of Physics: Conference Series **2606** (2023) 012008 doi:10.1088/1742-6596/2606/1/012008

Metrological perspective on the use of ultrasound physiotherapy equipment in Brazil

J G D Silva, A V Alvarenga and R P B Costa-Félix

National Institute of Metrology, Quality, and Technology, Av. Nossa Senhora das Graças, 50, Xerém. Duque de Caxias - RJ, Brazil

jessycag1@gmail.com; avalvarenga@inmetro.gov.br; rpfelix@inmetro.gov.br

Abstract. Despite several scientific proofs about the success of using therapeutic ultrasound for different therapies, there are also reports of failures and ineffectiveness in applied treatments. Much of this failure is due to the poor working conditions of the ETUS equipment used, which causes the reduction or absence of therapeutic effects. This survey asked professionals about the methods they use and the methods they know and are available on the market to test their equipment through a questionnaire. This study aimed to discuss, for the metrological perspective, the use of ultrasound physiotherapy equipment in Brazil. A problem for the metrological reliability of ETUS is related to the lack of metrological culture among users, who do not adopt procedures to assess their equipment, often because they are unaware of its importance and need. Periodic checks on equipment would give physiotherapists a resource with greater reliability and possibly more efficient therapies.

1. Introduction

The use of Ultrasound Therapy Equipment (ETUS) has become quite common in recent decades to support the treatment of various pathologies, especially those of the musculoskeletal system. It is among the most used by physiotherapy professionals due to its beneficial actions in biological tissues already being well established for different treatments [1].

However, just as there are several studies on the success of therapeutic ultrasound applications for different therapies, there are also reports of failures and ineffectiveness in applied treatments [2]. Much of this failure is due to the poor working conditions of the ETUS equipment used, which causes the reduction or absence of therapeutic effects. Several publications prove that equipment, in routine use, in clinics and physiotherapy departments are outside the current specifications, bringing disadvantages such as ineffective therapy, high-temperature injury, worsening of signs and symptoms of preexisting diseases, high costs and longer treatments, loss of time for the patient and professional and a drop in the credibility of treatments [1][2][3].

Manufacturers emphasize the importance of periodic calibration, which must be performed at least once every six months, annually, or when there is a problem or mechanical impact.

In Brazil, electrical equipment only has mandatory conformity assessment to technical standards to be launched in the market. When you need a record of National Health Surveillance Agency - ANVISA. During the period in which such equipment is in use in physical therapy establishments, there is no mandatory assessment, and consequently, there is no guarantee of metrological reliability.

There is no law or regulation that makes it compulsory to control other biomedical instruments to ensure metrological reliability, with calibrations tracked during their lifetime. The additional costs of tracked calibrations are very evident for healthcare management and administration. At the same time, the benefits such as better diagnostic and therapeutic outcomes for patients are difficult to quantify economically [4].

The objective of this study was to verify the periodicity of maintenance actions of the ultrasound physiotherapy equipment and to analyze the physiotherapist's training regarding general knowledge related to the verification/maintenance procedures used in clinical practice.

2. Materials and methods

A questionnaire was applied to verify the physical therapist's training regarding general knowledge related to the maintenance of the ETUS and to analyze the procedures used in practice.

A questionnaire is an advantageous technique as it allows reaching many respondents and in remote geographic areas. The primary data survey allowed a closer approximation of reality, as information was obtained directly from the target audience. We chose to use an online questionnaire due to its practicality and low application cost.

The purpose of the questionnaire was to know the experience of the physiotherapist in the application of ultrasound therapy, how the equipment is maintained, and whether the professional is informed about its conditions of use. With this, a panorama was defined with the profile of use and maintenance of the ETUS and of the professionals who use it in clinical practice. It was questioned whether the equipment's ultrasound emission is verified, which validation is carried out, which methodology is used, and the frequency of such proof.

The questionnaire provided a heterogeneous and significant sample as needed to extend the inferences to the population of physical therapy professionals who use ultrasound therapy in Brazil. Different locations in the country were reached, professionals with varying groups of ages, time of experience, and different degrees of training.

As a measurement instrument, the questionnaire had its reliability evaluated using the Cronbach's Alpha, which measures the correlation between the answers in a questionnaire by analyzing the profile of the answers given by the respondents as described in [5].

3. Results

3.1. Professional profile

Professionals from 21 states, including the Federal District, responded to the survey. Almost half of the professionals, 49%, who answered the questionnaire were between 26 and 35 years old, 25% were between 36 and 45 years old, 15% were between 18 and 25 years old, and 11% were 46 years old or more. Regarding the degree of academic training, 55% of the professionals who responded to the survey had a *lato sensu* postgraduate degree (specialization, MBA, or other), 21% were graduates or technologists, 14% had a master's degree, 10% were doctors. Regarding the relationship with the equipment, 83% declared to be an operator: the professional who uses the equipment on the patient, and 17% declared to be a supervisor, that is, to be the professional who manages or supervises the team of operators. Regarding the length of experience with the use of ultrasound for therapy, 41% of them responded that they had between 3 and 8 years of experience, 26% answered that they had less than two years of experience, 21% that they had more than 15 years of experience and 12% reported who use ultrasound therapy from 9 to 15 years. The sample comprised 44% by liberal and independent professionals or MEI (individual entrepreneur), 34% by professionals working in private practice, 14% working in teaching and research, and 8% working in public health units. Of the equipment that professionals have access to, 52% said they have access to only 1, while 44% said they have access to 2 to 5 pieces of equipment, and 4% have access to more than five equipment.

3.2. Importance and performance of equipment verification

Regarding the importance or need to perform the verification of ultrasound therapy equipment regarding its ultrasound output, 91% of professionals considered it important to perform this verification periodically, 8% considered it important to perform the verification without a defined frequency, and 1% considered it essential to complete the check only after the equipment has been repaired. No professional deemed it not necessary to carry out this verification.

The verification shows that 49% carry out the confirmation but without a predefined schedule, and 24% carry out the proof according to a predefined schedule. Already 19% of professionals do not perform or never performed this verification, and 8% only perform the confirmation after the equipment is repaired or undergoing maintenance.

Among those who carry out the verification, in most cases, 54%, this verification is carried out by an external service provider. The professional carried out validation in 39% of cases and by a department at the professional's workplace in 7% of cases.

Among those who carry out the verification, 58% of professionals responded that they do not know the method used, 34% use the method based on Cavitation or Mist Method, 7% described calibration, and 1% described other methods.

Among the physiotherapists who carry out the verification of ETUS, 26% have an undefined frequency for this verification, and 17% do not know the frequency. 16% of professionals perform the verification every six months, 12% annually, 10% check the equipment every day, 10% once a month, and 9% once a week.

Although 91% of professionals consider it necessary to periodically check the ETUS and more than 80% of professionals say that they perform the verification of the ultrasound emission by the equipment, whether regularly or not, the method used is not adequate or not known to.

3.3. Verification method

Professionals presented their VIEW õ es regarding them is anyone who uses, and m is all they know and are ã the available e LEVELS market. Noting that most professionals, 58%, responded not knowing the method or not to used any method for evaluation ETUS. Among those who know the method, 38%, the most used is cavitation. Of these, 25% consider they use methods reliable, that they ensure the proper functioning of equipment,19% consider methods they are a expensive and 42% consider methods them all simple and the equipment operator it self can perform.

Of those who know, the views are divided on the reliability, and 17% consider the existence method is all no confident and 18% say there is all confident will levels ensure the proper functioning of the equipment. On the cost of methods existing, the majority opined that believes that them methods is all cheap and is simple can perform.

About the insertion the a new method, 93% considered it to be necessary. Of these, 45% want a method is all simple, inexpensive and confiable. Already will 32% are the concerned only with the reliability of the new method proposed. 17% no the mind the cost and is interested in a method simple and confident. And 6% waiting a method inexpensive and confident, being indifferent to the complexity thereof.

4. Discussion

The data show that when the professionals themselves carry out the verification, the method used in 80% of the cases is cavitation, which does not allow quantification of the transmitted intensity or the characterization of the radiated acoustic field. We also note that when verification is performed by a workplace department or an outside service provider, the method employed is unknown to most professionals in both cases.

Most of the sample, 83%, was composed of equipment operators, professionals who use the equipment on patients, being in direct proximity to the questions raised in this research. The other part of the sample, 17%, was composed of professionals who manage or supervise a team of operators, which is also vital to analyze management behavior. Of the supervisors, 50% work in teaching and research, and among the operators, the majority, 47%, is a self-employed professional or MEI. It is noted that the behavior is similar in some aspects and significantly differs in others: 26.3% of supervisors do not carry out the verification compared to 18.3% of operators. Of those familiar with the method employed, supervisors use calibration and cavitation more, while operators use cavitation more. In both cases, the majority carry out the verification using an external service provider without a defined frequency.

It was possible to notice that 81% of the professionals improved through a postgraduate degree, either Latu Senso, Masters or Doctorate. It shows that professionals seek more important technical qualifications to meet increasingly demanding users. However, 18% of these professionals with a postgraduate degree (Latu Senso, Masters or Doctorate) declared not to carry out the equipment verification. Among those who carry out the equipment verification, 43% are unaware of the method used. Among professionals with a PhD, 81% working in teaching and research all declare to carry out the verification of the equipment, and 50% use calibration as a method.

In comparison, 30% are unaware of the method used. Among professionals with a master's degree, 68.8% perform the equipment verification, but 63.6% are unaware of the method used. Among professionals with a Latu Sensu postgraduate degree, 82.7% carry out the confirmation of the equipment, and 57.4% are unaware of the method used. Among graduates or technologists, 73.9% check the equipment, but 70% do not know the method used.

Answers were obtained from professionals who work in different segments, making it possible to analyze other realities. Among self-employed and self-employed professionals or MEI, 16.3% declared that they do not perform or never performed their verification. The method used is unknown by the majority who carry out the proof, 55.8%, and 41.9% use cavitation. Among the professionals who work in private clinics, 21.1% stated that they do not perform or never performed the verification of their equipment. Among the professionals who work in public health units, 22.2% said that they do not or never carry out the confirmation of their equipment. Among those who do the verification, 33.3% use calibration. Among those who work with Teaching and Research, 21.4% stated that they do not perform verification and 25% use calibration.

Professionals working in public health units and teaching and research are the ones who most use calibration to verify the ultrasound emission by ETUS. And in all sectors, verification is mainly carried out by external service providers, except for self-employed professionals or MEI where they perform verification in 47.5% of cases, using the cavitation method.

Analyzing by the time of experience with the equipment, only professionals with more than nine years of experience use calibration to verify the ultrasound emission by the therapy equipment, but without defined periodicity in most scenarios. Most use the services of external providers, except for professionals with less than two years of experience where they verify the ETUS as the primary method used is cavitation.

5. Conclusions

Given the results of studies on the reliability of biomedical devices in use, it can be concluded that legislative initiatives that make the metrological control of this equipment compulsory during its lifetime are incredibly urgent. Still, a challenge is evident: the lack of a network of capable laboratories for the calibration of biomedical equipment sufficient to meet the demands in the country [4].

To be effective, treatment with ultrasound therapy must be performed safely ç the accuracy and ã o, depending on the intensity, frequency, and duration of application. For this is considerable will level the physical therapy professionals can rely on the values indicated by the equipment during the treatment in patients.

Whereas, according to data from the Federal Council of Physical Therapy and Occupational Therapy (COFFITO) on ú number of physiotherapists in Brazil exceeds 200,000 professionals, which leads us to conclude that s t and problem reaches a size ã the alarming, since ETUS s ã the one of the equipment most commonly used in clinical practice.

This study aimed to discuss, from the metrological perspective, the use of ultrasound physiotherapy equipment in Brazil, showing that the other problem for the metrological reliability of ETUS is related to lack of metrological culture among users, who do not adopt procedures to evaluate their equipment, often unaware of its importance and need. Periodic checks on equipment would give physiotherapists a resource with greater reliability and possibly more efficient therapies.

References

[1] ZAUHAR, G., RADOJCIC, D.S., DOBRAVAC, D., JURKOVIC, S, 2015, Quantitative testing of physiotherapy ultrasound beam patterns within a clinical environment using a thermochromic tile, Ultrasonics, v.58, n.4, pp.6–10.
[2] FERRARI, C.B., ANDRADE, M.A.B., ADAMOWSKI, J.C., GUIRRO, R.R.J., 2010,
[3] ISHIKAWA N.M., ALVARENGA A.V., PAES L.F.C., PEREIRA WCA, MACHADO JC, 2002, Performance analysis of ultrasound equipment for physiotherapy, operating in the city of Rio de Janeiro, according to ABNT NBR IEC 61689. Brazil J Phys Ther 2002;6:63–69 [In Portuguese.].
[4] DRUMMOND, A. M. C. 2007, Confiabilidade Metrológica de Equipamentos Eletromédicos a Laser e a Luz Intensa Pulsada. Pontifícia Universidade Católica do Rio de Janeiro - PUC-RIO.
[5] SILVA, J. G.; ALVARENGA, A. V.; COSTA-FELIX., R. P. . Evaluating the Reliability and Validity of a Questionnaire Used to collect data about the Periodic Evaluation of Ultrasound Therapy Equipment. 2018.

Acknowledgments

This study was financed in part by the Coordenação de Aperfeiçoamento de Pessoal de Nível Superior – Brasil (CAPES) – Finance Code 001. The authors thank the Brazilian National Council for Scientific and Technological Development (grants 312.501/2017-0 [RCF], 430.376/2018-9 [RCF], and 307.562/2020-4 [RCF]) and the Research Support Foundation of the State of Rio de Janeiro (grant E-26/201.563/2014 [RCF] and 260.671/2021 [RCF]).

Correlation between ultrasonic velocity and strength in the early ages of concrete from a metrological approach

T C Dourado[1], F S Santos[1], E W J Santos[1], R L Gaiofatto[2] R P B Costa-Felix[1]

[1]Laboratory of Ultrasound (LABUS), Directory of Scientific and Industrial Metrology (DIMCI), National Institute of Metrology, Quality, and Technology (INMETRO), Duque de Caxias – RJ, Brazil

[2]Concrete Laboratory, Engineering and Computing Center, Catholic University of Petrópolis, Rua Barão do Amazonas, 124 - Centro, Petrópolis – RJ, Brazil

tcdourado@inmetro.gov.br; rpfelix@inmetro.gov.br

Abstract. The correlation between ultrasonic pulse velocity (UPV) and the quality of concrete parts is a resource that has been explored for decades. However, the significant variability of production of these constructive materials tends to require more robust techniques for better reliability of results. This work adopts a metrological approach in the correlation of the ultrasonic pulse velocity and the respective compressive strength values through destructive tests, evaluated in the first ages (7, 14, 30, and 90 days). For this analysis, three different concrete mixes were moulded into cylindrical shapes, using different proportions of the inputs that make up this material. After undergoing statistical and metrological treatment, the results obtained allowed to identify a similar behaviour of UPV growth with the strengths of the concrete samples, especially when analysing the measurement uncertainty bars. However, it is noteworthy that it is necessary to expand the tests in a more significant number of concrete mixtures. It is expected that the improvement of these tests can help reduce costs and execution times of concrete structures where stripping depends on the strength achieved.

1. Introduction

Concrete is today the most demanded construction material on the planet. Its compressive strength is one of the most determining physical properties in its design as a structural element [1]. The most traditional and reliable compressive strength measurement method is performed through destructive testing, in which cylindrical specimens (most common), previously moulded, are broken by a hydraulic press that provides the force in which the disruption of each specimen occurred. In Brazil, the procedures for carrying out this test, regardless of destination, are established by ABNT NBR 5739:2018 (Concrete - Compression tests of cylindrical specimens) [1][2].

On the other hand, the propagation velocity of ultrasound longitudinal waves in concrete could give some information about the quality and homogeneity of a piece of concrete.

The measurement method more commonly used is to verify the propagation of ultrasonic pulse velocities (UPV) in concrete, is a non-destructive testing (NDT), performed by using equipment, which emits an ultrasonic pulse through a piece of concrete and measures the transit time. The procedures for carrying out this test are established by ABNT NBR 8802:2019 (Hardened concrete - Determination of ultrasonic wave velocity) [3]. Therefore, improving the correlation between UPV and compressive strength is essential for the construction industry, mainly related to quality control and structural integrity. However, the significant variability of mixtures and the almost always heterogeneous nature of concrete are obstacles in establishing correlations between the physical properties of materials and their compressive strength.

This work follows a metrological approach in the correlation of ultrasonic velocity and compressive strength measurement results in samples of concrete mixtures in their early ages.

2. Theoretical reference

The advantages of conducting non-destructive tests instead of destructive ones are countless. According to Neville (1997) [1], non-destructive testing increases safety and optimizes the construction schedule. Furthermore, according to Garbacz and Garboczi (2003) [4], the development of NDT methods is technically and economically significant, as it allows for loco analysis and evaluations, repeating the tests in the same regions as often as necessary. Thus, the search for this information through tests can provide benefits to the construction industry.

Several studies in the technical literature sought to correlate the ultrasonic pulse propagation velocity in concrete and its resistance. In most of these studies, it was observed that many factors affect these parameters. For example, according to Hernandez et al. (2002) [5], the water-cement ratio (W/C) is the main factor influencing the strength of concretes, increasing it with an increase in the amount of cement and reducing it with an increase in the amount of water.

In contrast, Del Río et al. (2004) [6] suggest that factors other than the W/C ratio, mainly the proportion and granulometry of aggregates, have a direct influence on the results of ultrasound propagation velocity measurements because of the dispersion of the signal during the measurements and the variety of components of the natural aggregates.

Several authors, over the years, point to an exponential relationship between strength and ultrasonic velocity. Panzera et al. (2010) [7] presented a small summary of equations developed in studies like this one, table 1, presented below, represents this summary. Note that table 1 f_c represents the compressive strength of concrete, and V_p represents the ultrasonic velocity in concrete mixtures [8].

Table 1. Summary of essential equations relating to resistance and ultrasonic velocity in concrete.

Equation	R^2	References
$f_c = 1{,}46 \times e^{0{,}77V_p}$	0,80	Turgut (2004) [9]
$f_c = 1{,}19 \times e^{0{,}715V_p}$	0,59	Nashn't *et al.* (2005) [10]
$f_c = 8{,}4 \times 10^{-9}\,(V_p \times 10^3)^{2{,}5921}$	0,42	Kheder (1999)[11]
$f_c = 1{,}2 \times 10^{-5}\,(V_p \times 10^3)^{1{,}7447}$	0,41	Kheder (1999) [11]
$f_c = e^{[(-3{,}3 \pm 1{,}8) + (0{,}0014 \pm 0{,}0004) \times (V_p \times 10^3)]}$	0,48	Del Rio *et al.* (2004) [12]

It is important to understand that the general use of these equations for any concrete composition will not return accurate estimates. They were obtained from the combination of tests results performed on specimens with specifics concrete compositions. More in-depth studies on the subject are necessary to define one equation or a set of equations for general use.

For the possible correlations between the two quantities studied here, using specifics compositions of concretes, according to Garbacz and Garboczi (2003) and Komlos et al. (1996), analyzes are usually performed from regressions curves models, with reasonable accuracy. These curves are also known as calibration curves [8].

3. Materials and methods

In this work, three different concrete mixtures were used. Specimens were moulded with metallic cylindrical moulds 100 mm diameters and 200 mm height. Table 1 shows the content of mixtures.

The specimens were tested by the two proposed measurement methods at four different ages: after 7, 14, 28, and 90 days of moulding.

Mixtures were composed of different types of gravel (gravel 0 and gravel 1), different proportions of water and cement, combining two types of cement (Portland Cement III – CP III-40 RS), and variations in the amounts of sand used. It was also used the superplasticizer additive Tec-Flow 9030, which ensures greater fluidity to the mixture, facilitating the densification. Besides the different proportions of the materials used, each mixture, they have an exact origin. Table 2 indicates the composition of each of the three mixtures used. The specimens were rectified (cut at both ends) to ensure a flat surface with adequate parallelism. After rectification of samples, three physical quantities, i.e., length, time and force were measured.

Specimens were divided into three groups (concrete lines), totalizing 36 cylindrical concrete pieces with 10 cm in diameter and 20 cm in height.

Table 2. Composition of concrete mixes for this work.

Materials	Mix 1	Mix 2	Mix 3
Cement (kg)	13,30	19,60	26,00
Water (kg)	4,55	4,78	4,18
Sand (kg)	40,70	38,20	35,70
Gravel (kg)	49,70	46,70	43,60
Aditive (kg)	0,08	0,12	0,16
Water/Cement (W/C)	0,34	0,24	0,27

From the propagation distance and the transit time the velocity with which this pulse propagated can be calculated. Thus, measurement of diameter and height of each specimen (concrete cylinder) were performed before the beginning of the tests, using a digital calliper, in four consecutive repetitions.

After carrying out the tests, the respective measurement uncertainties were calculated for all results obtained following the Guide to the Expression of Uncertainty in Measurement (ISO GUM, 2008) [13].

In the metrological treatment of data, type A measurement uncertainties (related to repetitions) and type B (related to measurement equipment, such as a calibration certificate) were combined to determine the combined uncertainty and this one for the expanded uncertainty.

3.1. Uncertainties assessment

In this work, type A uncertainty is derived from 30 repeated measurements of the time of flight (ToF) (t), 12 measurements of length (s), and one destructive test of breaking load (F_c) for each sample. It is noteworthy the more significant variability of results of temporal magnitude concerning dimensional. So, it demands many collections of the first one for improvement in statistical quality. It is evident in ultrasonic velocity measurements of heterogeneous materials. Type B uncertainties were extracted from the calibration certificate of the respective equipment.

The mathematical metrological equation used in this article followed the approach used in Santos et al. [14]. Thus, the calculation of the expanded uncertainty (U) starts with the determination of the combined standard uncertainty (u_z) as follows:

$$\mathbf{u_z^2} = \sum_{\mathbf{i=1}}^{\mathbf{N}} \left(\frac{\partial \mathbf{UPV}}{\partial \mathbf{x_i}}\right)^2 \mathbf{u_{xi}^2} \qquad \mathbf{(1)}$$

where *f* represents the dependent quantity 'ultrasonic velocity or UPV', and x_i as independent quantity 'space' (s) and 'time of flight' (t), thus:

$$u_{UPV} = \sqrt{\left(\frac{1}{t}\, u_{space}\right)^2 + \left(\frac{s}{t^2}\, u_{time}\right)^2} \qquad \mathbf{(2)}$$

For compression resistance, the uncertainty of area was defined as 1% of area, or 78.5 mm^2. Area uncertainty was arbitrarily defined by the authors.

Thus, the final combined standard uncertainty was assessed as following:

$$u_\sigma = \sqrt{\left(\frac{\partial \sigma_c}{\partial F} u_{Fc}\right)^2 + \left(\frac{\partial \sigma_c}{\partial A} u_A\right)^2} = \sqrt{\left(\frac{u_{Fc}}{A}\right)^2 + \left(\frac{F \times 78.5}{A^2}\right)^2} \qquad (3)$$

where σ_c represents the dependent quantity 'resitance compression', and x_i as independent quantity 'force' (F) and 'area' (A).

Lastly, the expanded uncertainty (U) can be calculated by:

$$U = u_z\, k \qquad (4)$$

where *k* is the coverage factor which corresponds to a level of confidence defined of 95%.

3.2. Ultrasonic Pulse Velocity

To determine the ultrasonic pulse velocity, the measurement of the pulse propagation distance and flight time is necessary.

The pulse propagation distance is the length between the ends of the cylinders (height). It was measured by a digital caliper, series 500, model Digital ABSOLUTE AOS (Mitutoyo, Japan).

The transit time of ultrasonic pulses was measured by an ultrasonic flaw detector, model EPOCH 600 (Olympus, EUA), and three pairs of circular section transducers. In the first pair, with a nominal frequency of 500 kHz, one of the transducers with a diameter of 32 mm (model V-389) and another with a diameter of 25 mm (model A-301S), both produced by Panametrics-Olympus NDT, USA. The second pair has a nominal frequency of 1000 kHz, both transducers of 12.5 mm and A-303S model (Panametrics-Olympus NDT, USA). In the third pair, with a nominal frequency of 54 kHz, both transducers of 20 mm and sold together to the 58-E4800 ultrasonic data collection equipment from CONTROLS, an Italian company specializing in construction products.

There are different acquisition arrangements for ultrasonic measurements. In figure 1a, the direct transmission acquisition arrangement is presented. This is the most recommended by the standards and commonly used in the determination of UPV. That is why it was used for this purpose in this work.

With the transducers positioned on the flat faces of the concrete cylinder, this research adopted border regions for the collection of measurements to meet the requirements of item 6 of ISO 16831:2012(E) *Non-destructive testing – Ultrasonic testing – Characterization and verification of ultrasonic thickness measuring equipment*. The standard recommends that the distance between the transducer edges and the face of the collection object is not less than one diameter. Jigs were developed and printed in a 3d printer, according to figure 1b, and used in 2b.

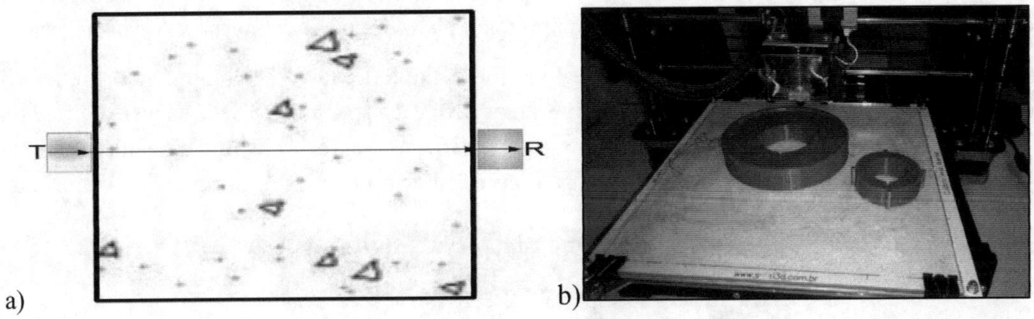

Figure 1. a) Array of acquisition direct transmission method, and b) Additive manufacturing of jigs.

These tests were performed at 7, 14, 28, and 90 days from the molding date of each line. For each of the specimens, four measurements of the height were collected, according to figure 2a. Then, for each frequency, five measurements of the ultrasonic pulse transit time were performed (according to schematic 2b) at five different points on the faces of the specimens. Each of these measurements originated from two consecutive readings, disconnecting, and reconnecting the transducers. To better control the quality of measurements, it was not allowed to diverge more than 0.5% between them. These measurement cycles were repeated once more on each specimen. In cases of divergence above the 0,5% limit, new readings were taken at the same point.

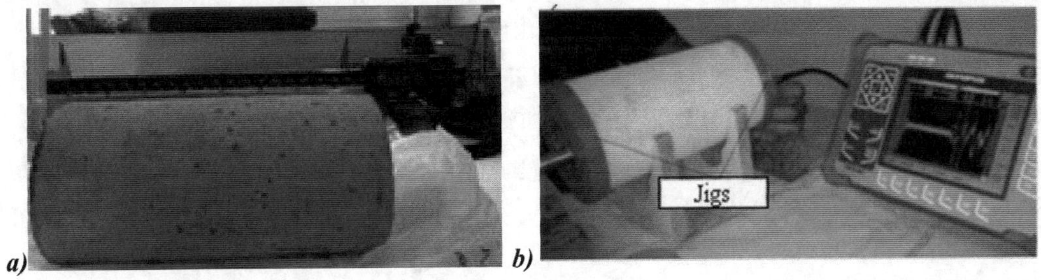

Figure 2. a) Caliper realizing dimensional measurements in the specimen; and b) Ultrasonic system Epoch 600, measured by the direct transmission method according to ABNT NBR 8802:2013.

3.3. Compressive Strength

To determine the compressive strength, it is necessary to measure the cracking force of the specimens and the contact area of the applied force.

Concrete compressive strength measurement systems are hydraulic presses. The press should apply progressively increased continuous force at fixed or adjustable rates (depending on the instrument). The used loading speed must be constant and, according to ABNT NBR 5739:2018 - Ensaio de Compressão de Corpos-de-Prova Cilíndricos, its value may be from 0.30 MPa.s-1 to 0.60 MPa.s-1. Loading is stopped when the instrument detects a drop in the applied force, which indicates the specimen rupture.

To carry out the test, it is necessary to measure the diameter (d) and height (h) of the specimens with an accuracy of ± 0.1 mm. Then, position them individually in the measurement system so that the axis of each specimen coincides with the axis of the compression plates (upper and lower) of the instrument.

After applying the load and measuring the breaking force, the test ends. The compressive strength of each concrete sample corresponds to the breaking strength of each specimen. The tests were carried out in a hydraulic press, model WAW-100-C-NS 042-NC 1167 A (Servo UTM, India), with a nominal capacity of 1,000 kN and a resolution of 0.01 kN. Figure 3 shows an image of the assay.

Figure 3. Specimen compression tests in the press.

The equipment undergoes calibration procedures periodically, with the last calibration being carried out in July 2018 installed in the Catholic University of Petrópolis.

The contact area was calculated from the specimen's diameters, measured also by a digital calliper, series 500, model Digital ABSOLUTE AOS (Mitutoyo, Japan) shown in figure 2a.

With the contact area of the specimen calculated, the stress with which each sample was broken is determined. This stress corresponds to concrete (f_c) compressive strength, usually expressed in megapascals (MPa).

Compressive strength measurements were performed on the specimens. The execution of these measurements, for the four ages (7, 14, 28, and 90 days) and sample quantities mentioned before, followed the criteria: faces of the compression plates and specimens were cleaned and dried; positioning of the specimens was carried out so that its axis coincided with the axis of the equipment's compression plates; carrying out the load with a constant increase in the applied load; and adequate recording of the results obtained.

As this is a destructive test, it was only possible to perform one measurement of compressive strength by the specimen.

4. Results and Discussions

4.1. Results

Table 3 presents the results of UPV and compressive strength by type of mixture tested in this research. Note that table 3 also contains the uncertainty values calculated according to the protocol exposed in the previous item.

Table 3. Measurement results for this work.

	Age [days]	Strength [MPa]	$U_{Strength}$ [MPa]	Velocity to F1=54 kHz [m.s^{-1}]	U_{F1} [m.s^{-1}]	Velocity to F2=500 kHz [m.s^{-1}]	U_{F2} [m.s^{-1}]	Velocity to F3=1.000 kHz [m.s^{-1}]	U_{F3} [m.s^{-1}]
Mix 1	7	18,56	0,20	3797	15	3982	20	-	-
	14	22,71	0,21	3824	18	3992	20	-	-
	28	27,29	0,19	3913	11	4111	95	4017	14
	90	38,42	0,59	4059	8	4261	47	4190	14
Mix 2	7	35,90	0,17	3946	16	4062	34	-	-
	14	40,92	0,20	3925	21	4071	9	4087	20
	28	46,78	0,53	4045	19	4187	8	4261	19
	90	67,64	0,48	4148	9	4309	11	4353	10
Mix 3	7	37,66	0,19	4016	34	4123	15	4130	13
	14	49,61	0,43	4081	24	4211	15	4358	13
	28	59,25	0,46	4222	11	4335	12	4444	18
	90	68,45	1,43	4223	19	4378	11	4439	62

Figures 4, 5 and 6 show the correlation graphs between compressive strength and ultrasonic velocities, by type of concrete mix. Note that the calibration curves were formed from the measured information and their respective uncertainties, where velocities are grouped by frequency of the ultrasonic emission signal.

Journal of Physics: Conference Series **2606** (2023) 012009 doi:10.1088/1742-6596/2606/1/012009

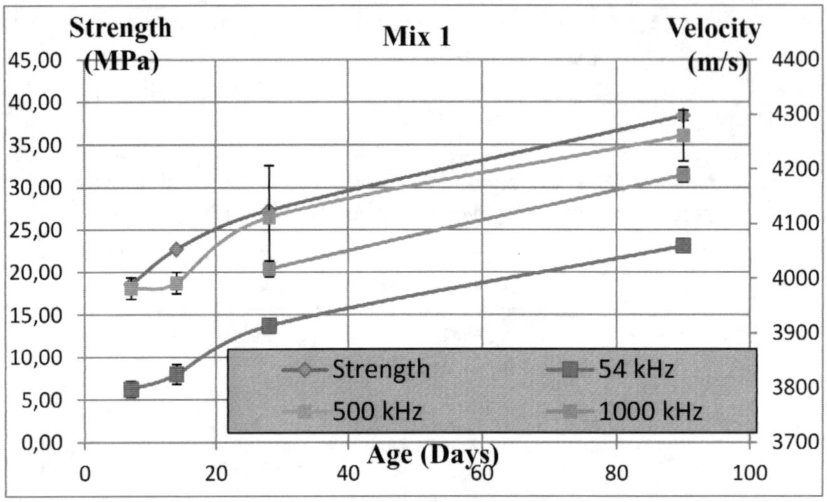

Figure 4. Concrete compression and ultrasonic velocity increase with time for mix 1.

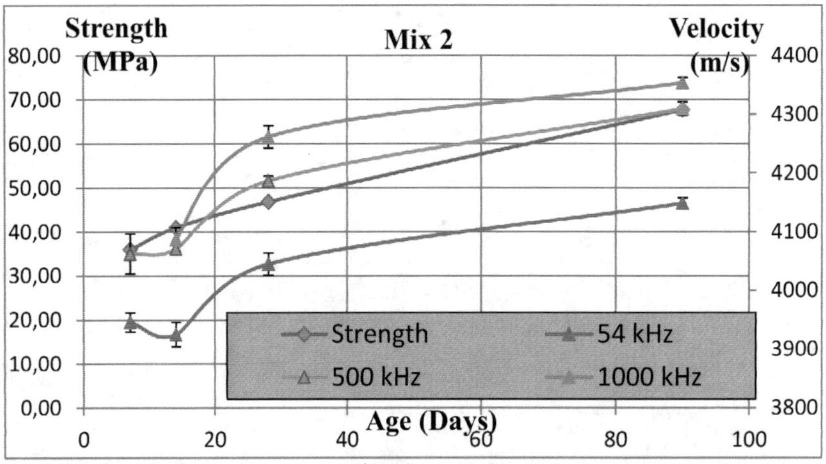

Figure 5. Concrete compression and ultrasonic velocity increase with time for mix 2.

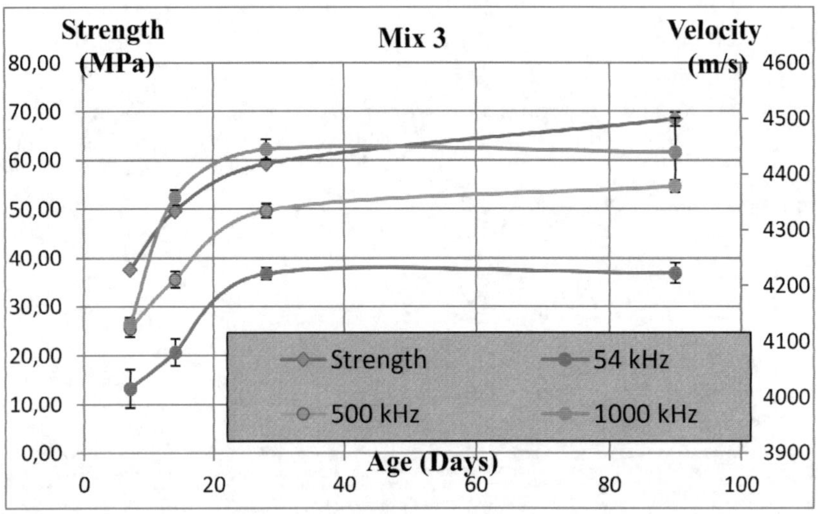

Figure 6. Concrete compression and ultrasonic velocity increase with time for mix 3.

4.2. Discussion

From the calculated values for compressive strength and respective uncertainties, one can note that they are different for all mixtures and ages.

Mixtures 1 and 2 presented very close UPV values for 54 kHz and 500 kHz at the ages of 7 and 14 days. However, there is a distinction in the UPV values between these ages and those of 30 and 90 days. Note that for 1000 kHz frequency, some UPV points were not presented due to data loss.

On the other hand, mixture 3 presented statistically distinct data for all frequencies of 7 and 14 days. Only the 500 kHz frequency managed to offer detailed UPV data between the ages of 30 and 90 days.

In general, by the behavior of the graphs presented, it is possible to see that the curves for all mixtures give an excellent correspondence between the growth of the ultrasonic velocity and the compressive strength of concrete during the first ages.

Likewise, this behavior is following results already identified in the literature on the subject.

5. Conclusions

From the data obtained and the composite graphs for this research, it was possible to observe the direct relationship between the increase in ultrasonic velocity and the compressive strength of concrete in the early ages.

The good correlations obtained represent a more significant deepening in this type of non-destructive analysis, which has been deepened in the literature for years.

It is possible to notice that one of the contributions of this work is the diffusion of metrological rigor for the measurements of compressive strength and UPV. And although not applied at this stage, additional features such as normalized error analysis can allow a quantitative comparison of results between different ages.

However, what was also observed was that the protocol used for data collection could be improved since not all resistance variations observed in the gold standard (destructive test) were observed through the UPV.

Finally, another nuance explored in this work, which may be helpful in the analysis of future outcomes, is the combination of calibration curves at different ultrasonic frequencies. This characteristic can be exploited due to the dispersive character of concrete, which usually presents different velocity depending on the frequency.

References

[1] Neville, A. M. Propriedades do concreto. 2a. ed. São Paulo: Pini, 1997.
[2] Associação Brasileira de Normas Técnicas, NBR 5739: Concreto - Ensaios de compressão de corpos-de-prova cilíndricos, Rio de Janeiro, 2018.
[3] Associação Brasileira de Normas Técnicas, NBR 8802: Concreto endurecido – Determinação da velocidade de onda ultrassônica, Rio de Janeiro, 2019.
[4] Garbacz A, Garboczi E. J., Ultrasonic evaluation methods applicable to polymer concrete composites. NIST Report, NISTIR 6975. Gaithersburg: National Institute of Standards and Technology; 2003.
[5] Hernández, M. G.; Anaya, J. J.; Izquierdo, M. A. G.; Ullate, L. G. Application of micromechanics to the characterization of mortar by ultrasound. Ultrasonics. Vol 40, 2002, p. 217 – 221.
[6] Del Río, L. M.; Jiménez, A.; López, F.; Rosa, F. J.; Rufo, M. M.; Paniagua, J. M. Characterization and hardening of concrete with ultrasonic testing. Ultrasonics. Vol 42, 2004, p. 527 – 530.

[7] Panzera, T. H., Christoforo, A. L., Cota, F. P., Bowen, C. R., Ultrasonic pulse velocity evaluation of cementious materials, Minas Gerais, 2011.

[8] Baesso, R.M., Oliveira, P.A., Morais, G.C., Alvarenga, A.V., Costa-Felix, R.P.B. Using ultrasonic velocity for monitoring and analysing biodiesel production. Fuel, 2018, 226, pp. 389–399.

[9] Turgut, P., Research into the Correlation between Concrete Strength and UPV Values, NDT.net 2004, 12.

[10] Nash't, S. A'bour, A. Sadoon Finding an unified relationship between crushing strength of concrete and non-destructive tests Middle East Nondestructive Testing Conference & Exhibition, Bahrain, Manama (2005).

[11] Kheder, G. A two-stage procedure for assessment of in situ concrete strength using combined non-destructive testing Mater. Struct., 32 (1999).

[12] Del Río, L. M. Jiménez, A.; López, F.; Rosa, F. J.; Rufo, M. M.; Paniagua, J. M. Characterization and hardening of concrete with ultrasonic testing. Ultrasonics. Vol 42, 2004, p. 527 – 530.

[13] JCGM, 2008, Evaluation of Measurement Data – Guide of the Expression of Uncertainty in Measurements, JCGM 100:2008.

[14] Santos, E. W. J., Pinto, E. S. F., Santos, F. D. G., Dourado, T. C., and Costa-Félix, R. P. B. 2021 J. Phys.: Conf. Ser. 1826 012007.

[15] Komlos, K., Popovicks, S., Nürbergeva, T., Babal, B., Popovicks, J. S., Ultrasonic pulse velocity test of concrete properties as specified in various standards, Cement and Concrete Composites, 1996, p.357.

[16] Associação Brasileira de Normas Técnicas, NBR 5739: Concreto - Ensaios de compressão de corpos-de-prova cilíndricos, Rio de Janeiro, 2018.

[17] ISO 1920-7:2004 Testing of concrete-- Part 7: Non-destructive tests on hardened concrete.

[18] ASTM-C597-16. Standard Test Method for Pulse Velocity Through Concrete. Standard, ASTM International, West Conshohocken, PA - USA, 2016.

[19] BS EN 12504-4:2004, Testing concrete. Determination of ultrasonic pulse velocity. Standard, British Standards Institution, London, UK, 2004.

[20] Prassianakis, N.; Giokas, P. Mechanical properties of old concrete using destructive and ultrasonic non-destructive testing methods. Magazine of Concrete Research. Vol 55, 2003, p. 171 – 176.

Acknowledgments

The authors thank the Brazilian National Council for Scientific and Technological Development (grants 312.501/2017-0 [RCF], 430.376/2018-9 [RCF], and 307.562/2020-4 [RCF]) and the Research Support Foundation of the State of Rio de Janeiro (grant E-26/201.563/2014 [RCF] and 260.671/2021 [RCF]).

Journal of Physics: Conference Series **2606** (2023) 012010 doi:10.1088/1742-6596/2606/1/012010

Proposition of water meter for buildings based on a thin-film resistive sensor: electromechanical and thermal characterizations

L S Gonçalves [1,4], K A R Medeiros [2] and C R H Barbosa [3]

[1] Pontifical Catholic University of Rio de Janeiro, Postgraduate Program in the Department of Chemical and Materials Engineering, Rio de Janeiro, Brazil

[2] Pontifical Catholic University of Rio de Janeiro, Mechanical Engineering Department, Optical Fiber Sensors Laboratory, Rio de Janeiro, Brazil

[3] Pontifical Catholic University of Rio de Janeiro, Postgraduate Program in Metrology, Rio de Janeiro, Brazil

[4] Corresponding author: laisgoncalves@aluno.puc-rio.br

Abstract. Notoriously, measurement proves to be essential to encourage water conservation. Thus, this work characterizes two thin-film resistive sensors (bend sensors) with different coatings aiming at their application to measure water consumption. For this, the individual water measurement system is presented and discussed briefly. Then, resistive sensor parameters and operating principles are detailed, as well as the mathematical formulation of the correlation of the sensor resistance with the flow velocity. Methodologically, the system developed for the electromechanical and thermal characterization of the sensor is presented. The results allowed obtaining a quadratic relationship between the sensor deflection angle and the resistance. Furthermore, it was observed that the polyester-coated sensor presented a low hysteresis value when subjected to temperature variation, obtaining a negative linear relationship between the sensor resistance and temperature. In turn, the polyimide-coated sensor did not show low hysteresis when exposed to temperature change and flexed at 50° and 70° angles. Finally, it is proposed for future work computer simulations and experimental tests to confirm the applicability of the sensor for water measurement.

1. Introduction

As a consequence of economic progress, population growth, and changes in consumption patterns, the use of freshwater has increased significantly in the last 100 years. This increase was six fold over this period and continues to increase – according to the United Nations Educational, Scientific and Cultural Organization (UNESCO), water consumption is expected to grow by approximately 25 % by 2030. In addition, other worrying aspects are the exponential drop in water quality and water stress, which affects more than 2 billion people when basically estimated by availability as a function of supply [1].

The so-called economic water scarcity means that water is physically available, but the infrastructure for accessing it is precarious, and many regions of the world are facing this situation. This precariousness is due to deficient regulation, lack of incentives, and/or lack of investment in the water sector [1]. However, some economic studies have found that, when investing US$ 1 in public hydrological monitoring systems, on average, US$ 4 is generated in social benefits [2], highlighting the socioeconomic value and the need for more studies focused on the technological advancement in the water sector.

Notoriously, metering is one of the most effective procedures for encouraging water conservation. However, in some countries – such as Brazil – water only begins to be valued and rationed when other sectors (for example, electric) face problems or when there are long periods of drought. That is, the concern, often, is not directly related to the water itself, but to other resources. Thus, a crisis is necessary for water consumption to attract attention. Furthermore, unfortunately, water is often taken for granted on Earth.

Therefore, considering the importance and proportion of the water sector, more studies must be carried out to contribute to the advancement of water meter technologies to overcome the limitations of those currently available and make these meters more modern and with higher reliability.

Thus, this study aims to contribute to the advancement of water meter technologies, specifically for measuring the volume of water drained in a pipeline of autonomous units in buildings. It presents electromechanical and thermal characterizations of a flexible type resistive sensor to evaluate its applicability in buildings water measurements, particularly inserted inside some hydraulic components (such as flow control valves), normally present in these installations.

2. Literature review

2.1. Individualized water measurement system

Conceptually, the Water Individualized Measurement System (IMS) corresponds to the installation of a water meter in the branch of each autonomous unit in buildings, which can be residential, commercial, or mixed. In general, this type of measurement allows for a fairer water bill, as each unit pays only for its expenditure, resulting in a more conscious consumption, with less waste. However, a large number of buildings have collective water metering. That is, the measurement of monthly water consumption is made for the entire building, not considering what each unit consumed, but what the entire building consumed. In addition, there are even more alarming cases, in which some buildings do not have any type of instrument to measure the volume of water supplied, with the consumption being estimated based on the number of rooms in the property, or based on the constructed area, in square meters (m^2).

To change this scenario and encourage individualized measurements to be widely used, in Brazil Federal Law 13.312/2016 was enacted, establishing national guidelines for basic sanitation. This law mandates the adoption of environmental sustainability standards, such as the individualized measurement of water consumption in new buildings [3]. Although this law only provided for a nationwide mandatory implementation of water SMI in new buildings in 2016, several places in the country were already enacting local laws before that, such as Distrito Federal (Law 3.557/ 2005), Niterói (Law 2.340/2006), Recife (Law 16.759/2002) and Curitiba (Law 10.785/2003) [4–7].

In addition, it is noteworthy that, although the Federal Law 13.312/2016 does not mandate the replacement of collective measurement by IMS in old buildings (built before the law went into effect), some case studies [8–10] indicate a significant improvement both in the bill and in the waste of water, which encourages the unit owners to implement the water SMI voluntarily. With this, further studies are essential to provide advances in meter technologies for the water sector, aiming to ensure the sustainability of water resources and to contribute to more conscious and sustainable consumption.

2.2. Thin-film resistive sensor

A bend sensor is a thin-film resistive sensor that changes its electrical conductivity as it is bent, as the sensor is formed from a single thin layer of a plastic substrate coated with a resistive material (paint based on carbon/polymer). By bending this type of sensor, the paint separates into several micro-cracks that, depending on the bending movement, determine an increase or a decrease in the electrical conductivity of the material [11], and the radius of curvature or angular deflection of the sensor determines the sensor electrical resistance.

The bend sensor has electromechanical properties that arouse the interest of many researchers. Thus, its application has spread to several areas. One of them is medical, which concentrates a large number of studies, with this type of sensor being used to measure the rotations of small joints in the body; for the assessment of trunk posture in spinal diseases and rehabilitation follow-up [12]; and for semi-automatic goniometric measurements of finger joint extension capabilities using a sensory glove [13]. Additionally, there are studies that show the bend sensor being used to monitor neck posture where the system immediately reminds the user to correct the neck posture [14]; to continuously monitor fetal health – sensors are connected to the pregnant woman and detect fetal movement [15]; to constitute a 3D printed prosthesis [16]; to recognize and categorize different health-related sitting postures in an office chair [17]; and for the approximate reconstruction of the surface geometry of the human system, which can be used to measure shape, record dynamic changes, or examine deliberate movement to assess the human body and performance [18].

In addition, the thin-film resistive sensor is present in applications aimed at ground monitoring, where the sensor is used to measure the soil settlement and horizontal movements of a sandy slope model due to continuous loading testing in a laboratory [19]; and to detect and locate road accidents in real-time [20].

Compared to other areas, the water sector does not present many studies using the bend sensor for flow measurement. However, it is possible to verify some studies that demonstrate the applicability of this type of sensor for this purpose. For example, Fan *et al.* [21] employ the sensor to monitor the airflow velocity in real-time, helping to save energy consumption in a wastewater treatment process; Xu *et al.* [22] report a way to monitor water flow in pipelines using a flexible resistive film; Srinivasan, Sen, Kumar and Sabebu [23] use the sensor as a cantilever beam for flow measurement for moderate flow applications; Stewart, Fox and Harnett [24] developed a velocity sensor using the bend sensor for application in streams; Gonçalves, Medeiros, and Barbosa [25] presented a study of the resistive sensor application to measure the flow of water through the comparison between the results of experimental electromechanical characterization with a computer simulation, and these same authors [26] also proposed an adaptation of the valves commonly present in residential hydraulic systems, using a bend sensor, and demonstrated a new methodology for measuring the volume of water consumed in autonomous units.

In short, it is observed that thin-film resistive sensors have numerous applications in the most diverse areas, such as medical, aeronautical, automotive, robotics, and even soil studies. In recent years, this type of sensor has been used increasingly for being able to vary its resistance when flexed and also being light, robust, and inexpensive. These sensors are easily adaptable in hydraulic systems of autonomous units (residential, commercial, or mixed) in buildings.

2.3. Bend sensor for measuring water consumption

A correlation of the sensor resistance with the flow velocity is necessary to use the bend sensor to measure the volume of water drained in a pipe. For this, the sensor is considered similar to a cantilever beam subjected to a uniformly distributed load q, in which the maximum deflection (δ_{max}) and the maximum rotation angle (θ_{max}) occur at the free end of the beam. For the calculation of θ_{max}, it is used

$$\theta_{máx} = \frac{qL^3}{6EI}, \tag{1}$$

where L is the beam length, and EI is the flexural rigidity.

As the proposal is to use the sensor inside a pipe, the loading intensity q is considered as a result of the drag force – equation (2) – experienced by the sensor immersed in the flowing water divided by the length (L) of the sensor.

$$F_A = \frac{1}{2} C_D \rho v^2 A, \tag{2}$$

where C_D is the drag coefficient, ρ is the mass density of the fluid, v is the flow velocity, and A is the area of the object perpendicular to the flow.

Then, by combining equations (1) and (2), the maximum deflection angle is calculated by

$$\theta_{máx} = \left(\frac{\frac{1}{2} C_D \rho v^2 A}{L}\right) \frac{L^3}{6EI}, \tag{3}$$

As the purpose is to correlate the sensor resistance with the flow velocity, through analytical calculations, the velocity is calculated by isolating the variable v^2 from equation (3). Furthermore, according to Xu *et al.* [22], the relative resistance increment of the flexible resistive ($\Delta R/R_0$) is proportionally related to the sensor maximum deflection angle, as

$$\frac{\Delta R}{R_0} = K\theta_{máx} , \quad (4)$$

where R_0 is the resting resistance of the bend sensor (when relaxed – without having a force applied), ΔR is the difference between the bend sensor resistance (after loading) and the resting resistance, K is the deflection coefficient and θ_{max} is the maximum deflection angle.

Then, to obtain the relationship between the relative change in the bend sensor resistance and the flow velocity, equation (3) is substituted into equation (4), obtaining

$$\frac{\Delta R}{R_0} = K'v^2 , \quad (5)$$

where K' is the adjusted deflection coefficient, in this study obtained through Excel® software, from a linear fit between the relative resistance increment and the square of the flow velocity.

3. Materials and methods

Two types of characterization of thin-film resistive sensors were carried out to better understand some bend sensor properties and their behavior upon bending at different angles within an exploratory study, which seeks to investigate the feasibility of using it to measure water consumption. The first characterization method searched the electromechanical behavior of the sensor, while the second analyzed its electrothermal characteristics. For this, a test setup was developed, used in both characterizations, containing a Flexpoint® bend sensor [27], a Keysight U1233A portable digital multimeter, a screw to flex the bend sensor, a fixed support with a welded nut to guide the screw, a protoboard as a base for the bend sensor and, finally, a protractor to measure the angle of the sensor curvature. This test bench is shown in Figure 1.

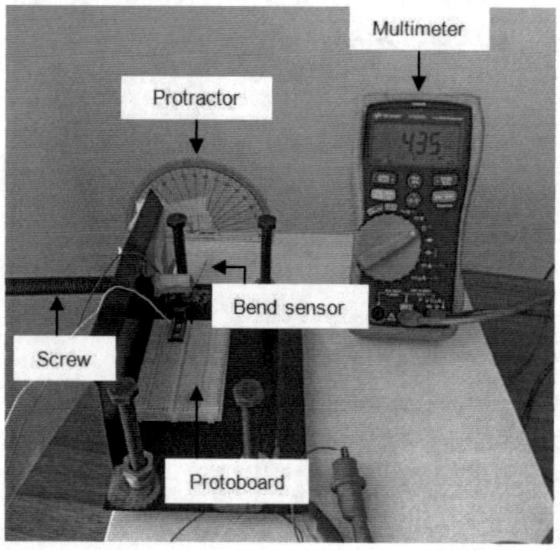

Figure 1. The test setup used in electromechanical and thermal characterizations.

Additionally, in both characterization methods, two Flexpoint® resistive sensors were used, with two inches in length (0.28" width and 0.005" thick) and different coatings, as shown in Figure 2, one with a coating of polyester (Figure 2a) and the other with polyimide coating (Figure 2b).

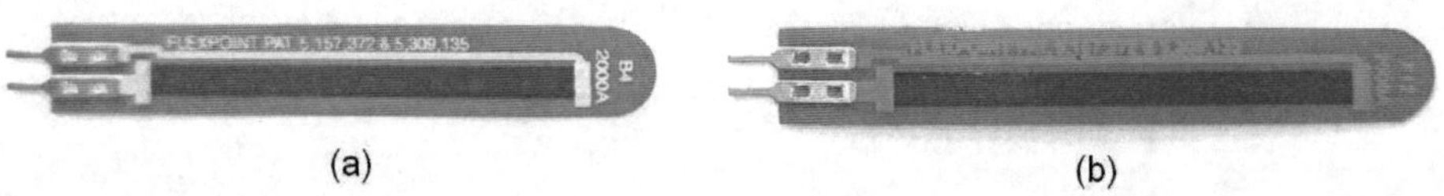

Figure 2. Bend sensors with two inches in length: (a) With polyester coating (b) With polyimide coating.

Furthermore, in this work, the values of the parameters needed to calculate v^2 are shown in Table 1.

Table 1. Bend sensors and flow parameters.

Parameters		
Density ρ at 20 °C (kg/m³)		998.2
Drag coefficient C_D [28]		2.05
Area A (m²)		3.61×10^{-4}
Length L (m)		0.0508
Moment of inertia I (m⁴)		1.16×10^{-15}
Elastic modulus E (GPa)	Polyester [29]	3.65
	Polyimide [30]	6.36

3.1. Bend sensor electromechanical characterization
Initially, to relate the resistance of the bend sensor with the angle of deflection, it was decided to execute a simplified characterization test, in which the bend sensor was flexed by the test bench screw, performing a two-step cycle, the first for loading of the sensor and the other for unloading. With this, it was also possible to obtain the sensor hysteresis curve.

In the first stage of the procedure, the angle was increasingly settling at different angles: 0° (resistance at rest), 10°, 30° (Figure 3a), 50°, and 70° (Figure 3b). In turn, in the second stage, the angle was varied decreasingly, from 70° until rest. In addition, in both stages at each angle variation, ten resistance measurements of the bend sensor were made to obtain the average value.

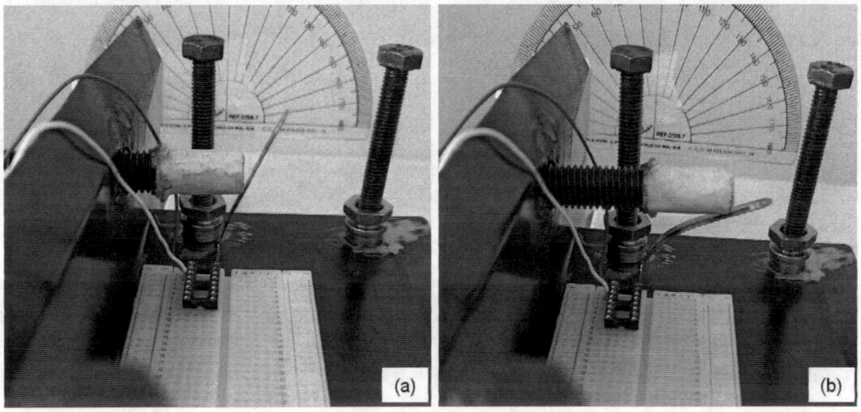

Figure 3. Deflection angle variation: (a) Sensor bent at 30°; (b) Sensor bent at 70°.

Finally, the relationship between the average bend sensor resistance and its inclination was determined. Thus, a regression equation and the coefficient of determination (R^2) were obtained via Excel® software.

3.2. Bend sensor electrothermal characterization
The second characterization was made to verify the influence of temperature on the bend sensor resistance and also to evaluate the sensor hysteresis concerning the effect of temperature.

In this characterization, it were used the test bench described in section 3 and a climatic chamber (THERMOTRON, model SM-8-82000), which was programmed to perform heating within a range of 20 °C to 40 °C with steps of 5 °C, and cooling from 40 °C to 20 °C with steps of 5 °C.

The test bench was placed inside the climatic chamber, and the sensor was bent with the screw at the angles of 0° (rest), 10°, 30° (Figure 4a), 50°, and 70° (Figure 4b) to obtain the resistance values. For each angle, a programmed cycle was performed, and as the target temperatures were reached, ten resistance measurements of the bend sensor were performed to obtain the average value.

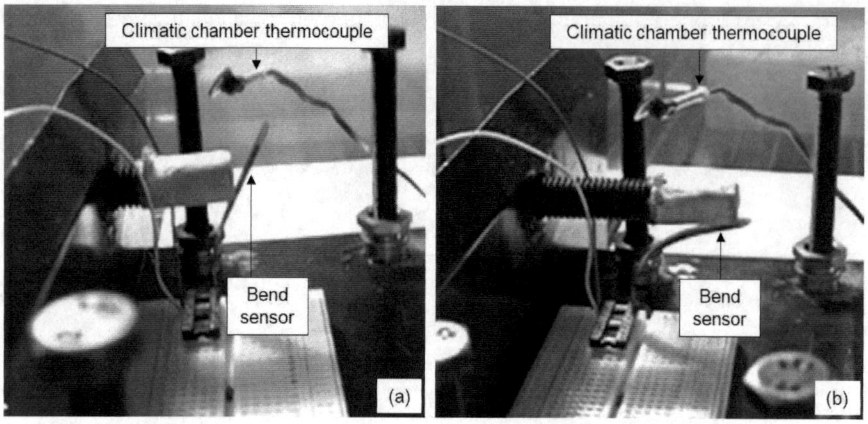

Figure 4. Deflection angle variation: (a) Sensor bent at 30°; (b) Sensor bent at 70°.

4. Results and discussion

In this section, the results of the electromechanical and thermal characterizations of the resistive sensors are presented. Furthermore, to perform a preliminary evaluation of the feasibility of applying the bend sensor to measure the water consumption flowing in a pipe, the results of the sensor resistance correlation with the flow velocity are presented.

4.1. Electromechanical characterization of bend sensor

The principal objective of this characterization was to know the sensor resistance variation characteristics about its angular deflection and, with that, to analyze its performance. In this characterization, two thin-film resistive sensors with 2-inch in length were evaluated: one with polyester coating and the other with polyimide, as mentioned previously.

The results for the bend sensor with polyester coating are shown in Figure 5. As the methodology made it possible to measure the resistances in the increments and decrements of the angle (Figure 5a), it was feasible to obtain the hysteresis curve of this bend sensor and to calculate its maximum hysteresis that, in this case, it was equal to 12 % of the full scale. Furthermore, this sensor presented a quadratic relationship between its resistance – average between the values in the increment and decrement of the angle – which varied between 3.85 kΩ and 5.25 kΩ, and its deflection. Thus, a regression equation and the coefficient of determination (R^2) were obtained via Excel® software (Figure 5b).

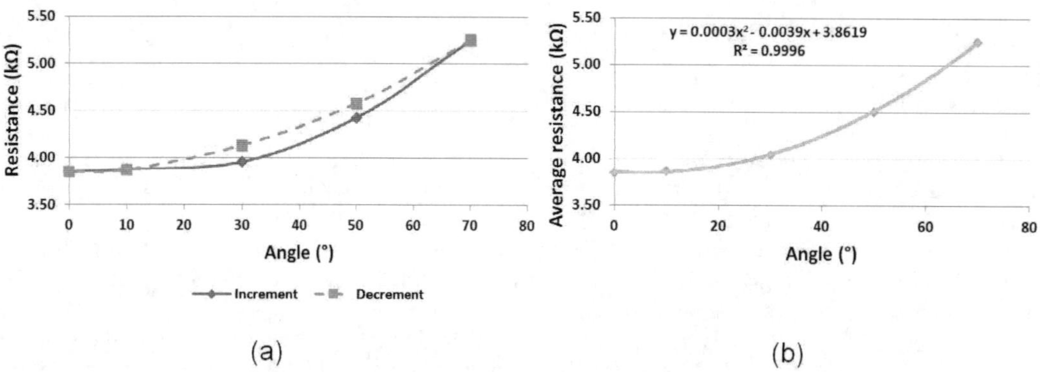

Figure 5. Resistance as a function of the bend sensor deflection angle with polyester coating: (a) Resistances in the angle increment and decrement; (b) Regression equation and coefficient of determination (R^2).

Regarding the bend sensor with polyimide coating, the results are shown in Figure 6. This sensor had a maximum hysteresis equal to 26 % of the full scale (Figure 6a) and, such as the sensor coated with polyester, a quadratic relationship between the average resistance, which ranged between 10.23 kΩ and 10.43 kΩ, and its deflection (Figure 6b). Similarly, a regression equation and the coefficient of determination (R^2) were also obtained through Excel® software (Figure 6b).

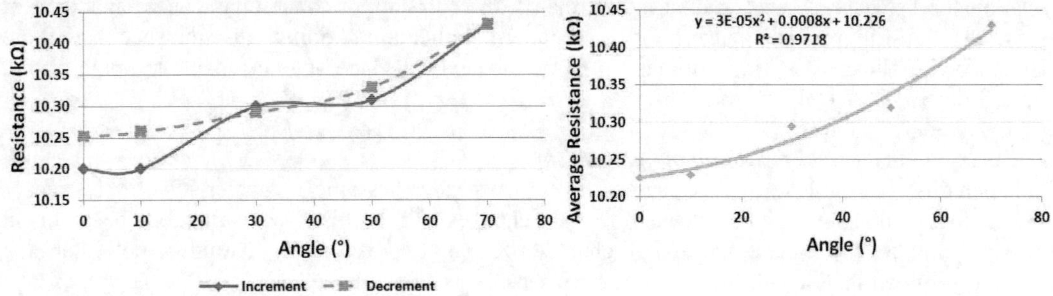

(a) (b)

Figure 6. Resistance as a function of the bend sensor deflection angle with polyimide coating: (a) Resistances in the angle increment and decrement; (b) Regression equation and coefficient of determination (R^2).

Considering the change in the flow velocity of water inside a pipe, it is interesting and necessary that the proposed water meter based on a thin-film resistive sensor has the smallest hysteresis value as possible, that is, the highest possible agreement between the "loading" and "unloading" curves. Thus, the two characterized sensors demonstrate promising applicability for measuring water consumption. Therefore, when comparing sensors with different coatings, in this case, the polyester one presents lower hysteresis.

As mentioned before, to previously analyze the application of the bend sensor to measure the volume of water flowing in a pipe, it was necessary to find a correlation between the sensor resistance and the flow velocity. For this, using the results of the bend sensor electromechanical characterization and the analytical calculations explained in section 2.3, it was possible to obtain this relationship. For each angle to which the bend sensor was submitted in this characterization, was obtained the average resistance – thus being able to obtain $\Delta R/R_0$ – and the variable v^2 – obtained from equation (3), in which its parameters are found in Table 1. Therefore, the relationship between the relative change in the bend sensor resistance and the flow velocity of equation (5) was established, as shown in Figure 7.

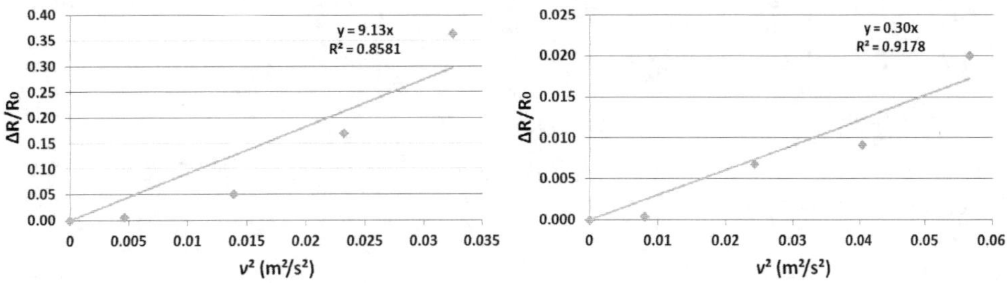

(a) (b)

Figure 7. Relative resistance increment response of the bend sensor: (a) With polyester coating; (b) With polyimide coating.

As a result of this correlation between sensor resistance and flow velocity, through statistical regression analysis, the polyester-coated bend sensor presented an adjusted deflection coefficient (K') equal to 9.13 s^2/m^2 and a linear model that explains approximately 86 % of the variance of the variable v^2 from the variable $\Delta R/R_0$. In turn, the bend sensor with polyimide coating presented K' equal to 0.30 s^2/m^2 and a coefficient of determination (R^2) of 0.92. In addition, for both correlations, the regression statistical significance was tested, and a result was obtained indicating statistical evidence that there is a relationship between the analyzed variables, that is, the correlations found characterize an excellent response of thin-film resistive sensors for the proposed application. Additionally, the speeds used are relevant for installing piping in buildings. In Brazil, for example, the velocity inside the pipe cannot exceed 3 m/s [31].

4.2. Electrothermal characterization of bend sensor

Fundamentally, the purpose of this characterization was to assess the characteristics of the sensor resistance variation when subjected to temperature changes and, in this way, evaluate its behavior. In the same way as the electromechanical characterization, two types of coating – polyester and polyimide – of two-inch length thin-film resistive sensors were evaluated.

This methodology allowed measuring the sensor resistances in the increase and decrease of temperature with the fixed curvature angle, consequently, it was possible to calculate and evaluate the hysteresis concerning the temperature. The first bend sensor characterized was the polyester-coated (the same that was used in the electromechanical characterization), which, when subjected to heating and cooling of the climate chamber, showed the same behavior for all angles tested - as the temperature increased, the sensor resistance decreased, and when the temperature decreased, the resistance increased (Figure 8) – thus demonstrating a good agreement between the "loading" and "unloading" curves of the sensor.

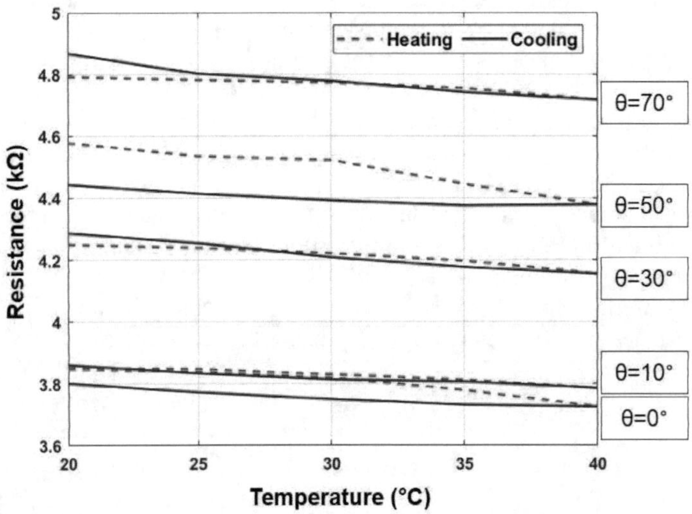

Figure 8. Resistance response of the polyester-coated bend sensor to temperature variation for different deflection angles.

In turn, the second sensor characterized was the one with a polyimide coating, the same that was used in the electromechanical characterization. When subjected to the cycle of the climatic chamber, this sensor presented a different behavior between the tested angles (Figure 9). When heating at angles of 0° (sensor at rest) and 10°, the resistance values decreased, and during cooling, these values increased (Figure 9a), as occurred in the polyester-coated sensor and were expected. However, for the angles of 30°, 50°, and 70°, this behavior was not maintained. At 30° angle, the resistance decreased with increasing temperature up to 35 °C, from 35 °C to 40 °C the resistance increased, and then, during cooling, the resistance continued to increase (Figure 9a). At angles of 50° and 70°, as the temperature increased, the resistance of the sensor also increased, and during cooling, with the decrease in temperature, the sensor resistance increased considerably (Figure 9b) – thus denoting a behavior irregular between the sensor "load" and "unload" curves.

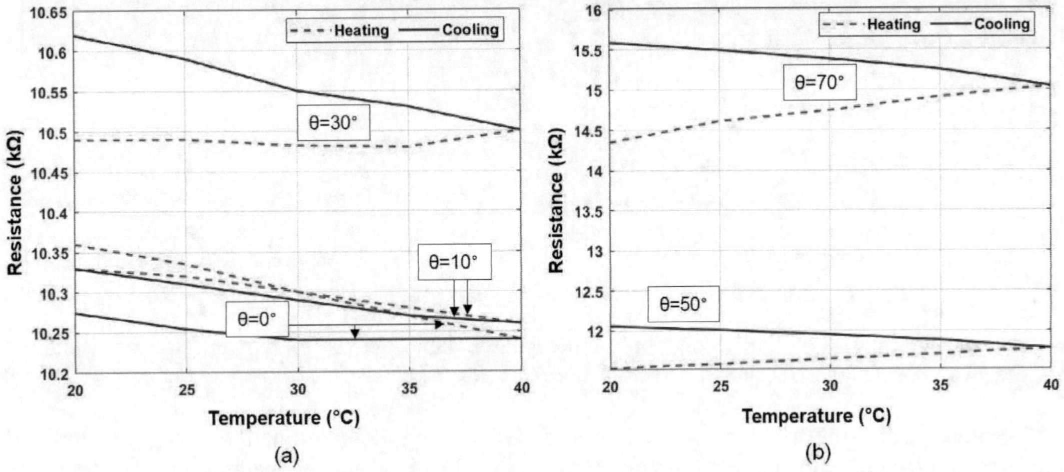

Figure 9. Resistance of the first polyimide coated bend sensor to temperature variation for deflection angles: (a) 0°, 10°, and 30°; (b) 50° and 70°.

In general, the bend sensor behavior with polyimide coating verified in this electrothermal characterization can be described as weird, as this polymer is part of the group of super engineering plastics, that is, they are ultra-performance polymers. Polyimide has significant and particular characteristics, such as, chemical and thermo-oxidative stability at high temperatures, resistance to solvents and radiation, low flammability and smoke emission, superior semiconductor properties, and high thermal stability. The latter is one of the most outstanding qualities of polyimide, enabling this material to be used for long periods at temperatures up to 200 °C and, on the other hand, for the short term at temperatures up to 480 °C without affecting its mechanical behavior [31–32].

As mentioned, this type of plastic has excellent electrical, chemical, and physical-mechanical properties, raising doubts regarding the results obtained in this characterization. For this reason, a third bend sensor – also coated with polyimide and two inches in length – was subjected to the identical electrothermal characterization to confirm the observed anomalous behavior.

Similar to the first bend sensor with a polyimide coating, this sensor showed irregular results (Figure 10). For the angles of 0°, 10°, and 30°, it presented a behavior with good hysteresis (Figure 10a); however, for the 50° and 70° angles, it demonstrated a disagreement between the "loading" and "unloading" temperature curves (Figure 10b). Consequently, this behavior makes it impossible to use the thin-film resistive sensor with polyimide coating for measuring water consumption in pipes of autonomous units, since, according to the specifications of the National Institute of Metrology, Quality, and Technology (INMETRO), water meters must be constructed of materials resistant to temperature variations between +1 °C and + 40 °C [34].

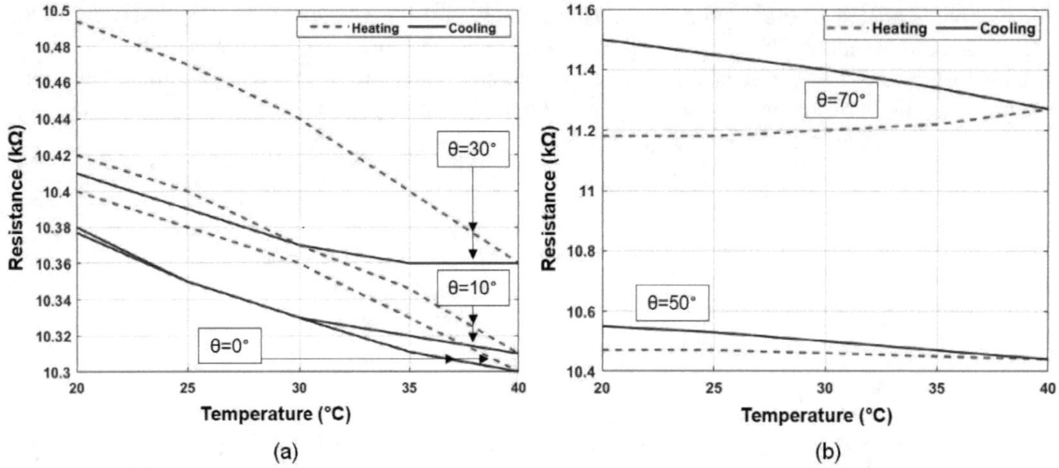

Figure 10. Resistance of the second polyimide coated bend sensor to temperature variation for deflection angles: (a) 0°, 10°, and 30°; (b) 50° and 70°.

Therefore, since the bend sensor with polyester coating demonstrated a more regular and satisfactory behavior, with better conformity between the "loading" and "unloading" curves, Figure 11 presents the results obtained indicating the measured quantities of two different ways: in Figure 11(a) the resistances average between the values in the increment and decrement of the temperature for each angle and Figure 11(b) the equal quantities related utilizing a color map.

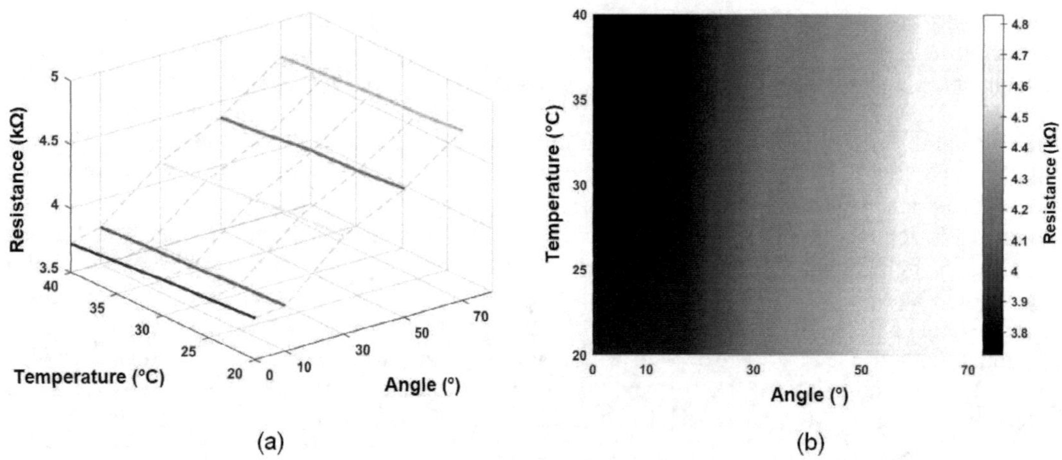

Figure 11. Response of the polyester-coated bend sensor about the average resistance between the temperature increment and decrement values for each angle, through: (a) Three-dimensional measured quantities graph; (b) Color map.

As previously exposed, the polyester-coated bend sensor had a negative linear relationship between the sensor resistance and the temperature variation, that is, the resistance decreases when the temperature increases. In this sense, the resistance response of the polyester-coated bend sensor can be seen in Figure 12(a) and Figure 12(b).

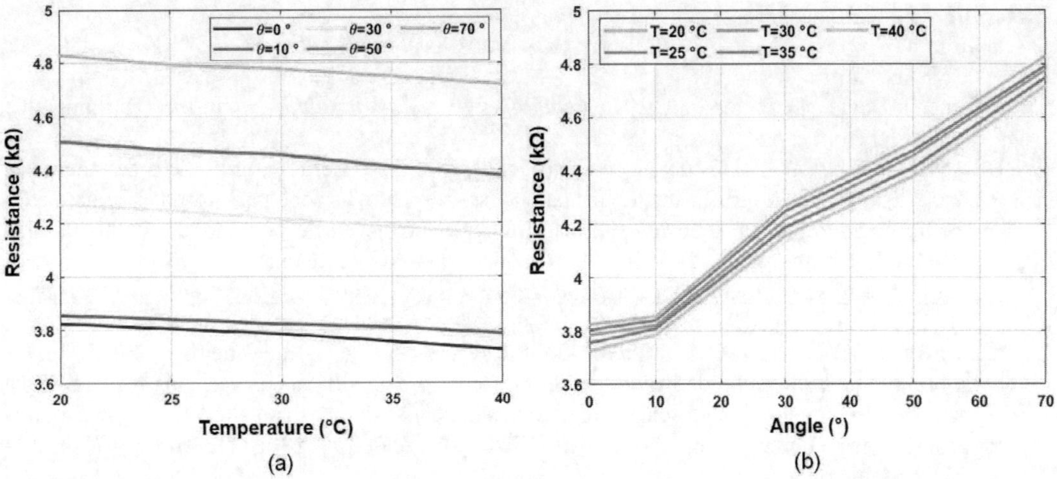

Figure 12. Polyester-coated bend sensor resistance as a function of: (a) Temperature; (b) Angle of deflection.

Although thin-film resistive sensors present, in the electromechanical characterization, an appropriate quadratic behavior between the sensor resistance and the deflection angle variation, it is noteworthy that this characterization does not consider the temperature effect. Therefore, for the proposed water meter, the electrothermal characterization test is more significant, as it is more similar to the reality of building pipes. Therefore, because the bend sensor with polyester coating has demonstrated a more stable behavior compared to polyimide, as well as a lower hysteresis in electrothermal characterization, it stands out as more viable in terms of applicability for measuring water consumption in pipes of autonomous units of buildings.

5. Conclusions

In this work, two resistive thin-film sensors (bend sensor) with two inches in length and with different coatings - one of polyester and the other of polyimide - were characterized about the angular deflection and the effect of temperature to evaluate the applicability of these sensors for measuring water consumption in the pipe of autonomous units in buildings.

For both sensors, the methodology used in the electromechanical characterization allowed to obtain the quadratic relationship between the sensor resistance and the angular variation and the linear correlation between the resistance and the flow velocity. Additionally, to consider that this linear correlation is acceptable, through the F-test, the regression statistical significance was tested.

In turn, the electrothermal characterization methodology allowed evaluating the resistance behavior when subjected to temperature variation. In this case, the polyester sensor demonstrated excellent performance – with increasing temperature, the resistance decreased. In contrast, the polyimide-coated sensor showed an unstable operation. Even though this type of polymer is excellent for applications with temperature variation, the behavior of resistance change with temperature change was different for the sensor deflection angles, making it necessary to characterize another bend sensor with polyimide coating for ratification purposes, which also showed the same irregular results.

Therefore, as the intended application of the proposed hydrometer is for measuring water consumption in building pipes, the polyester bend sensor demonstrated better feasibility to be used as a transducer for this meter. It can be easily inserted inside any hydraulic component (such as flow control valves), normally present in these installations, even avoiding major inconveniences with usual works in cases of installation of conventional water meters. Thus, as a suggestion for future work, it is expected to carry out computational simulations and tests in a hydraulic section to experimentally corroborate the potential of the bend sensor with polyester coating for measuring water consumption in autonomous building units.

6. References

[1] Nations U 2021 *The United Nations World Water Development Report 2021: Valuing Water.* UNESCO, Paris

[2] Gardner J, Doyle M e Patterson L 2017 *Estimating the value of public water data* (Durham, NC: Duke University) p 16

[3] BRASIL 2016 Lei nº 13.312, de 12 de julho de 2016 Altera a Lei nº 11.445, de 5 de janeiro de 2007 Estabelece diretrizes nacionais para o saneamento básico, para tornar obrigatória a medição individualizada do consumo hídrico nas novas edificações condominiais. **Diário Oficial da União**: seção 1 Brasília, DF n 132-A p 1-4 Available on: <http://pesquisa.in.gov.br/imprensa/jsp/visualiza/index.jsp?jornal=1000&pagina=1&data=12/07/2016>. Accesseed 2 April 2020

[4] GOVERNO DO DISTRITO FEDERAL 2005 Lei nº 3.557, de 18 de janeiro de 2005 Dispõe sobre a individualização de instalação de hidrômetro nas edificações verticais residenciais e nas de uso misto e nos condomínios residenciais do Distrito Federal, e dá outras providências. **Diário Oficial do Distrito Federal**: seção 1 Brasília, DF ano 35 n 14 p 1-52 Available on: <http://www.buriti.df.gov.br/ftp/diariooficial/2005/01_Janeiro/DODF%20014%2020-01-2005/Se%C3%A7%C3%A3o01-%2014.pdf>. Accesseed: 20 November 2020

[5] NITERÓI, RJ 2006 Lei nº 2.340, de 6 de junho de 2006 Estabelece, para projetos de edificações coletivas, a exigência de localização de hidrômetro para medição do consumo de água de cada unidade autônoma. **Diário Oficial da Prefeitura Municipal de Niterói**: atos do prefeito, Niterói, RJ p 1-15 Available on: <http://pgm.niteroi.rj.gov.br/downloads/do/2006/06_Jun/07.pdf>. Accesseed: 26 November 2020

[6] RECIFE, PE 2002 Lei nº 16.759, de 17 de abril de 2002 Institui a obrigatoriedade da instalação de hidrômetros individuais nos edifícios. **Diário Oficial da Prefeitura do Recife**: poder executivo, Recife, PE Available on: <http://www.recife.pe.gov.br/diariooficial-acervo/exibemateria.php?cedicacodi=203&aedicaano=2002&ccadercodi=2&csecaocodi=2&cmatercodi=6&QP=Lei+16759&TP=>. Accesseed: 26 November 2020

[7] CURITIBA, PR 2003 Lei nº 10.785, de 18 de setembro de 2003 Cria no Município de Curitiba o Programa de Conservação e Uso Racional da Água nas Edificações - PURAE. **Diário Oficial do Município de Curitiba**: atos do poder executivo, Curitiba, PR ano 36 n 72 p 1-12 Available on: <https://legisladocexterno.curitiba.pr.gov.br/DiarioConsultaExterna_Download.aspx?Id=291>. Accesseed: 26 November 2020

[8] Lima B C, Yamaguchi J K, Kussaba L L e Ferreira A T 2006 Sistema de medição individualizada de água: estudo de caso de edifício comercial em São Paulo. *REEC - Revista Eletrônica de Engenharia Civil* **11** pp 56-66 DOI 10.5216/reec.V11i3.37331

[9] Chaib E B D, Faria D M C P e Nascimento N de O 2017 Avaliação do potencial de redução do consumo de água potável por meio da medição individualizada de água em edifícios residenciais *XXII Simp. Brasileiro de Recurcos Hídricos* (Florianópolis) **22**

[10] Sousa V, Silva C M e Meireles I 2019 Performance of water efficiency measures in commercial buildings *Resour. Conserv. Recycl.* **143** pp 251–9 DOI 10.1016/j.resconrec.2019.01.013

[11] Orengo G, Sbernini L, Lorenzo N Di, Lagati A e Saggio G 2013 Curvature Characterization of Flex Sensors for Human Posture Recognition *Univers. J. Biomed. Eng.* **1** pp 10–5 DOI 10.13189/ujbe.2013.010103

[12] Saggio G e Orengo G 2018 Flex sensor characterization against shape and curvature changes *Sensors Actuators, A Phys.* **273** pp 221–31 DOI 10.1016/j.sna.2018.02.035

[13] Saggio G 2014 A novel array of flex sensors for a goniometric glove *Sensors Actuators, A Phys.* **205** pp 119–25 DOI 10.1016/j.sna.2013.10.030

[14] Guo YR, Zhang XC e An N 2019 Monitoring neck posture with flex sensors *9th Int. Conf. Information Science Technology (Hulunbuir)* vol 7 pp 459–63 DOI 10.1109/ICIST.2019.8836806

[15] Borges L M, Barroca N, Velez F J e Lebres A S 2009 Smart-clothing wireless flex sensor belt network for foetal health monitoring *3rd Int. Conf. Pervasive Computing Technologies for Healthcare* pp 1–4

[16] Jabin J, Chowdhury A M, Adnan M E, Islam M R e Mahmud S S 2019 Low cost 3D printed prosthetic for congenital amputation using flex sensor *5th Int. Conf. On Advances in Electrical Engineering (Dhaka, Bangladesh)* pp 821–25 DOI 10.1109/ICAEE48663.2019.8975415

[17] Hu Q, Tang X e Tang W 2020 A smart chair sitting posture recognition system using flex sensors and FPGA implemented Artificial Neural Network *IEEE Sensors Journal* **20** pp 8007–16 DOI 10.1109/JSEN.2020.2980207

[18] Starck J R, Murray G, Lawford P V e Hose D R 1999 An inexpensive sensor for measuring surface geometry *Medical Engineering and Physics* **21** p 725–29 DOI 10.1016/S1350-4533(99)00097-1

[19] Yao C, Hong C, Su D, Zhang Y e Yin Z 2020 Design and verification of a wireless sensing system for monitoring large-range ground movement *Sensors Actuators, A Phys.* **303** p 1-7 DOI 10.1016/j.sna.2019.111745

[20] Ramesh M V, Vidya P T e Pradeep P 2013 Context aware wireless sensor system integrated with participatory sensing for real time road accident detection *Int. Conf. on Wireless and Optical Communications Networks* pp 1–5 DOI 10.1109/WOCN.2013.6616203

[21] Fan Y, Dai Z, Xu Z, Zheng S, Dehkordy F M, Bagtzoglou A, Zhang W, Gao P e Li B 2019 High resolution air flow velocity monitoring using air flow resistance-type sensor film (AFRSF) *Sensors Actuators, A Phys.* **297** pp 1–10 DOI 10.1016/j.sna.2019.111562

[22] Xu Z *et al.* 2019 Towards high resolution monitoring of water flow velocity using flat flexible thin mm-sized resistance-typed sensor film (MRSF) *Water Research X* **4** pp 1–10 DOI 10.1016/j.wroa.2019.100028

[23] Srinivasan CR, Sen S, Kumar A e Saibabu C 2014 Measurement of flow using bend sensor Proc 2014 *Int. Conf. on Advances in Energy Conversion Technologies (Manipal, India)* pp 27–30 DOI 10.1109/ICAECT.2014.6757056

[24] Stewart R L, Fox J F e Harnett C K 2013 Time-average velocity and turbulence measurement using wireless bend sensors in an open channel with a rough bed *Journal of Hydraulic Engineering* **139** pp 696–706 DOI 10.1061/(asce)hy.1943-7900.0000725

[25] Gonçalves L dos S, Medeiros K A R, Barbosa C R H 2021 Electromechanical characterization of resistive sensor for application in water metering *XXIII IMEKO World Congress (Yokohama, Japan)* pp 1–4

[26] Gonçalves L dos S, Medeiros K A R, Barbosa C R H 2023 Hydrometer design based on thin-film resistive sensor for water measurement in residential buildings *Water* **15** 1045 DOI 10.3390/w15061045

[27] Flexpoint *Standard bend sensor* Available on: https://media.digikey.com/pdf/Data%20Sheets/FlexPoint%20PDF's/Bend_Sensor_FLXT.pdf Accesseed 20 August 2020

[28] Kreith F 1999 *Fluid Mechanics* CRC Press LLC Florida

[29] Overview of materials for Polyester Film *MatWeb: Material Property Data* Available on: <http://www.matweb.com/search/DataSheet.aspx?MatGUID=40559706b4fd4aa0a43f5739799728f5&ckck=1>. Accesseed 2 February 2021

[30] Overview of materials for Polyimide *MatWeb: Material Property Data* Available on: <http://www.matweb.com/search/DataSheet.aspx?MatGUID=ab35b368ab9c40848f545c35bdf1a672>. Accesseed 2 February 2021

[31] ASSOCIAÇÃO BRASILEIRA DE NORMAS TÉCNICAS – ABNT 1998 NBR 5626 Instalação predial de água fria Rio de Janeiro pp 1–41

[32] Liaw DJ, Wang KL, Huang YC, Lee KR, Lai JY e Ha CS 2012 Advanced polyimide materials: syntheses, physical properties and applications *Progress in Polymer Science* **37** pp 907–74 DOI 10.1016/j.progpolymsci.2012.02.005

[33] Sroog C E 1991 Polyimides *Progress in Polymer Science* **16** pp 561–94 DOI 10.1016/0079-

6700(91)90010-I
[34] INSTITUTO NACIONAL DE METROLOGIA NORMALIZAÇÃO E QUALIDADE – INMETRO 2000 Portaria nº 246 de 17 de outubro de 2000 pp 1–12

Acknowledgments
The authors are grateful for the financial support of the development agencies CNPq, CAPES, FINEP, and FAPERJ and for the logistical support of the Centro Tecnológico do Exército – Brasil (CTEx). This work was financed in part by the Coordenação de Aperfeiçoamento de Pessoal de Nível Superior – Brasil (CAPES) – Finance Code 001.

Study on the reproducibility of the ice point realization at the Thermometry Laboratory of Inmetro

M A Pereira Neto[1,3], P H F Diniz[1], A G Torres [2], A R Oliveira[1] and K N Quelhas[1]

[1] Thermometry Laboratory, National Institute of Metrology, Quality and Technology (Inmetro), Duque de Caxias, RJ, Brazil
[2] Fluid Measurement Sector, National Institute of Metrology, Quality and Technology (Inmetro), Duque de Caxias, RJ, Brazil
[3] Author to whom any correspondence should be addressed.

maneto@inmetro.gov.br

Abstract. Despite being classified today as a secondary reference point of temperature, the ice point (0.0 °C) is still very important in many thermometry applications due to its simplicity, effectiveness, reproducibility and low cost. Uncertainties as low as some hundreds of microkelvin can be achieved, but it depends on the level of care of the experimental procedure practiced by the laboratory. The Thermometry Laboratory of Inmetro (Later) has adopted for a long time a conservative value available in literature for the uncertainty of the ice point. In order to reach more realistic results, in line with the specific equipment and experimental procedure of the lab, a study on the reproducibility of the ice point realization was conducted, enabling a significant reduction of the uncertainty previously adopted. In addition, from the uncertainty analysis performed, it was possible to show the great importance of the standardization of the ice bath preparation procedure for the achievement of better results.

1. Introduction

The ice point, as it is traditionally known, refers to the equilibrium temperature of the air-saturated mixture of ice and water, which occurs at 0.0 °C, at a pressure of 101.325 kPa [1]. Its importance in the field of thermometry is undeniable, since it was used for a long time as the defining fixed point for many temperature scales before being replaced by the triple-point of water in the International Practical Temperature Scale of 1948, Amended Edition of 1960 (IPTS-48) [2,3,4].

Despite being downgraded to the status of a secondary fixed-point at this time, the ice point still remains very important for industry and the accredited calibration laboratories, because of its simplicity, effectiveness, reproducibility and low cost. In the Thermometry Laboratory of Inmetro (Later), it is used as a thermal medium for the reference junction of thermocouples and for the measurement of the 0 °C point in the calibration of liquid-in-glass thermometers, digital thermometers and industrial platinum resistance thermometers.

To achieve the best level of uncertainty, it is essential to follow the recommended procedures to realize the ice point [1,5]. According to the literature, when the ice point is realized adopting a very careful experimental procedure, uncertainties of the order of 100 μK can be achieved [6,7]. In spite of it, Later has adopted for a long time a conservative value available in literature for the uncertainty of the ice point [5,6]. It has been for decades characterized by a rectangular probability distribution, with a half-width of 0.005 °C. In order to evaluate a realistic uncertainty for the ice point realized at Later, in line with the specific equipment and experimental procedure of the lab, a more rigorous study on the reproducibility of the ice point temperature was conducted.

2. Experimental procedure

In this study, the measurements were performed by four different technicians, using the same equipment and following the same procedure. The technicians, designated in this study as T1, T2, T3 and T4, in addition to measuring, were responsible for preparing the ice bath and for any other actions necessary for the experimental setup.

Each technician performed five sets of measurements, at different days, so that it could reflect possible changes related to the experimental conditions, most of them inherent to the preparation of the ice point: ice/water ratio, aeration of the mixture, level of chemical impurities, thermometer immersion, waiting time for stabilization, etc. Each set was made up of three measurements in sequence: one at the triple point of water (*TPW1*), one at the ice point (*IP*) and another one again at the triple point of water (*TPW2*). Each measurement represented the mean of 100 readings (approximately 3 min), acquired after detecting temperature stabilization.

The measurement system was composed by a long stem, metal-sheathed, 25 Ω standard platinum resistance thermometer (SPRT), connected to a high accuracy digital thermometer set to perform measurements in electrical resistance instead of temperature. Regarding the ice point, it was prepared according to the *slush method*, proposed by the Consultative Committee for Thermometry (CCT). In this method, a slush made of shaved ice and cold liquid water must be well mixed and aerated. There must be sufficient ice in the slush to avoid having a layer of water containing no ice at the bottom of the container [1]. Shaved ice was obtained from deionized water using a food-grade ice machine and a Dewar vacuum flask was used to prepare and maintain the ice bath. In all measurements of the ice point, the thermometer immersion was approximately 350 mm.

Figure 1 shows the main items that composed the experimental setup of this study.

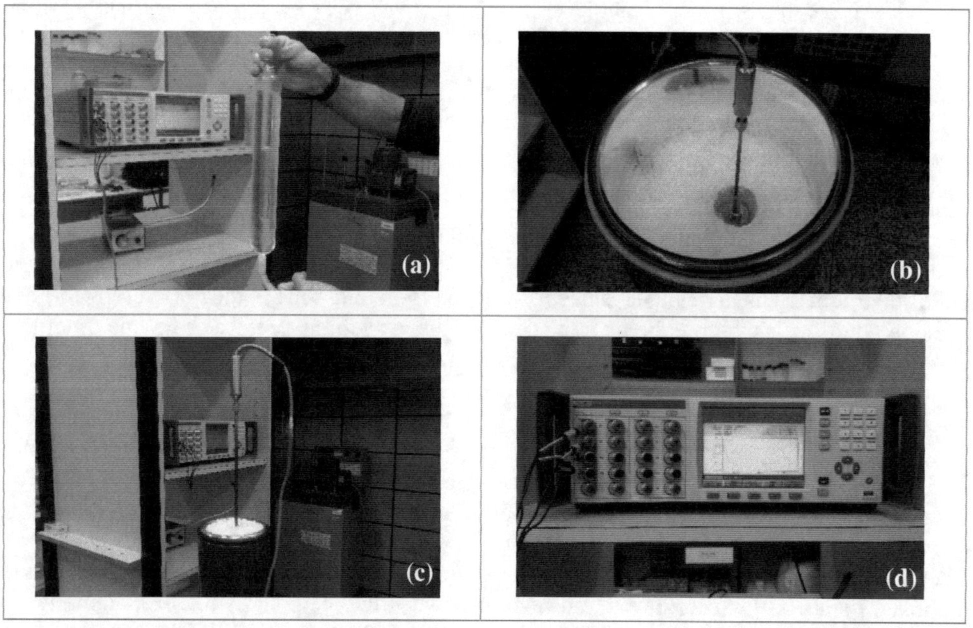

Figure 1. Triple point of water cell (a), SPRT during TPW cell measurements (b), SPRT during ice point measurements (c) and digital thermometer used for the resistance measurements (d).

3. Measurement model and uncertainties analysis

For the development of the measurement model used in this study, it was initially considered a simplified model, in which the difference between the temperatures of the triple point of water and the ice point ($T_{TPW} - T_{IP}$) is obtained from the corresponding difference between the thermometer signals in electrical resistance ($\bar{R}_{TPW} - \bar{R}_{IP}$), multiplied by its sensitivity around the ice point (dT/dR). In this model, represented in equation (1), the temperature of the ice point, considered as the measurand, is isolated and the resistance measured at the triple point of water is expressed as the arithmetic mean between the values of the first and second measurements.

$$\begin{aligned} T_{TPW} - T_{IP} = [\bar{R}_{TPW} - \bar{R}_{IP}]\frac{dT}{dR} \rightarrow T_{IP} = T_{TPW} - [\bar{R}_{TPW} - \bar{R}_{IP}]\frac{dT}{dR} \rightarrow \\ \rightarrow T_{IP} = T_{TPW} - \left[\frac{1}{2}(\bar{R}_{TPW1} + \bar{R}_{TPW2}) - \bar{R}_{IP}\right]\frac{dT}{dR} \rightarrow \\ \rightarrow T_{IP} = T_{TPW} - \frac{1}{2}\frac{dT}{dR}\bar{R}_{TPW1} - \frac{1}{2}\frac{dT}{dR}\bar{R}_{TPW2} + \frac{dT}{dR}\bar{R}_{IP} \end{aligned} \tag{1}$$

Based on this simplified model, the corrections for each input quantity are applied according to equation (2), whose symbols are described in table 1. It is important to highlight that this model was developed to determine the ice point temperature measured by each technician and, therefore, was applied separately for each one of them. A joint analysis of the results of the four technicians will be presented in section 4 (*Results and discussion*).

$$\begin{aligned} T_{IP} = (T_{TPW} + \delta T_{TPW}) - \frac{1}{2}\frac{dT}{dR}\left(\bar{R}_{TPW1} + \delta R_{TPW1,repr} + \delta R_{TPW1,res} + \delta R_{TPW1,hd}\right) \\ - \frac{1}{2}\frac{dT}{dR}\left(\bar{R}_{TPW2} + \delta R_{TPW2,repr} + \delta R_{TPW2,res} + \delta R_{TPW2,hd}\right) + \frac{dT}{dR}\left(\bar{R}_{IP} + \delta R_{IP,repr} + \delta R_{IP,res}\right) \end{aligned} \tag{2}$$

Table 1. Variables involved in the measurement model of the ice point temperature. Index *i* represents the first (*i* = 1) or second (*i* = 2) measurement at the TPW cell.

Symbol	Description
T_{IP}	Temperature measured at the ice point.
T_{TPW}	Temperature of the triple point of water according to ITS-90 (0.01 °C).
δT_{TPW}	Triple point of water cell temperature correction provided by its calibration certificate.
$\bar{R}_{TPWi}$	Mean electrical resistance of the SPRT in the *i*-th measurement at the TPW cell.
$\delta R_{TPWi,repr}$	Correction, in resistance, due to the reproducibility in the *i*-th measurement at the TPW cell, considering the five sets of measurements.
$\delta R_{TPWi,res}$	Correction, in resistance, due to the limited resolution of the digital thermometer in the *i*-th measurement at the TPW cell.
$\delta R_{TPWi,hd}$	Correction, in resistance, due to the effect of the hydrostatic head in the *i*-th measurement at the TPW cell.
$\bar{R}_{IP}$	Mean electrical resistance of the SPRT at the ice point.
$\delta R_{IP,repr}$	Correction, in resistance, due to the reproducibility in the measurement at the ice point, considering the five sets of measurements.
$\delta R_{IP,res}$	Correction, in resistance, due to the limited resolution of the digital thermometer in the measurement at the ice point.
dT/dR	Sensitivity of the temperature in relation to the electrical resistance of the SPRT, around 0 °C.

It should be noted that, with the exception of δT_{TPW} and $\delta R_{TPWi,hd}$, all other corrections are considered null. However, they must still be included in the model, as the uncertainties associated with each one of them are non-zero.

Regarding the evaluation of $\bar{R}_{TPWi}$ and $\bar{R}_{IP}$, it was initially calculated the mean of the 100 readings of each measurement set, and then the average of the results obtained for the five sets of measurements was computed.

Finally, to calculate the corrections arising from the effect of the hydrostatic head of the TPW cell, equation (3) was used.

$$\delta R_{TPWi,hd} = \frac{dT}{dh}(h_{col} - h_{sens})\frac{dR}{dT} \tag{3}$$

In this equation, dT/dh is the sensitivity of the temperature in relation to the height of the liquid column above the measurement point in the TPW cell and has a fixed value of -0.00073 °C/m. h_{col} and h_{sens} are, respectively, the height of the liquid column above the thermowell bottom and the distance between the end of the thermometer sheath and the midpoint of the sensing element, both in meters. Finally, dR/dT is the sensitivity of the electrical resistance of the SPRT in relation to the temperature, around 0 °C.

Since measurements i = 1 and i = 2 were performed using the same TPW cell and the same thermometer, and the effect caused by the hydrostatic head is systematic and dependent only on the characteristics of these two items, it is assumed that $\delta R_{TPW1,hd}$ and $\delta R_{TPW2,hd}$ are equal and totally correlated. To avoid the need to treat the correlation between these two input quantities in the uncertainty analysis, both will be treated as a single correction, expressed by $\delta R_{TPW,hd}$, so that the measurement model is reorganized in equation (4) as follows:

$$\begin{aligned} T_{IP} = (T_{TPW} + \delta T_{TPW}) - \frac{1}{2}\frac{dT}{dR}(\bar{R}_{TPW1} + \delta R_{TPW1,repr} + \delta R_{TPW1,res}) \\ - \frac{1}{2}\frac{dT}{dR}(\bar{R}_{TPW2} + \delta R_{TPW2,repr} + \delta R_{TPW2,res}) - \frac{dT}{dR}\delta R_{TPW,hd} + \frac{dT}{dR}(\bar{R}_{IP} + \delta R_{IP,repr} + \delta R_{IP,res}) \end{aligned} \tag{4}$$

Having made the previous considerations and defined the measurement model, the combined standard uncertainty of the ice point temperature measured by each technician is evaluated following the approach of the Guide for the Expression of Measurement Uncertainty – GUM 2008 [8]. Thus, the propagation of uncertainties is performed according to equation (5), in which u_j and c_j represent, respectively, the standard uncertainty of each input quantity and its respective sensitivity coefficient, which are listed in table 2.

$$u_c(T_{IP}) = \sqrt{\sum_j [c_j \times u_j]^2} \tag{5}$$

Table 2. Standard uncertainties, probability distributions and sensitivity coefficients considered for the evaluation of the combined standard uncertainty $u_c(T_{IP})$.

u_j	Description	Distribution	c_j
$u(\delta T_{TPW})$	Uncertainty due to the TPW cell temperature correction. It corresponds to the uncertainty value stated in its calibration certificate, divided by the respective coverage factor.	Normal	1
$u(\bar{R}_{TPWi})$	Uncertainty due to the repeatability of the resistance indications in the i-th measurement at the TPW cell. Given by the square root of the sum of the squares of the experimental standard deviations of the mean in each measurement set divided by five: $u(\bar{R}_{TPWi}) = \sqrt{(\sigma_1^2 + \sigma_2^2 + \sigma_3^2 + \sigma_4^2 + \sigma_5^2)/5}$.	Normal	$-\frac{1}{2}\frac{dT}{dR}$
$u(\delta R_{TPWi,repr})$	Uncertainty due to the reproducibility of the i-th measurement at the TPW cell. Given by the experimental standard deviation of the mean of the five resistance measurements taken in the first (i = 1) or second (i = 2) measurement at the TPW cell.	t-Student	$-\frac{1}{2}\frac{dT}{dR}$
$u(\delta R_{TPWi,res})$	Uncertainty due to the limited resolution of the digital thermometer in the i-th measurement at the TPW cell. It corresponds to half the resolution of 1×10^{-7} Ω, divided by $\sqrt{3}$.	Rectangular	$-\frac{1}{2}\frac{dT}{dR}$
$u(\delta R_{TPW,hd})$	Uncertainty due to the effect of the hydrostatic head on the first and second measurements at the TPW cell. It is considered the effect on resistance caused by a possible error of 5 mm when assessing the height of the hydrostatic head [replacing $h_{col} - h_{sen}$ by 0.005 m in equation (3)], divided by $\sqrt{3}$.	Rectangular	$-\frac{dT}{dR}$
$u(\bar{R}_{IP})$	Uncertainty due to the repeatability of the resistance indications in the measurement at the ice point. Given by the square root of the sum of the squares of the experimental standard deviations of the mean in each measurement set divided by five: $u(\bar{R}_{IP}) = \sqrt{(\sigma_1^2 + \sigma_2^2 + \sigma_3^2 + \sigma_4^2 + \sigma_5^2)/5}$.	Normal	$\frac{dT}{dR}$
$u(\delta R_{IP,repr})$	Uncertainty due to the reproducibility of the measurement at the ice point. Given by the experimental standard deviation of the mean of the five resistance measurements taken at the ice point.	t-Student	$\frac{dT}{dR}$
$u(\delta R_{IP,res})$	Uncertainty due to the limited resolution of the digital thermometer in the measurement at the ice point. It corresponds to half the resolution of 1×10^{-7} Ω, divided by $\sqrt{3}$.	Rectangular	$\frac{dT}{dR}$

4. Results and discussion

A summary of the results achieved by each technician is presented in table 3, including the ice point temperature T_{IP}, the combined standard uncertainty $u_c(T_{IP})$, the number of effective degrees of freedom η_{eff}, the coverage factor k (for a level of confidence of 95.45 %) and the corresponding expanded uncertainty $U(T_{IP})$.

Table 3. Summary of the results of each technician.

Technician	T_{IP} / °C	$u_c(T_{IP})$ / °C	η_{eff}	k	$U(T_{IP})$ / °C	w_k
T1	0.0008	0.00050	4.68	2.72	0.0014	0.029
T2	-0.00032	0.00014	11.96	2.23	0.00031	0.384
T3	-0.00038	0.00012	14.28	2.19	0.00027	0.492
T4	-0.00082	0.00028	4.80	2.69	0.00076	0.094

Table 3 also shows the weight related to each result. It is an inverse-variance weighting factor [9] used to determine a consensus value for the ice point temperature measured by the four technicians, being calculated according to equation (6). w_k and $u_{c,k}$ represent, respectively, the weight and the combined standard uncertainty for each technician.

$$w_k = \frac{\frac{1}{{u_{c,k}}^2}}{\frac{1}{{u_{c,1}}^2}+\frac{1}{{u_{c,2}}^2}+\frac{1}{{u_{c,3}}^2}+\frac{1}{{u_{c,4}}^2}} \tag{6}$$

In order to calculate the consensus value of the ice point temperature assessed in this study, it was used the Linear Pool method, implemented in the NIST Consensus Builder tool (NICOB) [10]. As input data, this method uses the main information about the result of each technician (ice point temperature, combined standard uncertainty, number of effective degrees of freedom and weight), assuming for each one a t-Student or a Gaussian probability distribution. Then, the probability distributions of the four technicians are combined to produce a resulting (consensus) distribution, formed from values drawn randomly from each input distribution. The probability of taking a value from a certain distribution is proportional to the corresponding weight w_k. The consensus value and its standard uncertainty are then given, respectively, by the mean and the standard deviation of the resulting probability distribution. From this, it is also possible to obtain the measurand lower and upper limits for a given level of confidence. The output of the results provided by NICOB tool is shown in figure 2.

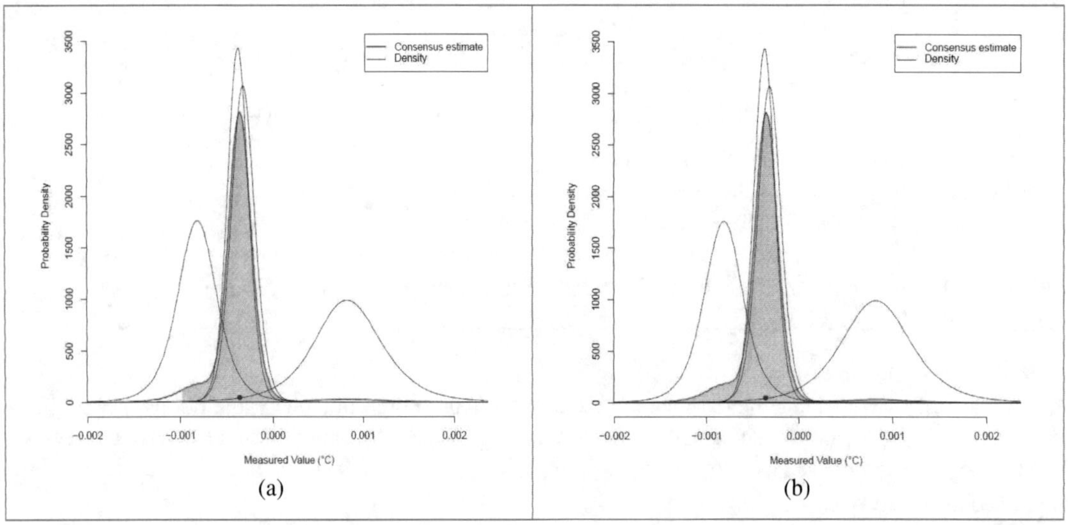

Figure 2. Output of the NIST Consensus Builder tool (NICOB), using the Linear Pool method and considering levels of confidence of 95.45 % (a) and 99.73 % (b).

In figure 2, the thinnest lines represent the probability density functions of the input distributions, while the thickest one represents the same function for the resulting distribution. The highlighted region represents the range corresponding to the considered level of confidence and the dot indicates the location of the consensus value. Table 4 presents a summary of the results generated by NICOB tool, using the Linear Pool method.

Table 4. Summary of the results generated by NICOB tool, using the Linear Pool method.

Consensus value / °C	Standard uncertainty / °C	Level of confidence of 95.45 %		Level of confidence of 99.73 %	
		Lower limit / °C	Upper limit / °C	Lower limit / °C	Upper limit / °C
-0.00036	0.00030	-0.00098	0.00051	-0.00148	0.00164

In figure 3, the ice point temperatures measured by the technicians, with the error bars representing their respective combined standard uncertainties, are displayed in a graph along with the lines indicating the consensus value and the limits of the intervals corresponding to the levels of confidence of 95.45 % and 99.73 %.

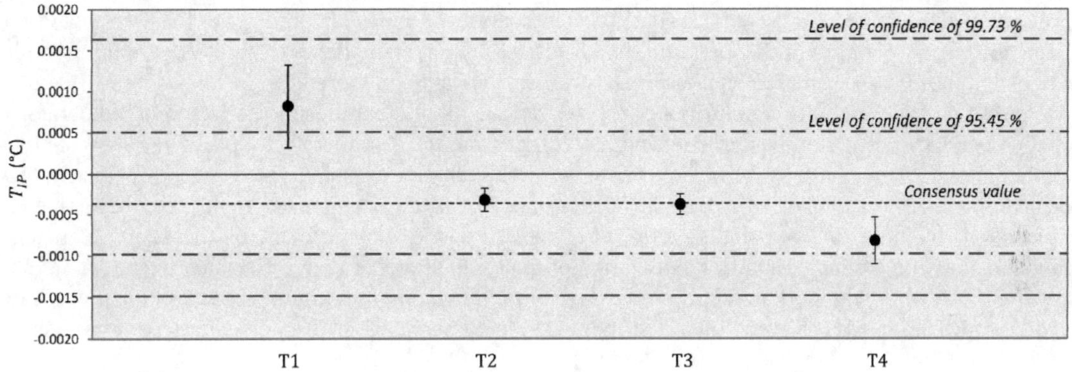

Figure 3. Measurement results of each technician (T1,…,T4), presented together with the consensus value and the limits corresponding to the levels of confidence of 95.45 % and 99.73 %.

The percentage contribution of each input quantity to the combined standard uncertainty was also evaluated. In this analysis, presented in figure 4, it is considered the ratio between the square of the standard uncertainty of each input quantity (in temperature units) and the square of the combined standard uncertainty of the ice point temperature measured by each technician (T1,…,T4).

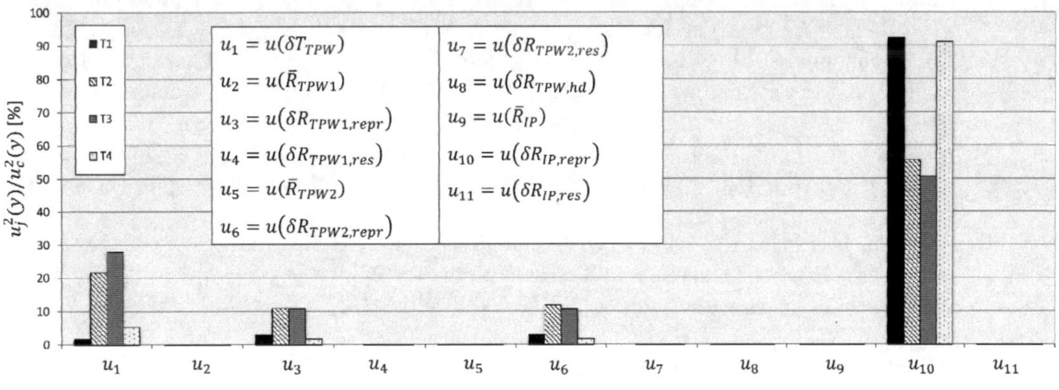

Figure 4. Percentage contribution of the standard uncertainty of each input quantity (u_j) to the combined standard uncertainty (u_c).

It can be seen that the greatest contribution to the combined standard uncertainty comes from the reproducibility of the five measurements of the SPRT resistance signal at the ice point, and then, to a lesser degree, from the calibration of the TPW cell and the reproducibility of the five measurements performed in the cell. Additionally, it can be seen that the other input quantities have only a negligible influence on the combined standard uncertainty.

5. Conclusion

Ice point temperature measurements performed by four different technicians resulted in a consensus value of -0.00036 °C. Considering the lower and upper limits of the intervals corresponding to the levels of confidence of 95.45 % or 99.73 %, it can be seen that such result is consistent with the definition value of this fixed point.

The standard uncertainty of 0.00030 °C, obtained for the probability distribution resulting from the calculation of the consensus value, is consistent with the level of uncertainty (of the order of 100 μK) understood as possible to be reached when the ice point is performed using a very careful experimental procedure. This result represented a significant reduction of the uncertainty previously adopted in Later for the temperature of the ice point, characterized by a rectangular probability distribution, with a half-width of 0.005 °C, based on conservative values available on literature.

Analysis of the percentage contribution of the various input quantities to the ice point temperature uncertainty indicates that the main contribution comes from the reproducibility of the measurement at the ice point. This may be associated with several factors that may vary slightly between the measurement sets: ice/water ratio, aeration of the mixture, level of chemical impurities, thermometer immersion, waiting time for stabilization, etc. Therefore, the improvement of the results depends fundamentally on the increasing standardization of the ice bath preparation procedure.

This study will be opportunely repeated as a way to assess possible improvements in the procedure and/or equipment used, or even converted into a routine assessment with an adequate pre-established frequency.

6. References

[1] Edler F, Kim Y G, Machin G, Pearce J and White D R 2018 *Guide on Secondary Thermometry: Specialized Fixed Points above 0 °C* (Paris: Consultative Committee for Thermometry) p 36.

[2] Stimson H 1961 International Practical Temperature Scale of 1948 - Text revision of 1960 *Journal of Research of The National Bureau Of Standard*, **65A**, n. 3, pp 139-145.

[3] Preston-Thomas H 1990 The International Temperature Scale of 1990 (ITS-90) *Metrologia* **27** pp 3-10 Erratum: *Metrologia* 27, p. 107.

[4] Hill K and Steele A 2014 The International Temperature Scale: Past, Present, and Future *NCSL International Measure*, The Journal of Measurement Science **9** 18.

[5] ASTM International 2011 ASTM E563-11 Standard Practice for Preparation and Use of an Ice-Point Bath as a Reference Temperature (West Conshohocken: ASTM International) pp 326-329.

[6] Mangum B W 1995 *Reproducibility of the temperature of the ice point in routine measurements*, TN-1411 (Gaithersburg: National Institute of Standards and Technology) p 17.

[7] Harvey A H, McLinden M O and Tew W L 2013 Thermodynamic Analysis and Experimental Study of the Effect of Atmospheric Pressure on the Ice Point *Temperature: Its Measurement and Control in Science and Industry* **8** AIP Conf. Series 1552 (New York: Ed. C Meyer and AIP) pp 221-226.

[8] Joint Committee for Guides in Metrology-Working Group 1 2008 JCGM 100:2008 Evaluation of measurement data: Guide to the expression of uncertainty in measurement (GUM), JCGM/WG 1.

[9] Steele A G and Douglas R J 2006 Extending chi-squared statistics for key comparisons in metrology *Journal of Computational and Applied Mathematics* **192** pp 51-58.

[10] Koepke A, Lafarge T, Possolo A and Blaza T 2020 *NIST Consensus Builder: User's Manual*, (Gaithersburg: National Institute of Standards and Technology) p 85.

The importance of proper storage of agar phantom on the water-glycerol-benzalkonium chloride mixture to maintain the ultrasonic group velocity

R M Souza[1], R P B Costa-Félix[2] and A V Alvarenga[2]

[1]Instituto Federal de Educação, Ciência e Tecnologia do Rio de Janeiro (IFRJ), Brasil
[2] Laboratory of Ultrasound (Labus), Directory of Scientific and Industrial Metrology (Dimci), National Institute of Metrology, Quality and Technology (Inmetro), Brasil

E-mail: raquel.ms.eng@gmail.com; rpfelix@inmetro.gov.br; adrotmail@gmail.com

Abstract. The present paper studied the metrological impact on the ultrasound group velocity in not storing the agar phantoms in the water-glycerol-benzalkonium chloride maintenance mixture. The ultrasonic velocity and the phantom's thickness were measured by the pulse-echo technique seven times within 44 days, whereas four replicates were carried out under repeatability conditions. The phantom's thickness did not differ statistically along the days since the normalized error was less than 1.0. Based on the results obtained, one can conclude that the ultrasound group velocity decreased in a logarithmic function along the days.

1. Introduction

The international standard IEC 60601-2-5:2015 [1] establishes basic safety and essential performance requirements for ultrasonic physiotherapy equipment. This particular standard recommends, among others, a test that aims to assess the heat generated on the surface of physiotherapy transducers. For this, a tissue-mimicking material (TMM) should be used. A phantom intended to inhibit cooling by convection and to model the acoustic properties of a specific tissue.

An agarose phantom's recipe is described, and the mixture components for its storage, preventing the soft TMM from degradation under the temperature between 18 °C to 25 °C [1]. The mixture comprises 88.1 % water weight, 11.9 % glycerol weight, and 0.5 % benzalkonium chloride weight. According to [2], the water-glycerol-benzalkonium chloride mixture adds to keep the long-term stability of ultrasonic properties of the phantom. In 2018, Souza et al. [2] assessed the long-term stability of the ultrasonic properties of a four years old TMM. In this 4-year gap, the phantom remained stored in a container filled with the water-glycerol-benzalkonium chloride mixture. The authors concluded that the TMM's shelf-life could be more extensive than two years if adequately preserved, as described in the standard [1].

The phantom's properties traceability is requisite to carry out reliable tests in evaluating the safety of ultrasonic physiotherapy equipment. There are several studies of the ultrasonic properties of agarose phantoms [2][3][4]. The TMM's ultrasonic properties measurements, such as ultrasonic velocity and attenuation coefficient, are usually carried out in the experimental setup underwater. According to [5], when an agar phantom is removed from the maintenance mixture and put underwater for a while repeatedly, a glycerol loss can be observed, impacting the ultrasonic group velocity.

Based on the preceding, the present paper intends to study the metrological impact on the ultrasound group velocity in not storing the agar phantoms in the water-glycerol-benzalkonium chloride maintenance mixture.

2. Materials and methods

2.1. Agar phantom

The reference value of the ultrasonic velocity of the standardized phantom [1] is 1,540 m.s-1 at the ambient temperature. One agar phantom (Ø 70 mm x 20 mm) was prepared according to the detailed recipe described in [1][6]. After that, it was stored in a tightly closed beaker containing a water-glycerol-benzalkonium chloride mixture in other to prevent it from drying out and possible degradation.

2.2. Experimental setup

The experimental setup (Figure 1) consists of a 5 MHz central frequency transducer (model A309S) (Olympus-NDT, USA) acting as transmitter and receiver. It is excited by a waveform generator model 33500B (Agilent, USA) with a sine burst of 20 cycles, 20 V peak-to-peak amplitude at 5 MHz. An oscilloscope model DSO-X 2014A (Agilent, USA) digitalizes the acquired signals. The experimental setup includes a stainless steel cylindrical reflector (Ø 63 mm x 10 mm). A calibrated thermo-hygrometer measures the temperatures of the distilled water inside the Ø 300 mm x 110 mm high water-bath model 557 (Fisatom, Brazil). A positioning system aligns the transducer. A similar experimental setup is described in detail in [5].

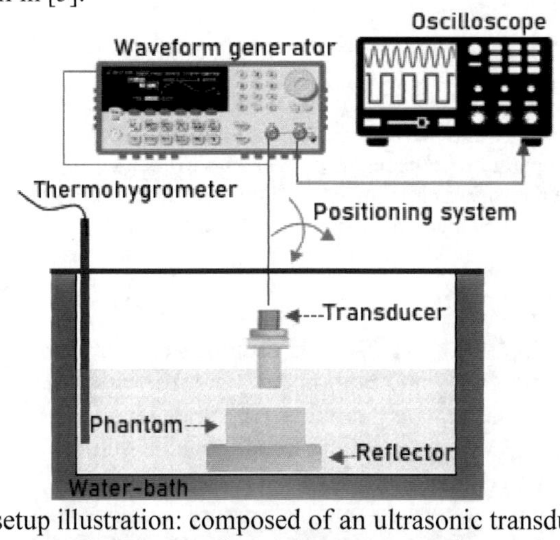

Figure 1. Experimental setup illustration: composed of an ultrasonic transducer, a stainless steel target (reflector), an oscilloscope and the waveform generator, a thermo-hygrometer to estimate the temperature of the water, and a positioning system for transducer alignment.

2.2.1. Measurement procedure. The phantom was stored in a closed beaker inside the water bath for 40 minutes before starting the measurements to obtain the thermal equilibrium. The beaker was removed during the measurements, and the phantom was kept inside the water bath filled with distilled water. On the first day, the ultrasonic velocity in the phantom, newly manufactured and until then preserved in the water-glycerol-benzalkonium chloride mixture, was measured. After that, the mixture was removed from the beaker. Then, the phantom was kept in the closed and dry beaker. Twenty-four hours later (Day 2), the ultrasonic velocity was measured under the same conditions. After that, the phantom remained stored in the dry and closed beaker. Measurements were carried out under the same conditions on days 4, 8, 16, 36, and 45. Four replicates of ultrasonic velocity measurements were performed each day. Regarding the pulse-echo technique used in the measurements, more details can be seen in [5]. It is essential to highlight that the measurements were carried out under repeatability conditions.

2.3. Measurement Uncertainty
The measurement uncertainty was estimated based on the BIPM/JCGM:100 (GUM) [7]. The mathematical model of the ultrasonic velocity can be found in [2]. The sources of uncertainty are well described in [5]. We previously presented Ishikawa's diagram to the ultrasound group velocity (Figure 2) in [5].

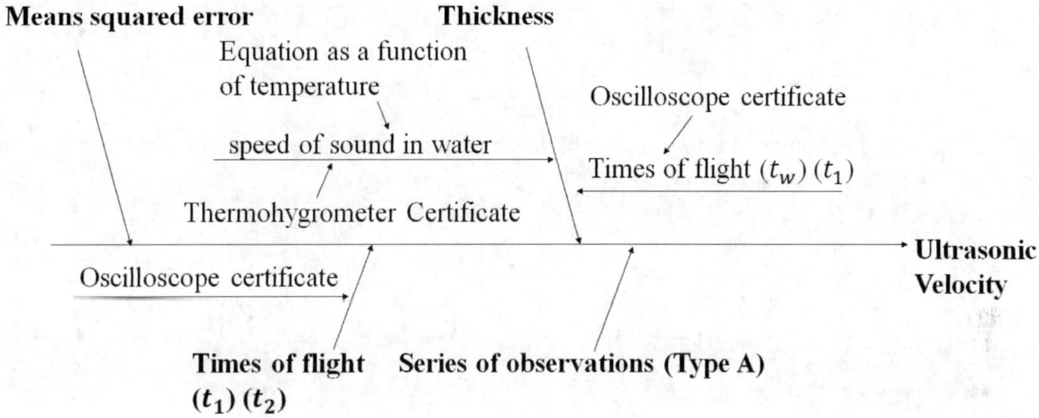

Figure 2. Ishikawa's diagram: sources of uncertainty for ultrasound group velocity [5].

The MSE's root calculated the uncertainty of the Mean Squared Error (MSE). This source was obtained from the regression line fit for all measurements.

3. Results
The phantom's thickness was measured using the pulse-echo technique under repeatability conditions (n = 4). Table 1 discloses the mean and the respective uncertainties to each day. Besides, Table 2 provides the results of the group velocity along the days, the standard uncertainties, the coverage factor, and the expanded uncertainties.

Table 1. The phantom's thickness and its respective expanded uncertainties (p = 0.95), together with Type A and Type B standard uncertainties and coverage factor (k).

Days	Thickness [mm]	Uncertainty Type A[mm]	Uncertainty Type B [mm]	Coverage Factor (k)	Expanded Uncertainty [mm]
1	19.956	0.004	0.014	1.96	0.029
2	19.946	0.0061	0.011	2.01	0.025
4	19.951	0.010	0.012	2.09	0.033
8	19.952	0.009	0.015	2.01	0.035
16	19.932	0.011	0.013	2.12	0.037
36	19.926	0.005	0.014	1.97	0.029
45	19.926	0.006	0.013	1.99	0.028

Table 2. The results of the group velocity measurements, its respective expanded uncertainties (p = 0.95), together with MSE, Type A and Type B standard uncertainties and coverage factor (k).

Days	Ultrasonic velocity [m s^{-1}]	Uncertainty Type A [m s^{-1}]	Uncertainty Type B [m s^{-1}]	MSE uncertainty [m s^{-1}]	Coverage Factor (k)	Expanded Uncertainty [m s^{-1}]
1	1539.2	0.56	1.48		1.97	3.6
2	1534.4	0.54	1.09		1.97	3.0
4	1528.1	0.50	1.26		1.97	3.2
8	1521.0	0.26	1.54	0.86	1.96	3.5
16	1514.2	0.24	1.32		1.96	3.1
36	1509.9	0.31	1.40		1.96	3.3
45	1506.8	0.26	1.26		1.96	3.0

The regression in Figure 3 establishes the mathematical relation between the ultrasonic velocity and the days in which the measurements were carried out.

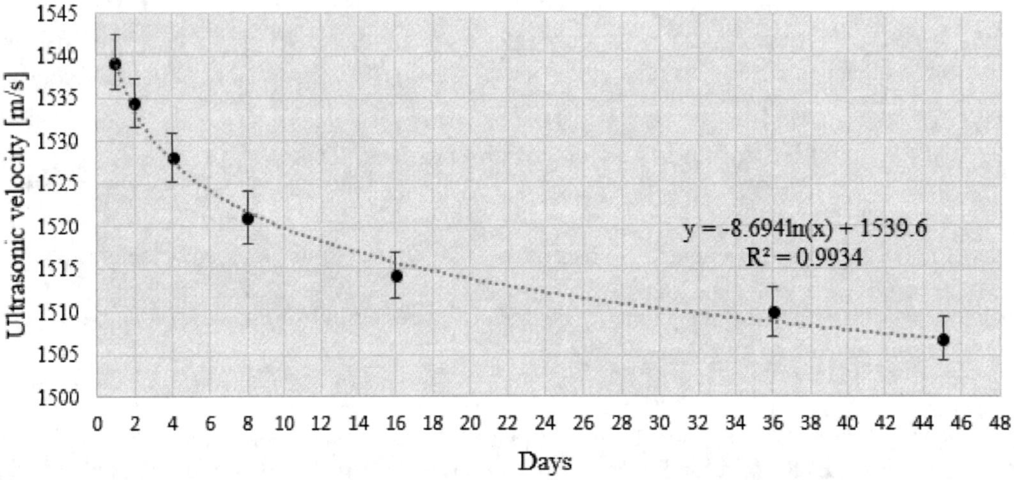

Figure 3. The results of the four replicates of group velocity measurements at each day, the uncertainty ranges and the regression line.

4. Discussions and Conclusion

Based on the result obtained (Figure 3), one can observe that the ultrasonic velocity decreases in a logarithmic function along the days. Besides that, the phantom's thickness (Table 1) did not differ statistically, in which the normalized error was less than 1.0. From the same graphic, we can infer that in just one week, the ultrasonic velocity decreased by 18.26 m s^{-1}.

The result presented here corroborates the ultrasonic velocity behavior of ultrasonic phantoms measured underwater, in which there is a loss of glycerol and a consequent reduction of the ultrasonic velocity [5][8].

It is essential to mention that this is the first controlled study that analyzes the impact of the absence of the water-glycerol-benzalkonium chloride mixing in standardized phantom storage.

Finally, one can conclude that not storing the agar phantoms in the water-glycerol-benzalkonium chloride mixture significantly impacts the ultrasonic velocity of the material along the time.

References

[1] Electrotechnical Commission. IEC 60601-2-5:2015: Medical electrical equipment Part 2-37: Particular requirements for the basic safety and essential performance of ultrasonic physiotherapy equipment 2.0 ed. Genova.

[2] Souza, R. M.; Monteiro, R. S.; Costa-Felix R. P. B.; Alvarenga, A. V. 2018 Ultrasonic properties of a four years old tissue-mimicking material. Journal of Physics: Conference Series **975**, 012025

[3] Santos, T. Q.; Alvarenga, A. V., Oliveira, D. P., Costa-Felix, R. P. B. 2017 Metrological Validation of a Measurement Procedure for the Characterization of a Biological Ultrasound Tissue-Mimicking Material. Ultrasound in Medicine & Biology **43**, pp 323-331

[4] Assis, M. K. M., Souza, R. M., Costa-Felix, R. P. B., Alvarenga, A. V. 2019 Assessment of ultrasonic properties of an agarose phantom at the frequency range 2.25 MHz to 10 MHz. In: Journal of Physics: Conference Series 1826 **1**:012005

[5] Souza, R. M., de Assis, M. K. M.; Costa-Félix, R. P. B., Alvarenga, A. V. 2020 Metrological Approach for Characterizing Ultrasonic Properties of Soft Tissue-Mimicking Material. In: Henriques J., Neves N., de Carvalho P. (Org.). IFMBE Proceedings. 1 ed.: Springer International Publishing. **76**, 1315-1328.

[6] Souza, R.M., Santos, T.Q., Oliveira, D.P., Souza, R.M., Alvarenga, A.V., Costa-Felix, R.P. B. 2016 Standard operating procedure to prepare agar phantoms. J. Phys. Conf. Ser. **733**, pp 012044

[7] BIPM, JCGM:100. Guide to the Expression of Uncertainty in Measurement (GUM). 2008 BIPM, Paris

[8] Alvarenga, A.V., Costa-Félix, R.P.B. Uncertainty assessment of effective radiating area and beam non-uniformity ratio of ultrasound transducers determined according to IEC 61689:2007. Metrologia, 2009, 46(3), pp. 367–374.

Acknowledgments

This study was financed in part by Pronametro (Inmetro). The authors thank the Brazilian National Council for Scientific and Technological Development (grants 312.501/2017-0 [RCF], 306.198/2017-7 [AVA], 430.376/2018-9 [RCF], 311.175/2020-1 [AVA], and 307.562/2020-4 [RCF]) and the Research Support Foundation of the State of Rio de Janeiro (grant E-26/201.563/2014 [RCF], and 260.671/2021 [RCF]).

Journal of Physics: Conference Series **2606** (2023) 012013 doi:10.1088/1742-6596/2606/1/012013

Basic procedures in the production of reference materials for ultrasonic non-destructive testing

E S F Pinto[1], W B Shiuan[1], E W J Santos[1], T C Dourado[1], R P B Costa-Félix[1]

[1]Laboratory of Ultrasound (LABUS), Directory of Scientific and Industrial Metrology (DIMCI), National Institute of Metrology, Quality, and Technology (INMETRO), Duque de Caxias – RJ, Brazil

tcdourado@inmetro.gov.br; rpfelix@inmetro.gov.br

Abstract. Conducting reliable measurements with traceability is a basic premise in commercial relations between nations. This role is performed by the national standardization institutes (NMI) and coordinated by the Bureau International des Poids et Mesures (BIPM). The use of Reference Materials (RM) and Certified Reference Materials (CRM) have contributed for decades, mainly to the pharmaceutical and biological products industry, providing high-quality measurements and adding value to the products. This research presents the primary standards and guidelines relevant to the definition of RM and CRM for the area of non-destructive testing (NDT) by ultrasound. The generation process of these materials must occur from short- and long-term stability and homogeneity tests following ISO Guide 35. Following a basic protocol defined for the experiments, the first results generated for the RM and CRM validation process are presented for the first results generated for a production process of a set of 24 carbon steel blocks for property calibration regarding ultrasonic velocity.

1. Introduction

The ability to perform measurements that follow metrological rigour plays an essential role in the economic development of nations. Measurement declarations and the respective uncertainties related to the physical quantities of the products must follow a credible international standardization that guarantees the fulfilment of the mutual agreements.

This standardization is guaranteed by the traceability of the physical properties measured and propagated by the National Metrology Institutes (NMI). Acting in the calibration of primary standards or distribution of Reference Materials (RM) and Certified Reference Materials (CRM), they have contributed for years in the production of high-quality measurements and adding value to products [1].

In the area of non-destructive testing (END) by ultrasound, it is no different. The RM and CRM generation process must occur from short- and long-term stability and homogeneity tests following ISO Guide 35 [2]. Additionally, its highest level of characterization is protected by the traceability chain that calculates the associated measurement uncertainty following the Evaluation of Measurement Data – Guide to the expression of uncertainty in measurement [3] and definitions contained in the International vocabulary of metrology (2012) [4].

The performance of ultrasonic measurements made using RM and previously calibrated devices guarantees the reliability of the generated results and allows for more informed decision-making. And in the case of CRM, since a calibration certificate backs these with international traceability, they allow adding even more quality to the results [1][5].

This work addresses the basic procedures for carrying out the feasibility study, planning and production of RM and CRM for metallic blocks of carbon steel, considering the physical property of the ultrasonic velocity of the longitudinal wave.

2. Theory

2.1 Ultrasound in END

Ultrasonic waves have frequencies between 20 kHz and 20 MHz, being above the limit of sounds detectable by the human ear. They correspond to mechanical waves that need a material medium to propagate, carrying acoustic energy. The passage of the ultrasonic wave through material causes disturbances in its atomic or molecular particles, removing the particles from their equilibrium position, which tend to return to their original positions due to the action of internal tensions [6].

Ultrasonic pulses generated in the available measuring instruments convert electrical signals into mechanical energy both on the way and on the way back [7]. These devices are typically composed of emission (emitted signal) and reception (received signal) transducers, both systems being integrated with different sensors or in one. This last case corresponds to the pulse-echo method and is shown in figure 1.

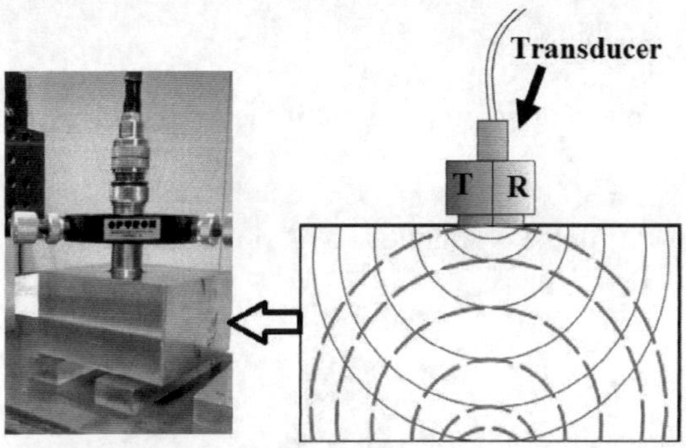

Figure 1. Schematic of the measurement system using the pulse-echo method.

Usually, devices that perform measurements in the field employ a standard block (used for checking the system) and a central processing unit, where an electrical impulse generator and a reading circuit are located.

These variations in signal energy are registered on the instrument's display and allow the evaluation of the times elapsed between the emission and reception of the ultrasonic signal, called the time of flight.

From collecting these propagated signals, it is possible to find discontinuities or interfaces between different materials. In other words, the occurrence of reflection and wave transmission phenomena allows indirect information such as location, depth, thickness, size, and orientation of the features of interest.

It is noteworthy that other losses of signal energy, such as acoustic attenuation of the medium and losses related to measuring instruments, must also be considered in the evaluations of the measured data. Specifically for END in ultrasound, the physical property it focuses on is the ultrasonic velocity of the longitudinal wave. In the context of RM and CRM, these arise to assist in the calibration of equipment when measuring materials with similar physical properties.

2.2 Standards for the production and development of RM and CRM

Both for the recognition of the basic nomenclature, as well as the preparation of the production planning of RM and CRM, it is essential to have access to the ISO GUIDE 35: 2017 [2], ISO GUIDE 30: 2016 [8], ISO GUIDE 33: 2019 standards [9], and NBR ISO 17034:2017 [10].

In Brazil, the standards NIE-DIMCI-30:2011 [11], NIE-DIMCI-31:2012 [12][11], NIE-DIMCI-32:2014 [13] as well as the documents DOQ-CGCRE-016:2016 [14] and DOQ-CGCRE-033:2012 [15], give support to interested parties.

2.3 Main requirements for RM and CRM

According to the definitions in ISO Guide 30:2016 [8], the reference material is a sufficiently homogeneous and stable material concerning one or more specified properties. The CRM is characterized as a metrologically valid procedure for one or more specified properties in the same document.

As already mentioned, what distinguishes the CRM from the RM is that the former is provided accompanied by a calibration certificate that presents the value of the measurand of the property of interest accompanied by its associated uncertainty, both with guaranteed measurement traceability [8].

2.4 Procedures in RM and CRM

Those interested in producing RM or CRM must have their quality system meeting the minimum requirements established by ISO/IEC 17025/2017 [16]. Essentially, this standard requires document control of the entire process, including a data protection system, validation of its measurement sheets, carrying out internal audits, carrying out preventive and corrective measures when applicable, in addition to valid calibrations for the systems directly involved in measurements.

In INMETRO, the standard that governs the planning and availability of these items is NIG-Dimci-001 - Planning and availability of reference materials (RM) and certificate of reference materials (CRM) [17], from planning to its availability.

According to the aforementioned NIG, the establishment of a property in a certain material comprises the study of all phases from planning to preparation of the report (in the case of RM) or the certificate (in the case of CRM).

Figure 2. presents requirements in the productions of RM and CRM.

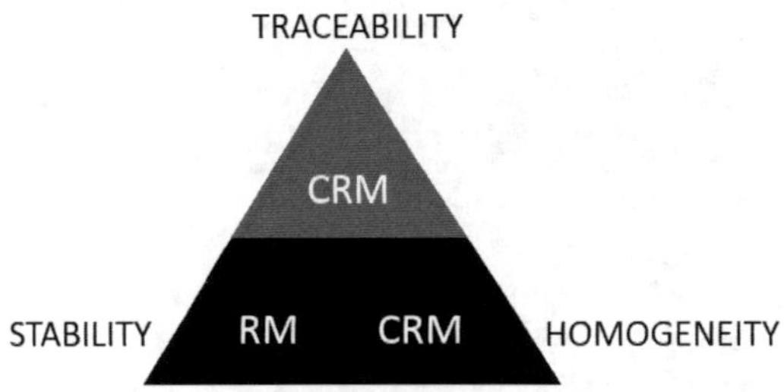

Figure 2. Pillars of RM and CRM production.

The flowchart outlined in figure 3 shows the main stages of the RM and CRM production process, from their selection to their availability, according to the ISO 17034:2017 standard [10].

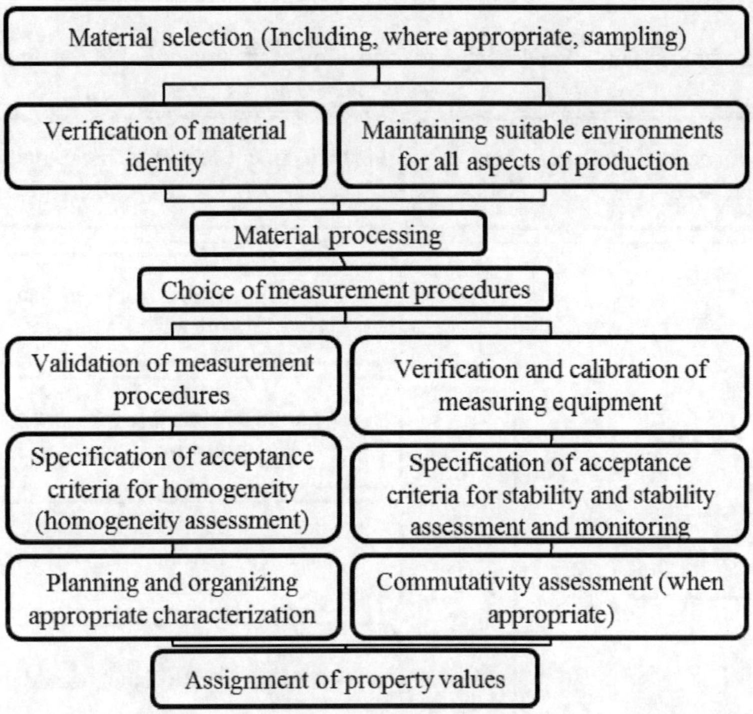

Figure 3. Flowchart with main stages in production of the RM and CRM.

From the selection of samples that would potentially meet the physical property in which the objective is to produce it as RM or CRM, the qualitative and quantitative assessment protocol must be carried out following the steps indicated in the flowchart in figure 4, created from main guide and standard mentioned.

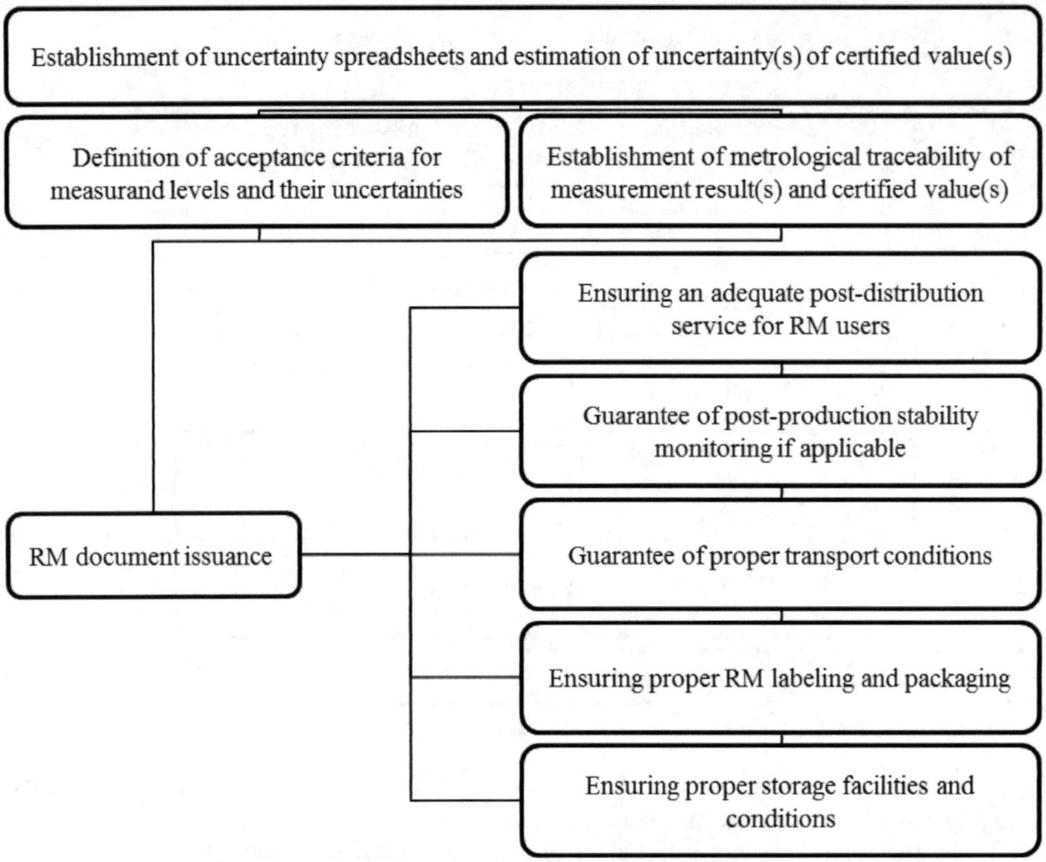

Figure 4. Flowchart with minimum metrological protocol to RM and CRM.

As already mentioned, this article discusses the basic planning guidelines to produce RM and CRM from metallic carbon steel blocks intend their subsequent availability for commercialization by INMETRO. The property selected for study and evaluation was the ultrasonic velocity of the longitudinal wave measured under the faces of these blocks.

3. Materials and methods

In this job, twelve 400 mm X 600 mm X 100 mm parallelepiped blocks were machined, named from here on from PB1 to PB12 and twelve stepped blocks with square faces of 400 cm X 400 mm and heights of 10 mm, 15 mm, 25 mm and 40 mm, referred to as SB1 to SB12.

Following the recommendation of ISO 16831: 2012 (E) Non-destructive testing - Ultrasonic testing - Characterization and verification of ultrasonic thickness measuring equipment [16], ultrasonic measurements of metal blocks must be performed with frequencies obtained from transducers above 5.0 MHz.

For the measurements performed in that planning, frequencies of 5.0 MHz, 7.5 MHz and 10.0 MHz were selected since they are commercially produced transducers and available in the ultrasound laboratory of INMETRO.

Figures 5a) and 6a) show images captured from each of the sets of said blocks. Likewise, figures 5b) and 6b) show the indication of the direction through a double arrow and respective nomenclatures related to the faces of dimensional measurements and ultrasonic signals.

The double blue arrows indicate the signals crossing the blocks and the green circles when the transducers were supported on this face.

The blocks also had their mass measured according to metrological rigour to assess possible losses resulting from the increasing and decreasing temperature concerning the ambient temperature.

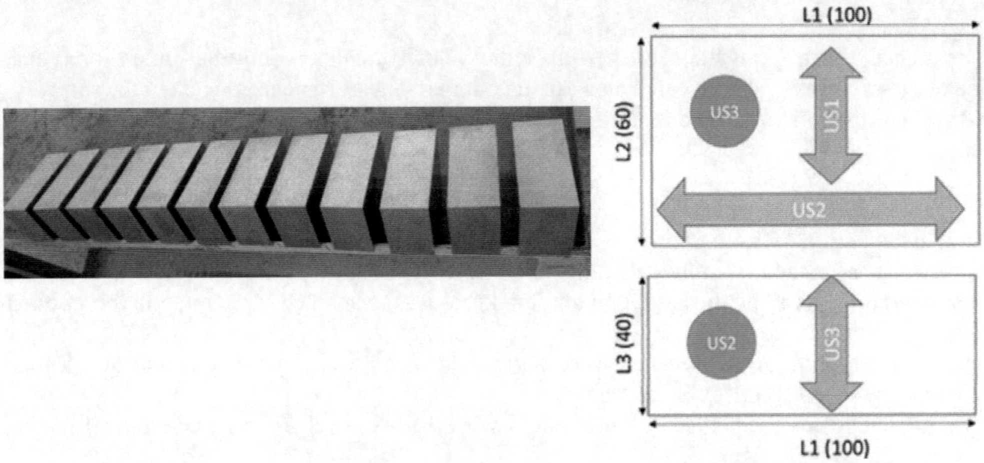

Figure 5. *Blocks B Nomenclatures of dimensions L1-L2-L3 and ultrasonic measurements US1-US2-US3 in lateral views of PB specimens.*

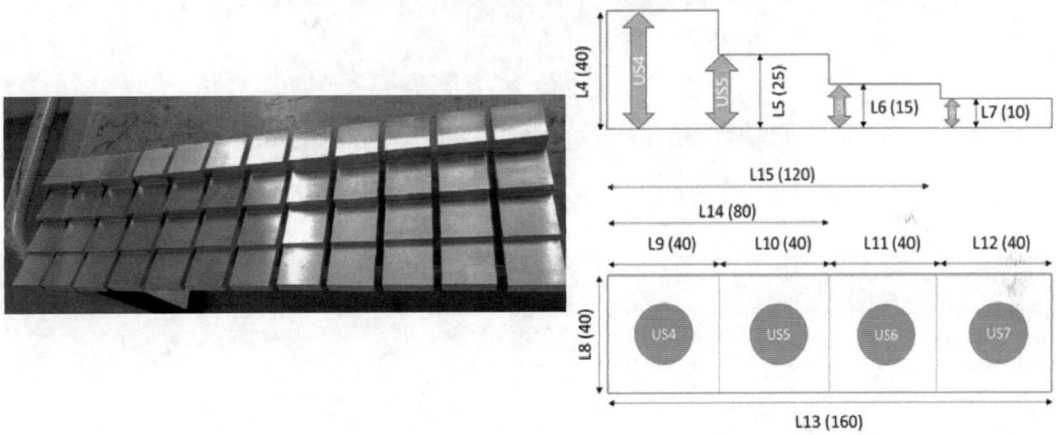

Figure 6. *Nomenclatures of dimensions L4-L5-L6-L7 and ultrasonic measurements US4-US5-US6-US7 in lateral (above) and upper (bellow) views of SB specimens.*

According to ISO GUIDE 30 [9], the measurement uncertainty to be declared in CRM is determined from the characterization, homogeneity and stability phases of the RM candidate samples.

3.1. Velocity ultrasonic quantity

From classical physics, the longitudinal ultrasonic velocity (LUV) can be determined from the relationship between the dimensional quantity and the propagation times of the wave within the blocks, that is, following the expression:

$$\boldsymbol{LUV} = \frac{2*L}{t} \qquad [1]$$

where:

LUV = longitudinal ultrasonic velocity;
L = block thickness, measured as described in 3.1.1;
t = double round trip time in the block;

Thus, as mentioned, the determination of the ultrasonic velocity requires only the dimensional and temporal quantities. However, this research also carried out mass measurements of each of the samples, aiming to determine their mass variations and allow to follow the evolution of the densities of the pieces throughout the study.

3.1.1. Dimensional measurement quantity

Each dimension was measured five times along its thickness. A 0-25 mm Mitutoyo micrometre (Japan) was used for measurements along the L1, L2 and L3 thicknesses, and another 25-50 mm micrometre for L4, L5 and L6 a calliper was used for L13, L14 and L15.
Dimensional measurements of the samples were performed using Maher 0-25 mm and 50-75 mm micrometres and an ABSOLUTE AOS digital calliper.
Note that dimensional measurements were performed with a spirit level for better alignment of the micrometer tips, according to figure 7.

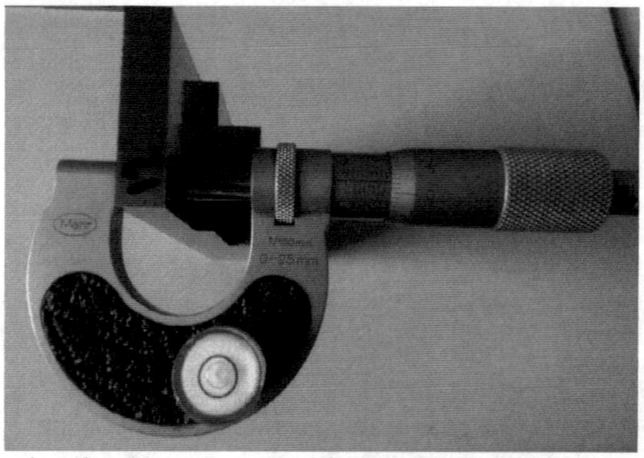

Figure 7. Dimensional measurements with support of micrometer and spirit level.

Transit time measurements of the longitudinal ultrasonic waves were carried out around the center of one of the faces of the blocks.

3.1.2. Time measurement quantity

The most general ultrasonic measurement procedure for longitudinal velocity calibration follows the pulse-echo method [19]. In this procedure, the emission and reception system are designed in a sensor that is coupled to the surface of the material to be tested, with or without the aid of a coupling fluid.

In this work, the pulse eco-method was used to determine and calibrate the ultrasonic velocity with the aid of a layer of mineral oil. The time-of-flight measurements were performed following the system shown in the figure 8. This is composed of an oscilloscope model DSO-X 3014A (Keysight, Malaysia), a function generator model 33250A (Agilent, Malaysia). The three ultrasonic transducers (central frequencies of nominal 5.0 MHz, 7.5 MHz, and 10.0 MHz) used had a diameter of 12.5 mm (model A-309S), manufactured by Olympus (USA).
Transit times were measured from the capture of the signals with the aid of the automation software LabView^{TM} (National Instruments, USA), according to the screen captured directly from that application and shown in figure 8.

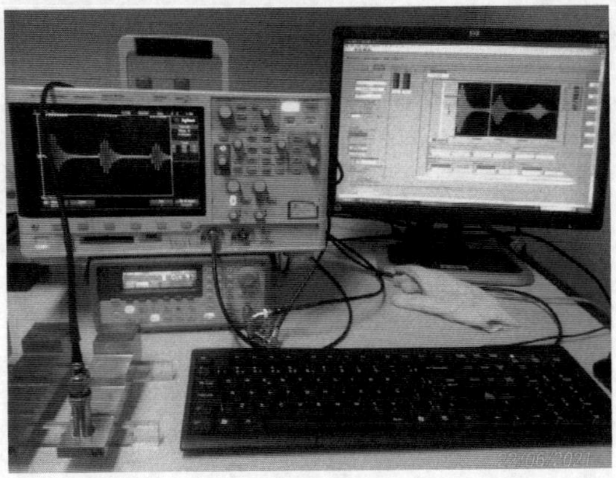

Figure 8. *Ultrasonic measurement system employed.*

3.2. Additional quantities

Mass measurements were carried out in each of the phases of temperature variation to which the pieces were submitted.

The determination protocol of each measurand was done by weighing 3 rounds of 5 weighing of each sample. With dimensional data of the parts and their respective masses, it is possible to measure and monitor the density of each of the samples used throughout the stability and homogeneity studies of the parts.

3.3. Stability study

According to ISO GUIDE 32:2017 [13], the short-term stability study must take place at different temperatures to evaluate each material property's effect. Sample temperatures may vary during transport from −50 °C to +70 °C, depending on the packaging and mode of transport. Shipping conditions can be defined based on the observed effects, and packing instructions are designed for any unwanted side effects. A short-term stability study typically takes 1 to 2 months but can be extended when optimal storage conditions must be determined simultaneously.

Also in the ISO GUIDE 32: 2017 [13], the stability test aims to determine the remaining degree of instability after preparation of the RR candidate or to confirm the stability of the material. However, even "stable" materials can exhibit instability for one or more property values. Therefore, a distinction is made between stability under specific conditions such as storage conditions and transport conditions.

In the specific case of this work, the stability study, for short, the temperatures of 2° C and 60° C were chosen to simulate possible extreme conditions. The study lasted two months, in which the samples were subjected to a temperature of 50 °C, where measurements were performed within a 30-day interval, as shown in the initial results in tables 1 and 2.

In other words, instability analysis mainly monitors the property value as a function of time and temperature variations to which the samples were submitted.

3.4. Homogeneity study

The homogeneity of a material represents the condition of uniformity of structure or composition with respect to one or more specific properties. An RM is said to be homogeneous for a given property if the value of the property, determined by tests on samples of a given size, is within the specified uncertainty limits.

The RM and CRM are primarily distributed in batches of units, so all units in the batch must be characterized by the same uncertainty for each property value. Therefore, an assessment of the homogeneity of candidate materials for RM and CRM is necessary.

The homogeneity study can cover the property variation between units of a lot of variation within the same unit. The evaluation of the variation between the same material units is always necessary, as during the intended use, it may be necessary to use a small portion of the material, in the case of solids and liquids or a small surface region as in the case of reference materials for calibration of devices used in ultrasound testing. This evaluation is of paramount importance. In the future, the necessary instructions are provided during its use, such as a minimum size of samples, care that must be taken with the material's surface, and how much to store it.
According to ISO GUIDE 35:2017 [2], the homogeneity assessment may include using previous evidence of the material's homogeneity, carrying out an experimental homogeneity study, or both.
There are exceptional cases where the evaluation can be discarded, as in the case of batches of a highly homogeneous material, where there are demonstrably previous experimental studies demonstrating that the packaging and storage do not affect its homogeneity
It is noteworthy that according to ISO GUIDE 31:2017 [12], for the effective validity of the uncertainty stated in the certificate of an CRM, a correct estimate of the effects due to both long-term stability and short-term stability is so important regarding the correct assessment of the inhomogeneity of the batch (ISO/IEC 17025: 2017 [14]; ISO GUIDE 35: 2017 [2]).

3.5. Temperatures at which samples are submitted
The temperatures to which the samples were subjected were 3 °C + 2 °C and 58 °C ± 2 °C. The samples were kept under this temperature for 3 weeks at each exposure, either in a freezer or in an oven. After each exposure period, new measurements were performed after the samples reached environmental equilibrium, between 23° to 25° and humidity around 50% to 75%.

3.6. Determination of metrological quantities
Although it is a determinant part of the RM and CRM analysis process for the quantities of interest, the focus of this work was restricted to presenting the basic procedures for the beginning of production and development of the studies necessary for their production.

In other words, the final step, which consists of the metrological analysis that consists of calculations, and which will define the metrological data of each sample, must be performed following the Evaluation of Measurement Data (2008) [3]. This document contains the step-by-step steps required to determine the measurand and its measurement uncertainties.
Briefly, the uncertainty parcels are divided into type A measurement uncertainties (related to testing repetitions) and type B (related to technical information about the test, such as a calibration certificate). After determining these portions, they are associated (combination of uncertainties, types A and B, for example) to determine the expanded uncertainty (a measure that serves as a reference for the uncertainty of any measurement.

4. Results
Based on the study and presentation of INMETRO's primary standards, guides and internal regulations necessary for selecting, measuring and producing reference materials, this research provides a minimum bibliographic framework to produce RM and RMC in the Brazilian context. Table 1 presents the results of the first round of measurements for BP and table 2 for BE. They present the calculated ultrasonic velocity values and their respective expanded uncertainties (U) for each block thicknesses, following the nomenclature presented in figures 5 and 6. The results presented were obtained from the combined standard uncertainty multiplied by the coverage coefficient k=1.96, which, for a t distribution, corresponds to a 95% coverage probability. This value was defined according to the publication Assessing Measurement Data - Guide to the Expression of Measurement Uncertainty - GUM 2008 [3].

Table 1. *Results of the first ultrasonic measurements for the 12 parallelepipedic blocks at a frequency of 5 MHz.*

	LUV1 [m s^{-1}] - 100 mm		LUV2 [m s^{-1}] - 60 mm		LUV3 [m s^{-1}] -40 mm	
Block identification	Value	U	Value	U	Value	U
BP 01	5880.2	3.9	5924.2	3.2	5895.4	3.8
BP 02	5914.3	3.0	5915.1	3.2	5903.3	3.7
BP 03	5880.2	3.0	5916.4	3.2	5907.2	3.7
BP 04	5882.4	2.9	5915.4	3.4	5907.5	3.7
BP 05	5879.2	4.0	5916.2	3.2	5898.1	3.7
BP 06	5887.6	3.0	5919.6	3.2	5906.9	3.7
BP 07	5879.7	3.0	5914.7	3.2	5900.3	3.7
BP 08	5913.4	3.0	5919.4	3.2	5904.7	3.7
BP 09	5880.6	3.0	5917.1	3.2	5904.4	3.7
BP 10	5881.7	3.0	5916.7	3.2	5902.7	3.7
BP 11	5881.1	3.0	5918.5	3.3	5920.5	3.9
BP 12	5881.1	3.0	5916.0	3.3	5899.0	3.7

Table 2. *Results of the first ultrasonic measurements for the 12 blocks scaled at the frequency of 5 MHz.*

	LUV4 [m s^{-1}] 40 mm		LUV5 [m s^{-1}] 25 mm		LUV6 [m s^{-1}] 15 mm		LUV7 [m s^{-1}] 10 mm	
Block identification	Value	U	Value	U	Value	U	Value	U
BE 01	5922.7	3.6	5923.0	4.1	5920.6	5.5	5933.8	7.6
BE 02	5923.3	3.7	5925.3	4.1	5921.5	5.4	5920.1	7.7
BE 03	5922.1	3.7	5925.1	4.0	5920.2	5.5	5928.4	7.6
BE 04	5922.8	3.7	5927.0	4.0	5916.7	5.6	5915.9	8.2
BE 05	5921.1	3.7	5917.2	4.2	5911.9	6.0	5912.0	8.1
BE 06	5915.0	3.7	5911.8	4.1	5911.2	5.6	5902.6	8.0
BE 07	5918.8	3.7	5917.7	4.0	5900.8	5.5	5893.6	7.9
BE 08	5911.3	3.7	5908.2	4.1	5905.9	5.5	5891.9	8.0
BE 09	5919.6	3.7	5911.4	4.0	5904.1	5.6	5893.8	7.9
BE 10	5921.2	3.7	5916.1	4.1	5909.3	5.5	5910.8	8.2
BE 11	5909.9	3.6	5906.9	4.0	5899.1	5.6	5887.9	8.0
BE 12	5921.8	3.6	5914.5	4.0	5907.6	5.6	5906.4	7.7

From the tables presented, the expanded uncertainty increases with decreasing thickness of the tested blocks. In other words, the process of evaluating the homogenization and stability of each of the blocks will take place from subsequent blocks, following the sequence protocol. Completing the comparisons between the results of this first round with the subsequent rounds will serve to analyse the stability and homogeneity of the results over time. Furthermore, it will be used to calculate an expanded uncertainty for each of the surfaces evaluated.

5. Conclusions

The discussions presented in this work seek to help those interested in the development of RM and CRM, specifically in the initial stages of production of standard blocks where the ultrasonic velocity is the reference quantity.

The main reference guides of the standards that serve as a basis for consulting the main stages and compliance with the basic requirements of feasibility studies, development and production of RM and CRM were presented. The initial results for the first round of ultrasonic measurements were presented, containing the speed and respective uncertainties expanded to the frequency of 5 MHz. to raise quality and add value.

References

[1] Maia TQS, Alvarenga AV, Souza RM, Costa-Félix RPB. Feasibility of Reference Material Certification for Speed of Sound and Attenuation Coefficient Based on Standard Tissue-Mimicking Material. Ultrasound Med Biol. 2021 Jul;47(7):1904-1919. doi: 10.1016/j.ultrasmedbio.2021.03.015. Epub 2021 Apr 23. PMID: 33896678.

[2] ISO GUIDE 35: Reference materials - General and statistical principles for certification, ed. 4, 2017.

[3] BIPM. JCGM:100 – Evaluation of Measurement Data – Guide to the expression of uncertainty in measurement. Genève: BIPM, 2008.

[4] BIPM JCGM. International vocabulary of metrology – Basic and general concepts and associated terms. 3rd Ed. – 2008 version with minor corrections. JCGM 200:2012. Paris: Bureau International de Pois et Measure. 2012. Miller, R. K., McIntre, P.: Acoustic Emission Testing (Nondestructive Testing Handbook, Volume 5), American Society for Nondestructive Testing, 1987.

[5] Alvarenga, A.V., Silva, C.E.R., Costa-Félix, R.P.B. Monte Carlo uncertainty assessment of ultrasonic beam parameters from immersion transducers used to non-destructive testing. Ultrasonics, 2016, 69, pp. 144–151.

[6] Miller, R. K., McIntre, P.: Acoustic Emission Testing (Nondestructive TestingHandbook, Volume 5), American Society for Nondestructive Testing, 1987.

[7] Maggi, L.E., Silva, C.E.R., Alvarenga, A.V., Costa-Felix, R.P.B. Ultrasonic calibration and certification of V1 and V2 type reference standard blocks for use in Non-Destructive Testing. Journal of Physics: Conference Series, 2011, 279(1), 012029

[8] ISO GUIDE 30:2015. Reference materials — Selected terms and definitions.

[9] ISO GUIDE 33:2015. Reference materials — Good practice in using reference materials.

[10] ISO 17034:2016 General requirements for the competence of reference material producers

[11] INMETRO - NIE-DIMCI- 30: Certificação de material de referência - Orientações sobre validação de métodos analíticos. 1ª revisão. Duque de Caxias, 2011.

[12] INMETRO - NIE-DIMCI- 31: Estudo de homogeneidade de material de referência. 2ª revisão. Duque de Caxias - RJ, 2011.

[13] INMETRO - NIE-DIMCI- 32: Estudo de estabilidade de material de referência. 4ª revisão. Duque de Caxias - RJ, 2014.

[14] INMETRO - DOQ-CGCRE-016:Orientação para a seleção e uso de materiais de referência. 2ª revisão. Duque de Caxias - RJ, 2010.

[15] INMETRO - DOQ-CGCRE-033:Orientações sobre análise crítica da documentação associada aos materiais de referência adquiridos. Duque de Caxias - RJ, 2012.

[16] ISO/IEC 17025: General requirements for the competence of testing and calibration laboratories, 2017.

[17] INMETRO – NIG-DIMCI-001: Planejamento e disponibilização de material de referência. Duque de Caxias, 2019.

[18] ISO 16831: 2012 (E) Non-destructive testing - Ultrasonic testing - Characterization and verification of ultrasonic thickness measuring equipment.

[19] Santos, T. Q., Alvarenga, A. V., Oliveira, D. P., & Costa-Felix, R. P. (2017). Metrological Validation of a Measurement Procedure for the Characterization of a Biological Ultrasound

Tissue-Mimicking Material. Ultrasound in medicine & Biology, 323-331. doi:10.1016/j.ultrasmedbio.2016.09.007.

Acknowledgments
The authors thank the Brazilian National Council for Scientific and Technological Development (grants 104.449/2021-7 [WBS], 312.501/2017-0 [RCF], 430.376/2018-9 [RCF], and 307.562/2020-4 [RCF]) and the Research Support Foundation of the State of Rio de Janeiro (grant E-26/201.563/2014 [RCF] and 260.671/2021 [RCF]).

Journal of Physics: Conference Series **2606** (2023) 012014 doi:10.1088/1742-6596/2606/1/012014

Polymer blocks homogeneity concerning the speed of sound for in vitro studies in implant dentistry

G C Morais[1], R C Mayworm[1], D T Giosa[2], F N G K Fontão[2], J M Granjeiro[1,2], R P B Costa-Felix-

[1]National Institute of Metrology, Quality and Technology, Duque de Caxias, Brazil.
[2] Faculdade ILAPEO.

E-mail: jmgranjeiro@inmetro.gov.br; rpfelix@inmetro.gov.br

Abstract. Dentistry implants methods are recursively enhancing the image localisation approach to improve patient point of intercession accuracy. Previous planning of the surgical act is on the cut edge of the technology. Materials that mimic the implant house, i.e., the maxillary bone, have been developed and improved material technology. Standard non-invasive and non-ionising radiation technologies are relevant tools to that approach. Ultrasound imaging techniques play a prominent role in that matter. Nevertheless, recent polymer materials specially developed to help the pre-implant studies had not been evaluated about their homogeneity of ultrasound parameters. Two specially developed polymer materials have their speed of sound analysed in a metrologically based methodology. At 1 MHz, a set of 10 blocks were considered statistically equivalent, unless one specimen. The speed of sound was 1020 m s^{-1} (expanded uncertainty for $p = 0.95$ equals 22 m s^{-1}) and 1676 m s^{-1} ($U0.95 = 10$ m s^{-1}). The ultrasound homogeneity is a vital outcome of the reliability of those materials used in dentistry implant pre-clinical studies.

1. Introduction

With the advances in digital flow in dentistry, added to the concept of reverse prosthetic planning, the dental market is beginning to migrate to implant systems that present solutions for accurately installing implants. Guided surgery has become popular among dentists and patients as surgical intervention becomes faster, minimally invasive, and more predictable [1]. Studies have shown that positioning errors occur more frequently in freehand techniques and can be minimised using surgical guides [2].

Virtual planning was an outstanding achievement in providing better 3D positioning, respecting anatomical and prosthetic parameters [3]. Furhauser et al. showed more excellent aesthetics and predictability in implants installed using the "flapless" technique guided in the anterior region of the maxilla due to the faithful transfer of the planned 3D positioning for the moment of surgery [4].

Laederach [5], in 2016, evaluated and compared the influence of the tolerance of different systems on the precision of guided surgery. This study was performed using digital images of acrylic specimens. Richard Koop [1], in 2012, tested the impact of different types of washer inserts and acrylic blocks and photographic digital recording. For these in vitro studies, there are different models of blocks. In addition to acrylic blocks, polyurethane blocks in the literature can present different densities similar to human bone.

This study aims to evaluate the homogeneity of the 10 polyurethane specimens, of double density (40PFC and 10 PFC), through ultrasonic tests, intending to validate them as methodologies of in vitro studies. The material of interest is similar enough to bone regarding densities. Ultrasound is the approach to define the homogeneity of the polyurethane blocks.

2. Materials and methods

2.1 Experimental setup

The ultrasound technique known as emission and reception was used. Two transducers are introduced, one being an ultrasonic pulse emitter, which passes through the sample, reaching, on its opposite side, the second transducer that works as a receiver. The study uses a transducer with a nominal frequency of 1 MHz and a diameter of 12.7mm as an emitter (model A-301S, NDT-Panametrics-Olympus Corporation, Japan) and as a receiver the transducer with a nominal frequency of 1 MHz and diameter of 25.4mm (model A-303S, NDT-Panametrics-Olympus Corporation, Japan), a waveform generator, 400 V amplitude (EPOCH 600, Olympus, USA), an oscilloscope (DSOX 1202A, Agilent Technologies, CA , USA) and a computer with a program developed in Python. Connecting the function generator output trigger with the oscilloscope input trigger is necessary to synchronise the equipment. To measure the temperature, a Thermohydrometer was used.

All bodies were submerged in water 5 days before starting the measurements to eliminate possible gaseous agglomerations inside the specimens, which could cause interference in the measurements. All measurements were performed underwater immersion to ensure good thermal stability in an acrylic tank.

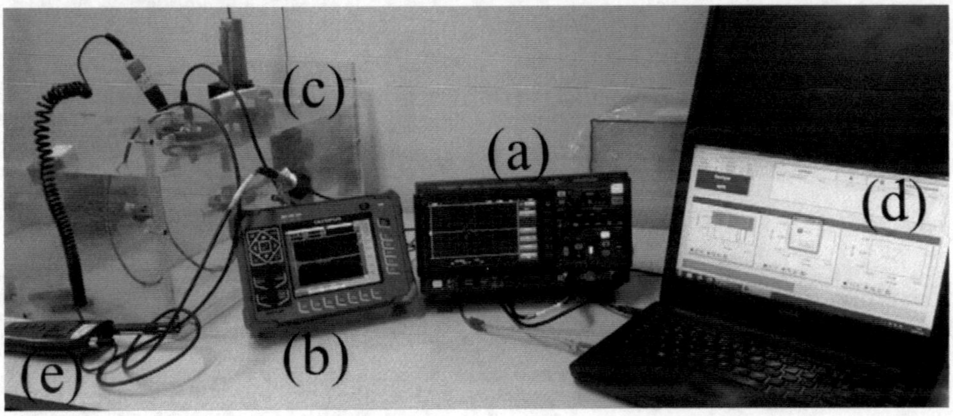

Figure 1. Figure with short caption (caption centred). Figure 1. Measurement system. (a) oscilloscope, (b) EPOCH 600 generator, (c) tank, (d) computer and (e) Thermohygrometer.

2.2 Measuring the speed of sound

Five repetitions were performed under repeatability conditions for each specimen, preceded by accurate alignment of the transducers. The 10 specimens were numbered (1 to 10). Due to their double density, their side were complementary labelled A and B. Finally, each side of the block was measured both vertically and horizontally, generating a total of 200 measurements. The temperature ranged from 19.5 °C to 21.5 °C during all measurements.

Figure 2. Example of a block to be measured.(a) Side A and (b) Side B.

The transducer alignment followed a different protocol from the traditional one. In this work, the transducers were adjusted according to the thickness of each body for each repetition. After the alignment of the system, the process of measuring the Flight Time of Side A (At) and Side B (Bt) was started.

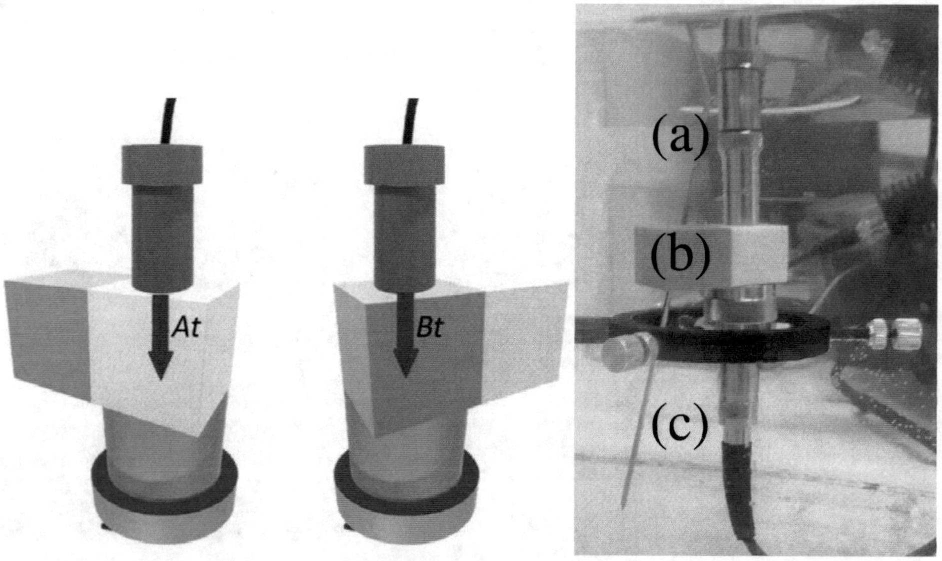

Figure 3. Submerged coupling system in the tank. (a) 1 MHz transducer with 12.7mm diameter, (b) Specimen and (c) 1 MHz transducer with 25.4mm diameter.

Body thickness ($\boldsymbol{x}$) was determined using a calliper, properly calibrated, and the velocity of each body was calculated using the equations:

$$vA=x \cdot (tA)^{-1} \quad (1)$$

$$vB=x \cdot (tB)^{-1} \quad (2)$$

To calculate the measurement uncertainty of the propagation velocity of the specimens, it was necessary to calculate the combined uncertainties of time and thickness used to calculate the velocity, considering the following sources of uncertainties: Type A uncertainty (μ_A) that is obtained by the standard deviation of the mean (3); Type B uncertainty are associated with the calibration of the calliper, which corresponds to 0.02 mm for k0.95 = 2 (μ_A), and the calibration certificate of the oscilloscope time base, used to determine the time of flight (0, 0022 % for $k_{0.95}$ = 1.96) – see (4).

$$\mu^A = \frac{S}{\sqrt{n}} \tag{3}$$

$$\mu_t^B = \left(\frac{0{,}0022\%}{2}\right) * t_{1max} \tag{4}$$

$$\mu_x^B = 0{,}02 \tag{5}$$

The combined uncertainty for specimen flight time (μ_t) in seconds was calculated (6). The combined uncertainty of the thickness on the specimen (μ_x) in seconds was calculated (7).

$$\mu_t^{\,2} = (\mu_t^A)^2 + (\mu_t^B)^2 \tag{6}$$

$$\mu_x^{\,2} = (\mu_x^A)^2 + (\mu_x^B)^2 \tag{7}$$

Based on equations (1) and (2), the following sensitivity coefficients related to thickness (Cx) (8) and time of flight on the specimen (Ct) (9) were determined:

$$Cx = \frac{\partial c}{\partial x} = \frac{1}{t} \tag{8}$$

$$Ct = \frac{\partial c}{\partial t} = -\frac{x}{t^2} \tag{9}$$

The combined uncertainty of the velocity of the specimens (μc) was calculated according to (10).

$$\mu c^2 = Cx^2 .\ \mu_x^{\,2} + Ct^2 .\ \mu_t^{\,2} \tag{10}$$

After calculating all these uncertainties, the final combined uncertainty of all bodies and the expanded uncertainty was calculated. As the uncertainty of type A(μ_v^A), obtained from mean velocity values, the uncertainty of type B is the highest value of the combined uncertainties between each body (μc) (10), and k.

$$\mu c_{final}^{\,2} = (\mu_v^A)^2 + (\mu_v^B)^2 \tag{11}$$

$$U_{exp} = k * \mu c_{final} \tag{12}$$

2.3 Communication between interfaces

The real-time Python approach was developed to optimise, streamline and remove possible human bias errors, using the platform both for oscilloscope control and for acquisition and processing of results. The application interface has a control screen, which through it allows the manipulation of oscilloscope data, also has a graphic area, where you can monitor the measured data, and finally, a table that at the end of each repetition presented the results, having the option to save the data in .doc. The following libraries were used; Tkinter, Matplotlib, Statistics, Pyvisa and Time.

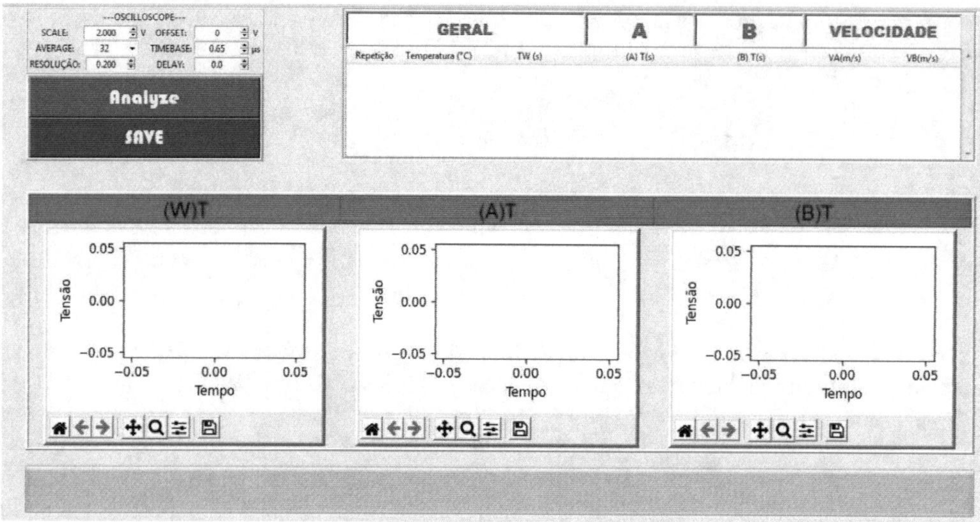

Figure 4. Application interface.

3. Results and discussion

The measurement result of the characterisation of the propagation velocity in the specimens and the respective combined uncertainties are presented in Table 1 for the two sets of measurements. The result of the expanded uncertainty of the mean value of all blocks is shown in Table 2. Due to the isotropic and homogeneous characteristics of the specimens, the values measured in the vertical and horizontal positions were considered equivalent.

Table 1. Average values of each body, with their respective combined uncertainties.

	Size A		Size B	
Specimen	Speed of sound (m s^{-1})	μc [m s^{-1}]	Speed of sound [m s^{-1}]	μc [m s^{-1}]
1	1031.33	1.31	1676.19	1.72
2	994.00	3.12	1675.53	1.57
3	1038.81	4.05	1678.80	2.28
4	1018.21	10.07	1672.62	1.78
5	1007.88	6.50	1672.53	1.80
6	1014.45	3.62	1682.01	2.27
7	1040.78	4.13	1677.51	3.31
8	1031.14	5.19	1680.07	1.89
9	1009.39	2.80	1676.59	2.61
10	1010.73	3.30	1672.32	5.02

Table 2. Value of all bodies and their respective expanded uncertainties.

Size A		Size B	
Speed of sound (m s^{-1})	U_{exp} (m s^{-1})	Speed of sound (m s^{-1})	U_{exp} (m s^{-1})
1020	22	1676	10

The results disclosed in Tables 1 and 2 reveals a quite reasonable homogeneity for all blocks. The normalised error applied to each block, and the average value for all blocks disclosed that only one specimen and side (number 2, Side A) was larger than 1. That analyses let to know that, concerning the ultrasonic quantity Speed of Sound, the homogeneity of the specimens are statistically assured.

Several in vitro studies evaluating the accuracy of guided surgery use acrylic blocks simulating human bone [1 and 5], distancing the dentist from the clinical reality. However, Calvert [6], obtained satisfactory results in mechanical and structural tests in polyurethane models. It is extremely important in studies of accuracy in guided surgery the use of a bone analogue that simulates the characteristics of human bone, providing truer results in in vitro research in dentistry.

The results obtained here support the use of polyurethane blocks in implant dentistry in vitro studies, since they simulate the characteristics of human bone with notable homogeneity in their composition. Thus, minimizing the bias factors in studies using bone mimetic materials to compare of the accuracy of guided implant systems.

4. Conclusion

The present work aimed to evaluate the homogeneity of polymer blocks, in two distinct densities, in relation to an ultrasound parameter, the Speed of Sound. With their homogeneity proven, these blocks can be considered an excellent material for in vitro study methodologies, which aim to assess the accuracy of guided surgery, since its structure is very close to the human bone structure.

References

[1] Koop R, Vercruyssen M, Vermeulen K, Quirynen M. Tolerance within the sleeve inserts of different surgical guides for guided implant surgery. Clin Oral Implants Res. 2013;24(6):630–4.

[2] Arısan V, Karabuda CZ, Mumcu E, Özdemir T. Implant Positioning Errors in Freehand and Computer-Aided Placement Methods: A Single-Blind Clinical Comparative Study. Int J Oral Maxillofac Implants. 2013;28(1):190–204.

[3] Schneider D, Marquardt P, Zwahlen M, Jung RE. A systematic review on the accuracy and the clinical outcome of computer-guided template-based implant dentistry. Clin Oral Implants Res. 2009;20(SUPPL. 4):73–86.

[4] Fürhauser R, Mailath-Pokorny G, Haas R, Busenlechner D, Watzek G, Pommer B. Esthetics of Flapless Single-Tooth Implants in the Anterior Maxilla Using Guided Surgery: Association of Three-Dimensional Accuracy and Pink Esthetic Score. Clin Implant Dent Relat Res. 2015;17:e427–33.

[5] Laederach V, Mukaddam K, Payer M, Filippi A, Ku€hl S. Deviations of different systems for guided implant surgery. Clin. Oral Impl. Res. 00, 2016, 1–5.

[6] Calvert KL, Trumble KP, Webster TJ, Kirkpatrick LA. Characterization of commercial rigid polyurethane foams used as bone analogs for implant testing. J Mater Sci Mater Med. 2010;21(5):1453-61. doi: 10.1007/s10856-010-4024-6.

Acknowledgements
The authors thank the National Program to Support the Development of Metrology, Quality, and Technology – Pronametro. Furthermore, the authors thank the Brazilian National Council for Scientific and Technological Development (grants 312.501/2017-0 [RCF], 430.376/2018-9 [RCF], 307.562/2020-4 [RCF], and 308156/2020-0 [JMG]) and the Research Support Foundation of the State of Rio de Janeiro (grant E-26/201.563/2014 [RCF], 260.671/2021 [RCF], SEI-260003/004495/2021 [JMG], and E-26/010.0020989/2014 [JMG]).

Ensuring the validity of measurement results of nitrite and nitrate in meat products

M C de Oliveira[1], M D de Almeida[2], A C P Osorio[2], R C de Sena[2], V Souza[3]

[1]Professional Master's Course in Metrology and Quality, Inmetro, Duque de Caxias, RJ, Brazil 25250-020; [2]Inorganic Analysis Laboratory, Chemical Metrology Division, Inmetro, Duque de Caxias, RJ, Brazil 25250-020; [3]Capacity Centre, Inmetro, Duque de Caxias, RJ, Brazil 25250-020

E-mail: marianacardoso41@hotmail.com

Abstract. Nitrate and nitrite measurement is a fundamental part of quality control in the meat products industries. The guarantee of technical validity of the results of these measurements can be achieved through the adoption of appropriate metrological tools such as method validation. In this study, the measurement of the mass fractions of nitrite and nitrate ions was performed by ion chromatography with UV detection. The method was linear ($r > 0.99$) over the working range. Limits of quantification values for measuring nitrite and nitrate in meat product were 8.0 mg kg^{-1} and 52.4 mg kg^{-1}, respectively. The mean values of recovery at all evaluated mass fraction levels were within the range of 88 % to 106 %. All the validation parameters resulted as satisfactory, and the method was considered adequate for the intended use.

1. Introduction

The production and consumption of meat products in the world has been expanding significantly in recent decades. With the increase in consumption, the use of food additives by the meat industries becomes increasingly essential so that their production levels can meet the market demands.

Nitrate and nitrite salts are widely used as curing agents for various types of meat products, due to the various technological advantages provided by them [1]. However, the indiscriminate ingestion of these salts can pose serious risks to human health [2]. Nitrite is considered toxic and can act on hemoglobin to form methemoglobinemia, which is unable to transport oxygen to tissues [3,4]. Furthermore, the reaction between nitrites and amines leads to the formation of nitrosamine compounds that are classified as possible carcinogenic for humans [5]. Nitrate must also be monitored because it can be converted into nitrite through the action of microorganisms present in the human body, mainly in the mouth and stomach [6]. Due to possible adverse health effects, several countries established legal limits for the addition of these curing salts in food, including in the meat and meat products. In Brazil, the addition of food additives to industrialized meat and meat products is officially regulated by Resolution of the Collegiate Director (RDC) n° 272 of the National Sanitary Vigilance Agency (ANVISA) [7].

The measurement of the mass fractions of nitrite and nitrate is a mandatory part of quality control in the industrial production of meat products. These measurements can be performed using a variety of analytical techniques, including ion chromatography with UV detection (IC-UV) [8]. The results obtained from these measurements must be reliable, in addition to complying with current legislative limits, to ensure the protection of the population's health. For this, it is necessary to adopt metrological tools to ensure the technical validity of the results generated in these measurements, in order to guarantee the quality and safety of these foods.

The validation of measurement procedures is one of the main metrological tools proposed by the ISO/IEC 17025 [9] to guarantee the technical validity of the measurement results. The aim of the present work was to validate the measurement procedure of the nitrate mass fraction and verify some performance parameters of the measurement procedure for nitrite in meat product by IC-UV. This evaluation was conducted as part of a technical feasibility study to produce a nitrite and nitrate reference material in a meat product.

2. Experimental

2.1. Reagents

The standard solutions were prepared using ultrapure water obtained by the Milli-Q system, model Elix 10 (Millipore®, USA). The eluent solution was prepared using potassium hydroxide reagent grade (Merck, Germany). Commercial Standard Reference Material of nitrate SRM 3185 lot 70309 (NIST, EUA) and sodium nitrite solution from the Key Comparison CCQM-K59 promoted by the National Institute of Metrology of Slovakia of 1 g kg^{-1} [10] were used for the fortification of thc mcat product samples and for the preparation of standard solutions.

2.2. Instrumentation

All standards, solutions and samples were weighted in a calibrated analytical balance (Sartorius, Germany), with 230 g capacity and 0.01 mg resolution. Meat product samples were homogenized using a basic analytical mill model A11B (IKA, Germany). A model 6-15H centrifuge (Petrotest®, Germany) was used for all centrifugation procedures. The measurement of the mass fractions of nitrite and nitrate was performed using the ICS-1000 ion chromatograph (Thermo Scientific™ Dionex™, USA) with an AD25 absorbance detection module (Thermo Scientific™ Dionex™, USA). Nitrate and nitrite were separated using an IonPac AS18 analytical column and AG18 guard column (Thermo Scientific™ Dionex™, USA).

2.3. Measurement of nitrite and nitrate mass fractions in meat product

The meat product samples were homogenized by a blender for at least 1 min. A 2.5 g portion of the homogenized sample was placed in a 50 mL Falcon tube and 12.5 mL of pre-warmed ultrapure water (50 – 60 °C) was added. The mixture was homogenized in a blender for 1 minute at 3,800 rpm. After cooling to room temperature, ultrapure water was added until a final mixture mass of 50 g was obtained. The mixture was centrifuged at 3,000 rpm for 30 min and the supernatant was removed and filtered using a 0.22 μm pore size PVDF syringe filter (Nova Analítica, Brazil). The filtrate was collected for chromatographic analysis.

The measurement of the nitrite and nitrate was performed by IC-UV. Nitrate and nitrite were separated using isocratic conditions. A solution of KOH 39 mM was used as eluent for the chromatographic analysis. The injection volume was 25 μL and eluent flow-rate was 1.0 mL min^{-1}. Analytes were detected using UV detection at 225 nm.

2.4. Evaluation of the performance level of the measurement procedure by IC-UV

The evaluation of the performance level of the nitrite and nitrate measurement procedure was conducted as part of a technical feasibility study to produce a nitrite and nitrate reference material (RM) in a meat product. This evaluation was carried out in order to demonstrate the suitability of this measurement procedure to carry out the measurements required for the studies to assess the homogeneity and stability under the conditions of transport of the batch of material produced in the feasibility study. For this, the measurement procedure must present an adequate level of performance and meet the criteria required by each of the studies involved in the production of RM.

The evaluation of the performance level of the nitrate measurement procedure was conducted through the complete validation process, with the standard solutions being prepared from the certified nitrate reference material SRM 3185 (NIST, USA). For nitrite measurement, only some performance parameters were evaluated as there was no certified reference material available to establish the metrological traceability of the results obtained. Therefore, the nitrite calibration solution from the key comparison conducted by the Slovak Institute of Metrology was used.

These assessments were conducted in compliance with the guidelines of National Institute of Metrology, Quality and Technology (INMETRO) [11] and the Ministry of Agriculture, Cattle and Supplying (MAPA) [12]. The performance parameters evaluated were selectivity, linearity, quantification limit and detection limit, trueness and precision. The definition of acceptance criteria was carried out so that the measurement procedure presented an adequate level of performance to be used in the feasibility study. For this, the acceptance criteria for accuracy and trueness were defined following the guidelines of ISO Guide 35 [13]. All statistical tests applied to treat the measurement results obtained were performed considering a confidence level of 95 %.

3. Results and discussion

3.1. Selectivity

Selectivity is an extremely important parameter so that the measurement procedure can be used to perform the measurements required in studies to produce a RM [13].

Two calibration curves were prepared for each analyte of interest in the absence and in the presence of meat matrix, extract in the proportion of 2.5 g of meat product to 50 mL of water. The standard solutions of the calibration curves were prepared gravimetrically at 5 levels of mass fractions of nitrite (0.5 to 2.5 mg kg^{-1}) and nitrate (5.0 to 25.0 mg kg^{-1}).

The slope of the curves was compared using the slope comparison t test. It was verified that the slopes of these curves were statistically equivalent (Table 1), demonstrating that there is no significant matrix effect in the conditions evaluated.

Table 1. Comparison of the slopes of the calibration curves for nitrite and nitrate in water (curve I) and in ham extract (curve II).

Parameter	Nitrite		Nitrate	
	Curve I	**Curve II**	**Curve I**	**Curve II**
slope	1.6×10^{-3}	1.6×10^{-3}	6.8×10^{-4}	6.7×10^{-4}
$S_{y/x}$[a]	2.5×10^{-5}	1.1×10^{-4}	6.6×10^{-5}	2.9×10^{-4}
S_x[b]	5.8×10^{-1}	5.1×10^{-1}	5.8×10	5.8×10
t test ($\alpha = 0.05$)	t_{calc} (1.945) < t_{crit} (2.056)		t_{calc} (1.418) < t_{crit} (2.056)	

[a]$S_{y/x}$ = linear regression standard error.
[b]S_x = sample variance x.

3.2. Linearity

The selected working ranges were from 0.5 to 2.5 mg kg^{-1} for nitrite and from 5.0 to 25 mg kg^{-1} for nitrate. The calibration curves were prepared gravimetrically at 5 levels of mass fractions of nitrite (0.5 to 2.5 mg kg^{-1}) and nitrate (5.0 to 25.0 mg kg^{-1}). The adequacy of the linear model was verified by determining the correlation (r) and determination (R^2) coefficients, whose values were all above 0.990, indicating a good adequacy of the linear model.

The linearity of these curves was also verified using the F test (F-Snedecor) in the regression analysis of variance. The linear model was fit for purpose ($F_{calc} > F_{crit}$) and the selected working range was considered linear for measuring the two analytes of interest.

3.3. Limit of detection and quantification

The limits of detection (LOD) and quantification (LOQ) were estimated from the parameters of the calibration curves for nitrite and for nitrate prepared in ham extract using Equations (1) and (2), where s_a is standard deviation of intercept and b is slope of regression line.

$$LOD = 3{,}3 \times \frac{s_a}{b} \tag{1}$$

$$LOQ = 10 \times \frac{s_a}{b} \tag{2}$$

The values of the detection and quantification limits of nitrite and nitrate in meat product are described in Table 2. The calculated values were multiplied for the dilution factor (20) for obtaining the LOD and LOQ values as analyte concentration in the matrix.

Table 2. Values of the detection and quantification limits for the measurement of nitrite and nitrate in meat product.

	LOD[a] (mg kg^{-1})	**LOQ[b] (mg kg^{-1})**
Nitrite	2.6	8.0
Nitrate	17.3	52.4

[a]Limit of detection.
[b]Limit of quantification.

3.4. Measurement trueness and precision

For nitrate measurement, the trueness was quantitatively expressed in terms of the measurement bias. As it has not been possible to find a suitable nitrate CRM in meat matrix, the measurement bias and the precision were evaluated in an analytical recovery study, using samples of ham fortified at two levels (22.0 and 25.0 mg kg^{-1}).

A recovery study was also conducted for nitrite measurement. The fortification of the ham samples was performed at three levels (0.5, 1.0 and 2.2 mg kg^{-1}) using the sodium nitrite solution from the Key Comparison CCQM-K59. Therefore, it was not possible to guarantee the metrological traceability of the values obtained in these measurements.

The precision was evaluated in terms of repeatability and intermediate measurement precision. Repeatability was assessed over the entire working range during the recovery study. Intermediate measurement precision was evaluated by comparing measurement results obtained in recovery studies conducted at different days.

The values of recovery and relative standard deviation of repeatability for each level of fortification studied are described in Tables 3 and 4. The recovery values satisfy both the acceptance criteria established by AOAC [14] and EC 657/02 [15]. However, stricter acceptance criteria for analytical recovery were established from 85 % to 110 % for analyte mass fraction values less than 10 mg kg^{-1} and from 95 % to 105 % for analyte mass fraction values greater than 10 mg kg^{-1}. The average recovery values obtained at the different fortification levels were satisfactory. It is concluded that the proposed measurement procedure did not show a trend and, therefore, was suitable for application in the feasibility study of the batch of material produced.

Table 3. Repeatability and recovery for the measurement of nitrate in the recovery study.

Level	**Study day 01**			**Study day 02**		
(mg kg^{-1})	**Rec[a] (%)**	**Mean (%)**	**RSD[b] (%)**	**Rec[a] (%)**	**Mean (%)**	**RSD[b] (%)**
	102.4			92.2		
22.0	108.9	104.9	3.5	100.2	95.9	4.1
	103.5			95.2		
25.0	99.8	102.6	2.8	101.3	101.0	2.7

	102.5			99.1		
	105.4			102.9		

[a] Recovery.
[b] Relative standard deviation.

Table 4. Repeatability and recovery for the measurement of nitrite in the recovery study.

Level	Study day 01			Study day 02		
(mg kg^{-1})	Rec[a] (%)	Mean (%)	RSD[b] (%)	Rec[a] (%)	Mean (%)	RSD[b] (%)
	85.1			103.7		
0.5	91.3	88.2	3.1	103.4	106.0	4.2
	88.2			110.8		
	98.3			103.0		
1.0	89.8	92.8	4.7	101.1	101.0	2.0
	90.4			98.9		
	104.2			95.9		
2.2	105.6	104.2	1.4	95.9	98.1	3.8
	102.7			102.4		

[a] Recovery.
[b] Relative standard deviation.

Another important parameter is precision [13]. The relative standard deviation of repeatability of the measurement procedure used to perform the measurements in the homogeneity study should ideally be less than the target measurement uncertainty of the value of the property of interest of the RM candidate [13], according to Equation (3), where: s_r represents repeatability standard deviation; n_{al} represents number of observations on each of n aliquots taken from each unit for measurement; and u_{trg} represents target measurement uncertainty, expressed as standard uncertainty, for the value of a property to be certified.

$$\frac{s_r}{\sqrt{n_{al}}} \leq \frac{u_{trg}}{3} \quad (3)$$

The repeatability standard deviation must meet the criteria established in accordance with Equation (3). The target uncertainty established for the MR properties of interest of nitrite and nitrate in meat product was 10 %. The acceptance criterion for the repeatability standard deviation of the measurement procedure was 4.7 %. The RSD values (%) obtained for both analytes were satisfactory and the proposed procedure proved to be suitable for application in the studies to assess the homogeneity and stability under the conditions of transport of the candidate batch for RM.

3.5. Measurement uncertainty

The evaluation of measurement uncertainty was performed as directed by the guide for the expression of measurement uncertainty [16], considering the following sources of uncertainty: uncertainty associated with the preparation of standard solutions, uncertainty associated with the preparation of the ham sample, uncertainty associated with the adjustment of the calibration curve and uncertainty in the repetition of the instrumental analysis of the sample.

The value of nitrate mass fraction in a cooked ham sample obtained was 202 mg kg^{-1} and the combined relative standard uncertainty value obtained was 3.1 %. The value of nitrite mass fraction was 25 mg kg^{-1} and the combined relative standard uncertainty value obtained was 3.6 %.

4. Conclusion

The procedure for measuring nitrite and nitrate in a meat product by IC-UV was validated according to INMETRO and MAPA guidelines, also observing the complementary guidelines of the ISO Guide 35. For both analytes, the calibration curves were considered fitted to the linear model in the working range studied, with correlation coefficients (r) greater than 0.990. The limits of quantification values for measuring nitrite and nitrate in meat product were 8.0 mg kg^{-1} and 52.4 mg kg^{-1}, respectively. The mean values of recovery at all evaluated mass fraction levels were within the range of 88 % to 106 %. The repeatability and intermediate precision were within the established criteria and the procedure was considered adequate to be used in the assessment of homogeneity in the proposed feasibility study. Thus, the measurement procedure has been validated and can be used within the scope of the technical feasibility study to produce a RM of nitrite and nitrate in meat product.

Acknowledgments
The authors warmly acknowledge the financial support from PRONAMETRO (Inmetro, Rio de Janeiro, Brazil).

References

[1] Binkerd E F, Kolari O E 1975 *Food and cosmetics toxicology* **13** 655–61
[2] Merino L, Ornemark U and Toldrá F 2017 *Advances in Food and Nutrition Research* **81** 65–107
[3] Swann P F 1975 *Journal of the Science of Food and Agriculture* **26** 1761-1770
[4] Fan A M & Steinberg V E 1996 *Regulatory toxicology and pharmacology* **23** 35-43
[5] Pegg R B & Honikel O K 2015 *Principles of curin* in *Handbook of fermented meat and poultry* ed JohnWiley & Sons p.19-30
[6] Merino L, Örnemark U & Toldrá F 2017 *Advances in food and nutrition research* **81** 65-107
[7] Agência Nacional de Vigilância Sanitária (Brasil). Resolução - RDC nº 272, de 14 de março de 2019. Estabelece os aditivos alimentares autorizados para uso em carnes e produtos cárneos. *Diário Oficial da União*, Brasília, DF, 18 de março de 2019, Seção 1, p 194
[8] Siu D C & Henshall A 1998 *Journal of Chromatography A* **804** 157–160
[9] ISO/IEC 17025: *General requirements for the competence of testing and calibration laboratories*
[10] MÁRIÁSSY M *et al.* 2008 *Metrologia* **45** 1-20
[11] INMETRO 2020 *DOQ-CGCRE-008: Orientação sobre validação de métodos analíticos*
[12] MAPA 2011 *Manual de garantia da qualidade analítica: Resíduos e Contaminantes em Alimentos*
[13] ISO Guide 35: 2017 *Reference materials - General and statistical principles for certification*
[14] AOAC International 201 *Guidelines for Standard Method Performance Requirements (Appendix F)*
[15] EC Commission decision 2002/657/EC of 12 august 2002 implementing council directive 96/23/EC concerning the performance of analytical methods and the interpretation of results. Off. J. Eur. Commun.: L221 8-36
[16] INMETRO 2012 *Avaliação de dados de medição: Guia para a expressão de incerteza de medição – GUM 2008*

Validation of a Bovine Viral Diarrhea Virus (BVDV) absolute quantification method by digital droplet PCR (ddPCR).

M R F Henrique[1], J L P Ramos-Júnior[1], R B Flatschart[1], S P C Barroso[2], M B Heinemann[3,4], da Fonseca F G[5], Granjeiro J M[1], V de Souza[1], and A V Folgueras-Flatchart[1]

[1] Diretoria de Metrologia Aplicada às Ciências da Vida (DIMAV), Instituto Nacional de Metrologia, Qualidade e Tecnologia - INMETRO - Duque de Caxias-RJ, Brazil.
[2] Instituto de Pesquisas Biomédicas (IPB) do Hospital Naval Marcílio Dias (HNMD) – Rio de Janeiro-RJ, Brazil.
[3] Faculdade de Medicina Veterinária e Zootecnia (FMVZ), Universidade de São Paulo (USP) – São Paulo-SP, Brazil
[4] Escola de Veterinária (EV), Universidade Federal de Minas Gerais (UFMG) – Belo Horizonte-MG, Brazil
[5] Instituto de Ciências Biológicas (ICB), Universidade Federal de Minas Gerais (UFMG) – Belo Horizonte-MG, Brazil

avflatschart@inmetro.gov.br

Abstract. A increasingly common way for the diagnostics of viral agents is to carry out PCR from biological samples, which can detect their nucleic acid during the acute phase of the disease. Taking into account the need for quality control of PCR tests in laboratories accredited by ISO/IEC 17025:2017, we intend to carry out feasibility studies for the production of Reference Materials (MR) for these molecular tests, using BVDV in the matrix serum as a model. The first step for this goal was the validation of an analytical method for quantification of viral RNA to characterize the material under study. Primers and probe from a commercial qualitative assay kit designed for Real Time RT-PCR (VetMax Gold BVDV - Thermo Fisher) were used for quantification of BVDV by ddPCR. The technique was optimized and a Detection Limit of 13 copies/μL was determined, allowing a Quantification Limit of 38 copies/μL. The method showed to be linear over two orders of magnitude. The method will be used in the homogeneity study and long-term stability tests for the pilot batch of BVDV in Fetal Bovine Serum MR, a model for the production of MR intended for PCR of BVDV and other *Flaviviridae*.

1. Introduction

Reference materials from virus nucleic acid are tools that help laboratories perform more accurate assays to confirm the diagnosis of viral diseases [3]. This type of material can be used for quality control and for evaluation of reagents, kits or test methods, on research, for the provision of services and in the development of innovative products for laboratory diagnosis [1]. In addition, these materials can be used in processes to obtain and maintain laboratory accreditation by ISO/IEC 17025:2017 and should allow providers to organize proficiency tests and rounds of interlaboratory comparisons [11,12,13]. In a feasibility study for the production of flavivirus reference materials, we chose to use BVDV (Bovine Viral Diarrhea Virus) as a model. The reliable diagnosis of the presence of this virus has been a concern on our laboratory, in the quality control of cells and supplies for culture, as bovine serum is a common supplement in growth media [7]. This disease agent may be present in cattle sera and is capable to infect cells of a large number of animal species in vitro [2]. As it can easily go unnoticed under optical microscope, infection by BVDV may cause cell alterations and interfere with the reliability of assays, research and also contaminate products derived from cell culture. Commercial methods for detecting BVDV for diagnosing the disease in cattle are qualitative techniques that shows the absence or presence of the virus in the biological samples analyzed. This is also the case with the real-time molecular RT-PCR method that we have implemented in the laboratory, which detects the presence or absence of viral RNA. For the characterization of the reference material that we intend to produce, it is necessary to validate a quantitative method that allows us to characterize the batch units through homogeneity and stability tests [9,13]. Using the primers and probe from VetMax Gold BVDV PI Detection Kit (Thermo Fisher), we optimized and validated a viral RNA quantification method by ddPCR (Droplet Digital PCR) [6,10]. This technique, accepted as a reference method, will provide an absolute quantification of nucleic acids with precision, sensitivity and robustness, without the need of calibrants [10,14]. In ddPCR, the reaction volume containing the PCR mix and the sample is partitioned into thousands of oil-encased droplets, which undergo the usual PCR thermocycling. The molecules generated by amplification of the analyte will be fluorescently labeled and detected in the droplets at the end of the reaction (end-point) [8]. Counting the fluorescent droplets in relation to the total droplets generated, analyzed according to the Poisson distribution, allows the quantification of the analyte in copies/μL in the reaction and in the initial sample [4,5]. The validation of this methodology applied to the quantification of BVDV RNA will allow the identification of the linear working range of the assay, the definition of the Detection Limit and the Quantification Limit of the Method for future use in the feasibility study for production of a reference material of this type.

2. Materials and Methods

2.1 Positive Controls, BVDV and HoBi-like reference strains

Reference samples of BVDV (Family Flaviviridae, genus Pestivirus) were used to evaluate the selectivity of primers and probe and in the optimization and validation tests of the method: Singer strain (BVDV-1, pestivirus A) and 890 (BVDV-2, pestivirus B). In addition to these, the Hobi-Like bovine strain (formerly known as an atypical form of BVDV and currently classified as pestivirus H) was also used. All these strains were cultivated in MDBK cells at the Veterinary School of the Federal University of Minas Gerais (UFMG). Supernatants of these cultures were collected in vials that were stored at -80°C until use. In some assays, a Fetal Bovine Serum (SFB) batch from a naturally infected source (stored at -20°C) and the positive control from VetMax Gold BVDV PI Detection kit (Thermo Fisher Scientific, USA) were used as positive controls.

2.2 DENV-1 and MDBK culture samples.

For selectivity test, a strain of Dengue Virus type 1 (DENV-1, cultivated in insect cells at the Institute of Biological Sciences of UFMG) and supernatant from MDBK cell culture was evaluated for amplification. Vials of this material were stored at -80°C until the time of use.

2.3 RNA extraction

RNA extraction from different viral strains, FBS and MDBK culture was performed using PureLink Viral RNA/DNA kit (Thermo Fisher Scientific, USA), following the manufacturer's instructions. The extracted RNAs were stored at -80°C until use.

2.4 Optimization of droplet digital PCR reaction conditions
RNA from the BVDV Singer strain was used in the ddPCR optimization steps, performed according to the recommendations of the One-Step RT-ddPCR Advanced Kit for Probes (Bio-Rad, USA) and using the VetMax Gold BVDV PI primers and probe Detection Kit (Thermo Fisher Scientific, USA). For this purpose, temperatures of reverse transcription and annealing of primers were optimized, seeking the temperature at which the highest quantification would occur, with the best apparent separation between populations of positive and negative droplets. The preparation of the emulsion for each reaction was performed with the Auto-DG apparatus (Bio-Rad, USA). Thermocycling of the plates with temperature gradient was carried out in the C1000 equipment (Bio-Rad, USA). The ddPCR QX-200 system (Bio-Rad, USA) was used for fluorescent droplet data acquisition.

2.4.1 Reverse transcriptase temperature
Reverse transcription temperatures (RT) in the range 42°C to 50°C were evaluated, setting the annealing temperature of the primers in the amplification step at 60°C (annealing temperature suggested by the manufacturer of the primers and probe for real time RT-PCR). A template-less control (NTC) was included along the run, with the highest RT temperature reaction.

2.4.2 Primes annealing temperature
Primer annealing temperatures within the range 59°C to 62°C were evaluated with the optimal RT temperature as established above. A template-less control (NTC) reaction was included along the run with the highest annealing temperature.

2.5 Parameters for validation
For validation of the proposed ddPCR method, the optimized conditions were used, as determined above, and in accordance with the guidelines of the document DOC-CGCRE-008 version 9 of 2020 (Inmetro, 2020).

2.5.1 Seletectivity
In the selectivity test, the amplification specificity of the set of primers and probe was evaluated in ddPCR. The RNAs extracted from the different BVDV strains mentioned above were amplified. Additionally, the following controls were evaluated: RNA from DENV-1 (also a member of the Flaviviridae family), RNA extracted from a FBS batch naturally contaminated with BVDV and RNA extracted from the supernatant of MDBK cells.

2.5.2 Limit of Detection and Limit of Quantification determination
To determine the Linear Working Range and the Limits of Detection (LD) and Quantification (LQ) of the method, the positive control of the VetMAX™-Gold BVDV PI Detection kit (Thermo Fisher Scientific, USA) was serially diluted and submitted to ddPCR. Eight dilutions were tested in duplicate.

3. Results and Discurssion

3.1Optimization of droplet digital PCR reaction conditions

3.1.1 Reverse transcriptase temperature
According to Bio-Rad RT-ddPCR reagents manual, the temperature range to be tested in RT optimization should be within the 42 °C to 50 °C range. Considering the options available in the thermocycler used, the temperatures actually tested were: 42.1°C, 45°C, 47°C and 49.8°C. The latter was the one that presented the highest quantification of the target template (Figure 1). From this step on, the RT temperature adopted in the RT-ddPCR became 50°C.

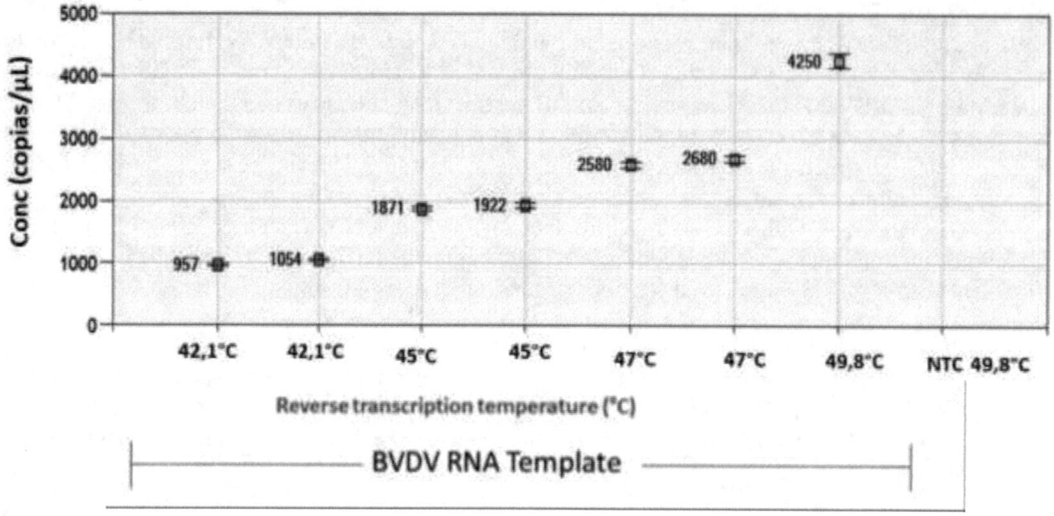

Figure 1: Figure 1: Results from the RT-ddPCR reverse transcription temperature optimization assays for absolute quantification of BVDV RNA. The reactions were performed using the same RNA from the BVDV-1 Singer viral strain, the primer and probe set from VetMAX™-Gold BVDV PI Detection kit and the reagents from One-Step RT ddPCR kit for probes. On a temperature gradient from 42.1°C to 47°C the samples were tested in duplicate. At the temperature of 49.8°C, RNA was evaluated in unicate along with a non-template control (NTC). This last temperature allowed the greater quantification of the template.

3.1.2 Primes annealing temperature

The best annealing temperature for the primers was selected from four annealing temperatures: 59.2°C, 60°C, 61°C and 61.8°C. There was no significant difference in template quantification under these different conditions (Figure 2 A). However, at a temperature of 60°C, the greatest distance was observed between the positive (blue) and negative (grey) droplet populations (Figure 2B).

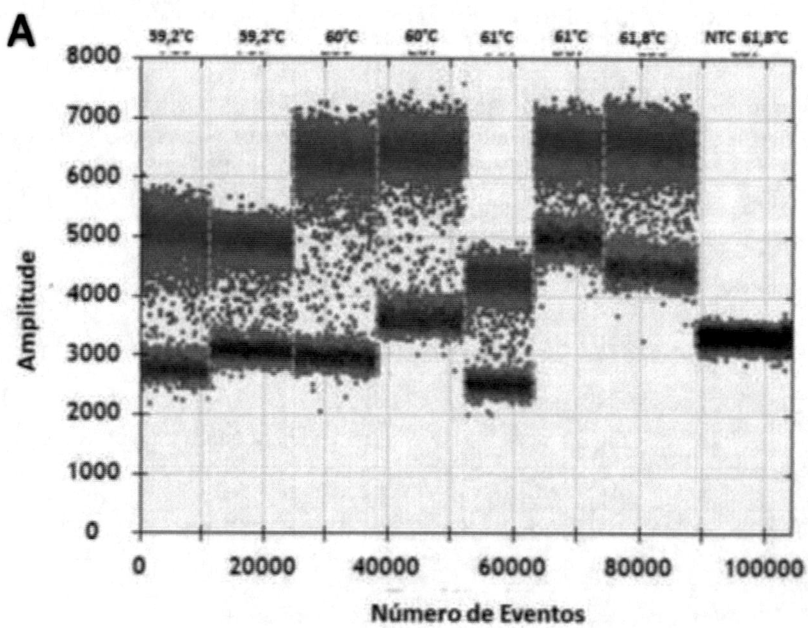

Journal of Physics: Conference Series **2606** (2023) 012016 doi:10.1088/1742-6596/2606/1/012016

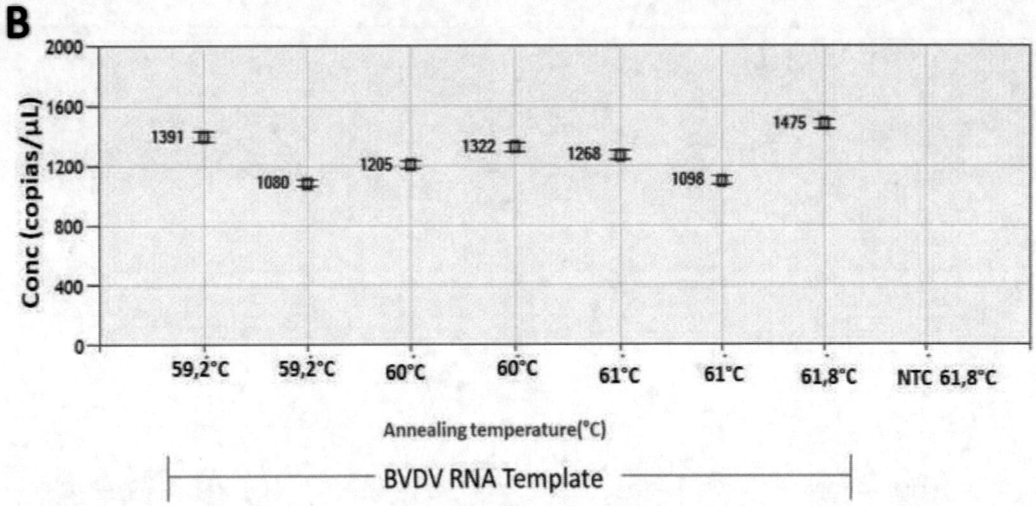

Figure 2: Results from RT-ddPCR annealing temperature optimization assays for absolute quantification of BVDV RNA. The reactions were performed with the same RNA from BVDV 1 Singer strain, the primer and probe set from VetMAX™-Gold BVDV PI Detection kit and reagents from One-Step RT ddPCR kit for probes. On a temperature gradient of 59.1°C to 61°C the samples were tested in duplicate. At 61.8°C, RNA was evaluated in unicate along with a non-template control (NTC). There was no significant difference between the quantification results between the different temperatures (B), but at 60°C (A) there was a greater separation between the positive (blue) and negative (grey) droplet populations.

3.2 *Parameters for validation*

3.2.1 Seletectivity

In selectivity tests, the amplification specificity of the set of primers and probe used (from VetMax Gold BVDV PI Detection kit) in the RT-ddPCR was evaluated.

There was amplification, as suggested by Thermo Fisher, in RNA samples from different BVDV strains (Singer/BVDV-1/pestivirus A and 890/BVDV-2/pestivirus B). In addition to what was predicted, the HoBi-like strain (formerly atypical BVDV, current pestvirus H) was amplified (Figure 3). The assay was specific for bovine pestiviruses. There was no amplification of DENV-1 (from the same Flaviviridae family, but from the flavivirus genus) and MDBK culture RNA.

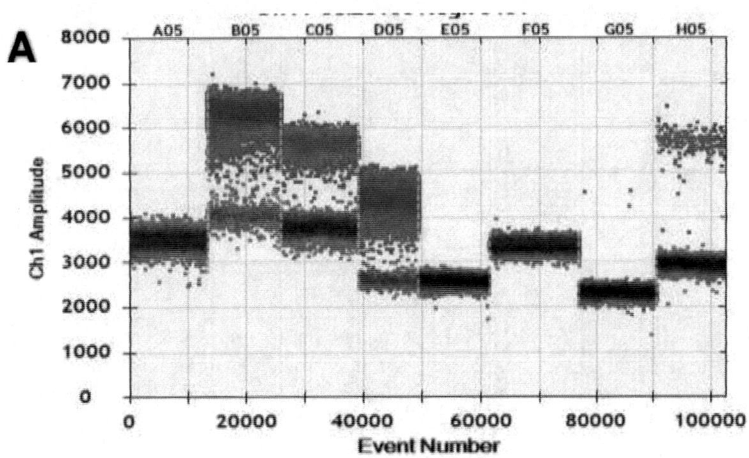

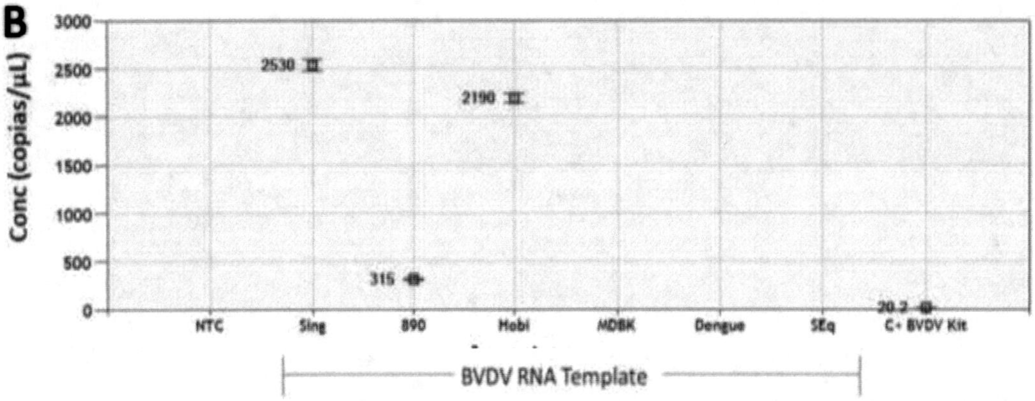

Figure 3: Results of primer and probe set selectivity assays from VetMAX™-Gold BVDV PI Detection kit used in ddPCR. In "B" the quantification of viral RNA samples (in RNA copies/µL) of the three pestivirus strains is observed: Singer (BVDV-1, pestivirus A), 890 (BVDV-2, pestivirus B) and HoBi- like (atypical BVDV, now pestivirus H). There is no quantification in samples with RNA from Dengue Virus (DENV-1, flavivirus), MDBK cells, and on no-template control (NTC). As positive controls, Fetal Bovine Serum (SFB) from a naturally infected batch and the positive control (C+) from VetMAX Gold BVDV PI Detection kit were used. In "A" the negative droplet populations (in gray) are distinguishable in all tested targets and the positive droplet populations (in blue) only in bovine pestiviruses.

3.2.2 Determination of detection limit and quantification of the method

The analytical curve of the proposed method for quantification of BVDV RNA by ddPCR was linear over the entire tested range. It was possible, by linear regression and statistical analysis, to calculate the Limit of Detection (LD) as 13 copies/µL and the Limit of Quantification (LQ) as 38 copies/µL. The linear working range is from 38 to 400 copies/µL (Figure 4).

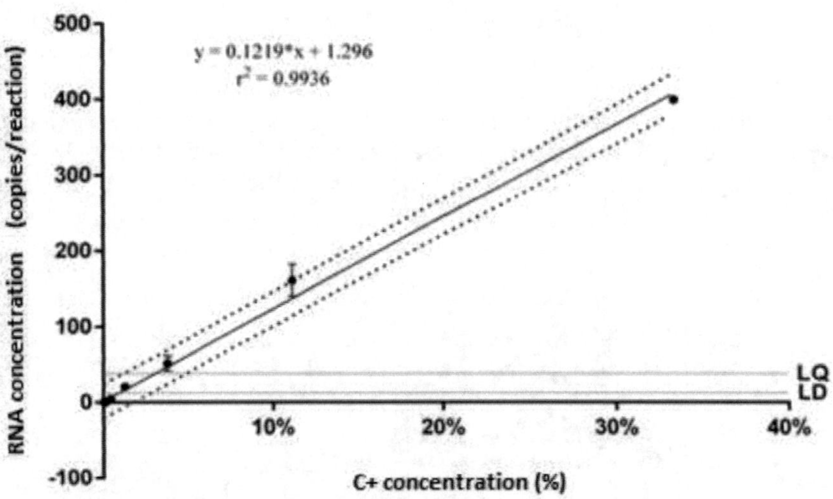

Figura 4: Analytical curve of the proposed BVDV RNA quantification assay by RT-ddPCR, defined by analysis of 8 concentrations of a serial dilution of VetMax Gold BVDV PI Detection Kit Positive Control (C+) in duplicate. The working range is linear over the entire range tested (r^2= 0.9936). The Detection Limit (LD=13

copies/μL) and Quantitation Limit (LQ=38 copies/μL) of the method were determined according to DOC - CGCRE-008 version 9.

4. Conclusion
The validation of this method for absolute quantification of BVDV RNA by ddPCR was possible using a set of primers and probe from a commercial qualitative kit (VetMax Gold BVDV from Thermo Fisher Scientific, USA). This validation will allow us to advance the feasibility study project for the production of RNA-based virus reference materials, using BVDV as a model. The proposed RT-ddPCR will be used in the homogeneity and stability tests of the first batch.

5. References
[1] Payne D A, Mamotte C D S, Gancberg D, Pazzagli M, van Schaik R H N, Schimmel H and Rousseau F 2010 Nucleic acid reference materials (NARMs): definitions and issues *Clinical Chemistry and Laboratory Medicine* **48**

[2] Glotov A G, Glotova T I and Koteneva S V 2018 Pestiviruses, which contaminate imported fetal bovine serum, may be a cause of the global spreading of viral diarrhea in cattle — a mini review *Sel'skokhozyaistvennaya Biologiya* **53** 248–57

[3] Jing R, Wang H, Ju S and Cui M 2018 Reference materials for molecular diagnostics: Current achievements and future strategies *Clinical Biochemistry* **56** 11–7

[4] Furuta-Hanawa B, Yamaguchi T and Uchida E 2019 Two-Dimensional Droplet Digital PCR as a Tool for Titration and Integrity Evaluation of Recombinant Adeno-Associated Viral Vectors *Human Gene Therapy Methods* **30** 127–36

[5] Nyaruaba R, Mwaliko C, Kering K K and Wei H 2019 Droplet digital PCR applications in the tuberculosis world *Tuberculosis* **117** 85–92

[6] Flatschart R B, Almeida D O, Heinemann M B, Medeiros M N, Granjeiro J M and Folgueras-Flatschart A V 2015 Absolute quantification of Bovine Viral Diarrhea Virus (BVDV) RNA by the digital PCR technique *Journal of Physics: Conference Series* **575** 012038

[7] Hafz E M, Salama A A A, Soliman A M, Ismail M E-B A, Farag G K and Ali A A E-S H 2017 Detection of Bovine Viral Diarrhea Virus in some Vaccines, Sera and Cell Cultures *Zagazig Veterinary Journal* **45** 314–22

[8] Quan P-L, Sauzade M and Brouzes E 2018 dPCR: A Technology Review *Sensors (Basel, Switzerland)* **18**

[9] Maurício A, Pereira M, Magalhães W and Cazarim L 2011 *Guia de validação e controle de qualidade analítica : fármacos em produtos para alimentação e medicamentos veterinários* (Brasilia: MINISTÉRIO DA AGRICULTURA, PECUÁRIA E ABASTECIMENTO)

[10] Bio-had 2020 Digital PCR *Bio-Rad Laboratories*

[11] ABNT 2017 NBR ISO 17034: Requisitos gerais para a competência de produtores de material de referência.

[12] ABNT 2017 NBR ISSO/IEC 17025:2017. Requisitos gerais para a competência de laboratórios de ensaio e calibração

[13] DOQ-CGCRE-008 2020 *ORIENTAÇÃO SOBRE VALIDAÇÃO DE MÉTODOS ANALÍTICOS* (Divisão de Acreditação de Laboratórios (Dicla))

[14] Pinheiro L and Emslie K R 2018 Basic Concepts and Validation of Digital PCR Measurements *Methods in Molecular Biology* 11–24.

Detection of Microbiological Contamination in Cell Cultures and their Raw Materials: a 6 years report of this activity.

J L P Ramos-Junior[1], O S Monteiro[2], N C dos Santos[2], R Correa[3], D P Cavalcanti[3], J Martins[3], L Boldrini[2], J M Granjeiro[2,4], A V Folgueras-Flatschart[2]

[1]Laboratory of Macromolecules (LAMAC), National Institute of Metrology, Quality and Technology (INMETRO), Duque de Caxias-RJ / Brazil.
[2]Laboratory of Tissue Bioengineering (LABIO), National Institute of Metrology, Quality and Technology (INMETRO), Duque de Caxias-RJ / Brazil.
[3]Laboratory of Microbiology (LAMIC), National Institute of Metrology, Quality and Technology (INMETRO), Duque de Caxias-RJ Brazil.
[4]Faculty Of Dentistry, Fluminense Federal University (UFF), Niterói-RJ / Brazil.

E-mail: avflatschart@inmetro.gov.br

Abstract. Contaminations in cell cultures are a source of concern for biotechnology laboratories, bio products manufacturers and cell banks as they involve economical loss, time loss and non-valid results. The main issues are fungi, bacteria (including mycoplasmas) and viruses. In this study, we present a retrospective of cell cultures quality control and their raw materials carried out at INMETRO throughout 6 years. From 2014 to 2019, different cell lines and raw materials derived from LABIO, from partner institutions of research projects, or from cooperation agreements were assessed. The detection of mycoplasmas were made using the biochemical method MycoAlert Plus. Contamination by bacteria and fungi were detected by a microbiological culturing according to the Brazilian Pharmacopoeia. Among the 944 assays carried out to detect bacteria and fungi; 5.6% presented contamination. Among 1143 assays of mycoplasmas detection; 10.8% presented contamination. Regarding raw materials assessment, 0.7% and 7.7% were contaminated with mycoplasmas and bacteria/fungi, respectively. In total, 8% of all samples presented contamination: 5.6% by mycoplasmas, 2.4% by bacteria/fungi and 0.1% by both. The dissemination of Good Practices on Cells Culture, the availability of these assays and the support of the implementation given to laboratories which are interested have been our contribution to improve control quality in the activity of cell cultures.

1. Introduction

Using cell cultures in Biotechnology and as a research tool have been of great relevance since the development of the technique in the beginning of the 20th century [1]. Manipulated under controlled environment and conditions, cell lines can be cultivated endlessly, keeping morphology similar to cells of tissues and organs of origin. Due to these characteristics, cell lines have been used to manufacture pharmacotherapics, bioproducts and as models of *in vitro* conditions and *in vivo* cell phenotypes [2, 3, 4].

The cell lines microbiological contaminants which concern culturing laboratories around the world are fungi, bacteria (including mycoplasmas), and viruses [5], derived from different sources as air, operator, raw materials or cells source tissues. Contamination by fungi and by the most part of bacteria are easily detected because they cause changes in color and turbidity of medium culture, they are visible to a common optical microscope and cause changes to the morphology of cell cultures [6]. In

order to avoid contaminations, many laboratories adopt antibiotics in cell culture routine, although some studies report that this action can cloak insidious bacteria instead of eliminating them [7].
On the other hand, contamination by mycoplasmas has been drawing the attention, as it is subtle, being undetected even using a common optical microscope frequently used in a laboratory routine. Due to their small size (which varies from 0.2 μm to 0.8 μm), and to the fact they do not have a rigid cell wall (only cell membrane), they are able to pass through the filters used to sterilize the medium culture [8]. They are not affected by antibiotics that are active in the cell wall synthesis, as penicillin and streptomycin.
Even though cell cultures contaminated by bacteria are far easily detected in general, there are reports of it being detected in up to 10% of the cultures and different cell lines from a laboratory were found to be contaminated by only one bacteria species [9, 10]. Regarding contamination by mycoplasmas, which are harder to detect, this pattern was observed from 32% to 85% of the cells cultivated in laboratories [11, 12, 13, 14]. In a data research on line, it was identified the presence of mycoplasmas sequences in 11% among data of gene expression of rodents and primate cells kept in the Genbank (NCBI), this data was published in high impact scientific journals, as Nature, Cell and Genome Research [15].
The contamination of cell cultures has been pointed out as one of the possible causes of *in vitro* assays results irreproducibility [16], being the assessment of possible contaminations pre-requisite of quality in product assays and trustful services created from cells. In 2004, LANGDON already stated that contaminations between cell lines and culture contaminations by microorganisms could not be neglected [17]. As they cannot be totally avoided, it is necessary to monitor and control the frequency of these events. Recently in 2017, FUSENIG and partners pointed out the changes that have been happening in recognized scientific journals, such as the *International Journal of Cancer*, where the cell authenticity assays and assays to detect mycoplasmas are turning into a mandatory requisite to submit articles [18]. In order to mitigate a potential risk of contamination, regulatory agencies of many countries are demanding for assays that prove the absence of mycoplasmas in specific phases of manufacturing bio products derived from cells, assuring their safety and quality [19].
Aiming to assure the quality of its cell cultures and *in vitro* assays, the Laboratory of Tissue Bioengineering (LABIO), implemented almost a decade ago, cell authenticity assays for human cells, detection assays of microbiological contaminants (bacteria and fungi) and mycoplasma detection [20]. Once these assays were available in our laboratory routine, they started to be available to external clients (research project partners or partnership agreements).
The dissemination of Good Practices on Cells Culture and the adherence of laboratories to the microbiological quality control of their cultures and raw materials is one of the consequences of this study. We shall present the history of these analyses from 2014 to 2019 as it follows bellow.

2. Methodology

2.1 Samples Analyzed

Samples of clarified supernatant of cell cultures and samples of raw materials of cell culture activity (culture medium, supplements and solutions) were sent for analyses by INMETRO laboratories (from 2014 to 2019) or external institutions like universities, research centers, service providers laboratories of *in vitro* assays (research projects partners or partnership agreements).
As a routine, cell culture samples were always collected for this quality control in the end and in the beginning of *in vitro* assays, during preparation, when unfreezing cell stocks (*Master Banks*) or when there was any suspicious of contamination due to changes in morphology, viability or doubling time.
Raw materials were always tested after getting to the laboratory or after its preparation, before clearance to be used and in case of a suspected contamination.

2.2 Microbiological Assays (to detect fungi and bacteria)

Fungi and bacteria detection assays follow the procedure of products sterility assays available in the Brazilian Pharmacopoeia [21]. Soybean Casein Digest Broth (TSB) was applied to grow up fungi and

thioglycollate fluid to grow up bacteria. Each one of the batches of these prepared media were previously tested regarding their sterility and capability to promote microbiological growth, using reference microorganisms.
During the assay, 1 mL of culture supernatant or raw material was inoculated in a aseptic enviroment, directly to test tubes; each one with 10 mL of TSB and thioglycollate media. Then, the TSB and thioglycollate tubes were incubated at 22.5 °C and 32.5 °C, respectively, under aerobic conditions for 14 days. By the end of this period, the positive result was acquired observing the turbidity in the inoculated tubes and the negative with the absence of turbidity. Negative and positive controls were included in each assay.
In order to minimize false negative results in samples with antibiotics, microbiological assays were carried out only after diluting the sample, looking for concentrations in which their effects were no longer observed.

2.3 Bioluminescence Testing (to detect mycoplasmas)
To detect mycoplasma contaminants of cell cultures and their raw materials, it was used a commercial biochemical method by bioluminescence, MycoALert PLUS Mycoplasma Detection Kit (Lonza, USA), according to the manufacturer's instructions. In summary, 100 uL of each sample (clarified supernatant or cell cultures raw material) were transferred to an appropriate tube on a Lucetta luminometer (Lonza, USA), 100 uL of the reagent provided in the kit were added to each tube. After 5 minutes, reading A was done. Later on, 100 uL of the substrate provided in the kit were added to each tube. After 10 minutes, reading B was done. The assay result was calculated by reason Reading B / Reading A. The samples that showed reason >1.2 were considered contaminated. The reason <1 indicated the absence of contamination by mycoplasmas. The samples with reason between 1 and 1.2 were considered borderline, suggesting a new sample test.
During the assay, the luminometer was set up to do the optical readings for 1 minute.
Considering that using antibiotics can restrict the growth of mycoplasmas, as it leads to false negative results, samples with antibiotics were not assessed. For the analyses of cell lines, they were cultivated without antibiotics for at least 2 weeks, and before the collection, they were not manipulated for at least 4 days. At least 1.5 mL of culture supernatant were collected (with 80-100% of confluence) and centrifuged in a Falcon tube at 200 RCF for 5 minutes. Then, 200 μL of this clarified supernatant were stored at -80 °C until the test.
For the analyses of raw materials, a 200 μL aliquot of each material was stored at -80 °C until the test.

3 Results and Discussions
The results of fungi, bacteria and mycoplasmas assays detection in cell cultures and in their raw materials carried out at LABIO as quality control mechanisms are showed in Table 1 and Figure 1.

Table 1: Result of samples checked by Microbiological Test (to detect fungi and bacteria) and by Bioluminescence test (to detect mycoplasmas) from 2014 to 2019.

			Microbiology Assays (fungi and bacteria detection)			Bioluminecent Assays (mycoplasmas detection)		
	Samples		Number of assays	Positive results	Contamination rate (%)	Number of assays	Positive results	Contamination rate (%)
Cell Lineages /	Jurkat	Human	2	0	0	2	0	0
	IEC-6	Rat	2	0	0	2	0	0
	J774 1.6	Mouse	2	0	0	3	0	0
	CHO-K1	Hamster	2	0	0	5	0	0
	HOS	Human	2	0	0	7	0	0
	HCC 1954	Human	3	0	0	3	0	0
	HFF-1	Human	3	0	0	5	0	0

	CCD 1059 SK	Human	4	0	0	3	0	0
	HACAT	Human	7	0	0	8	0	0
	HT-29	Human	7	0	0	11	0	0
	C2C12	Mouse	4	2	50	3	0	0
	Vero	Monkey	1	0	0	4	1	25
	RAJI	Human	2	0	0	2	1	50
	MC3T3	Mouse	2	0	0	6	2	33.3
	DU-145	Human	3	0	0	9	1	11.1
	V79-4	Hamster	9	0	0	12	1	8.3
	CACO-2	Human	12	0	0	14	1	7.1
	MCF-7	Human	13	0	0	16	3	18.7
	THP-1	Human	13	0	0	49	2	4.1
	MRC-5	Human	14	0	0	15	2	13.3
	HepG2	Human	31	0	0	40	3	7.5
	MDBK	Bovine	11	1	9.1	25	1	40
	FGH	Human	18	3	16.7	10	6	60
	L-929	Mouse	20	1	5	22	7	31.8
	U-87 MG	Human	22	4	18.2	14	3	21.4
	A549	Human	60	3	5	88	15	17
	Balb 3T3 cA31	Mouse	70	5	7.1	57	8	14
	Other cells	diverse	213	6	2.8	332	57	17.2
	SubTotal		**552**	**25**	**4.5**	**767**	**114**	**14.9**
Supplies	Culture media		149	10	6.7	154	2	1.3
	Suplements		91	3	3.3	89	0	0
	Solutions		152	17	11.2	133	0	0
	SubTotal		**392**	**30**	**7.7**	**376**	**2**	**0.5**
	Total		**944**	**55**	**5.8**	**1143**	**116**	**10.1**

The cell lines which was not observed any contaminations (by fungi, bacteria and mycoplasmas) was tested in a scarce volume during the period, indicating that its usage was seldom and they were less manipulated. However, some of the cells seldom assessed (less than 10) did not present contamination by fungi or bacteria, but high rates of contaminations by mycoplasmas: Raji (50%), MC3T3 (33,3%) and Vero (25%). More than just pointing out problems caused by careless manipulation of cultures, this fact draws the attention to possible contamination by mycoplasmas of cell stocks; on the contrary of the cell line C2C12, with 50% of contamination by fungi or bacteria and absence of mycoplasmas contamination.

From historical occurrences [22], it is known that raw materials used in cell culture as components of media culture, solutions and supplements are recognized source for the introduction of adventitious contaminants in bioprocesses. In general, the contamination rate of raw material by mycoplasmas (culture media, supplements and solutions) was very low (0.5%). Contaminant mycoplasmas were not found in products from animal source as supplements (for instance, bovine fetal serum) or solutions (for instance, trypsin). It demonstrates that the quality control of raw material suppliers has been adequate regarding the absence of this adventitious agent. Mycoplasmas were detected in work aliquots of culture media already handled by laboratory technicians in their cell cultures activities. Therefore, as the result indicates contamination, culture media are discarded, as well as the cells in which they were used. For this purpose, traceability of this information is essential. The contamination rate by fungi and/or bacteria of raw materials analyzed was 7.7%, which happened in individual work aliquots, exception given to a contamination accident of a stock solution prepared by an apprentice that did not follow the standard operational procedure adequately in the laboratory. Adequate and constant training for the staff is essential for this activity.

Regarding cell cultures analyzed, the contamination rate was less with bacteria and fungi (4.5%) than with mycoplasmas (14.9%). It can be related to the fact that contamination by bacteria and fungi are easier to be identified by the operator, who normally discards them without the need of a laboratory assay to confirm. As mycoplasma contaminations can be unnoticed during routine care of cell cultures, it is quite common to find cells contaminated by these agents only. The presence of only mycoplasmas was confirmed in different rates, for example, in samples of lines MCF-7 (18.7%), MRC-5 (13.3%), U-145 (11.1%), V79-4 (8.3%), HepG2 (7.5%), CACO-2 (7.1%) and THP-1 (4.1%).

Some of the lines tested, showed within this period, contamination by fungi and bacteria as well as by mycoplasmas. As this testing is carried out as a routine in the beginning and in the end of *in vitro* assays, also to form cell stocks (*Master Banks*), and when there is some suspicious of microorganism presence, the highest amount of analyses performed per lines is an indication of its higher manipulation in the laboratory. Then, it would be expected that the more manipulated they are, the higher the risk of contamination would be. Surprisingly, cell lines that were more tested within the period, A549 and Balb 3T3 clone A31, were not the ones that showed the higher rates of contamination by fungi or bacteria (5% and 7.1%, respectively) neither by mycoplasmas (17% and 14%, respectively). The line FGH, quite less tested, presented much higher rates of contamination by fungi or bacteria (16.7%) and by mycoplasmas (60%). The causes for this difference must be better assessed in the cell culture laboratory, so contaminations can be avoided, but without doubt, the level of training in Good Practices on Cells Culture of all operators who manipulate them are directly related to this result.

Analyzing the totality of assays within this 6 years period, of the 944 assays of fungi and bacteria detection and 1143 mycoplasmas detection assays, we had the absence of any contaminant in 86% of the cases. It shows that assays have been made routinely (as recommended) and not only to confirm a possible contamination (Figure 1A). Considering 8% of analyses that identified the presence of contaminants, more than two thirds (68.5%) were mycoplasmas, 29.8% were fungi and / or bacteria, and only 1.7% were observed mixed contaminations (with the presence of fungi and/or bacteria; and mycoplasmas) (Figure 1B). These results are corroborated by a study that searches multiple contaminations in cell cultures [23]. In this study, the author reinforces that contamination by mycoplasmas is a significant risk for any cell culture laboratory and emphasizes the relevance of a routine to detect theses agents.

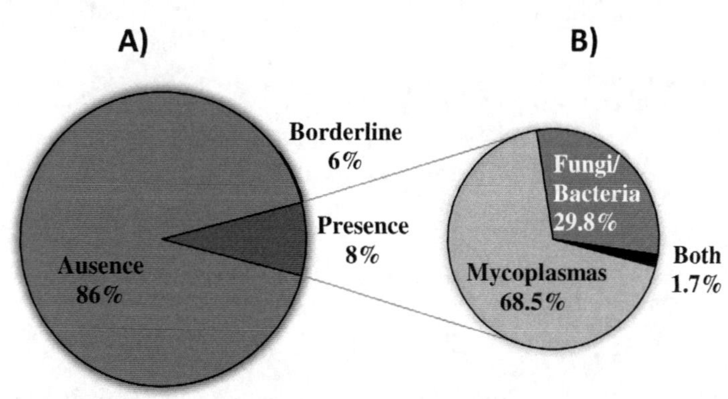

Figure 1: Contamination rate found in analyses made in cell cultures and their raw materials from 2014 to 2019. In A) "Ausence" indicates the absence of fungi, bacteria and mycoplasmas, "Presence" indicates the presence of fungi, bacteria and mycoplasmas, and "Borderline" refers to the adjoining results found in mycoplasmas detection assays. In B), considering the results in which some contamination was detected in the samples analyzed; we here present the percentages of contamination by fungi or bacteria, by mycoplasmas or both.

4 Conclusion

The risk of contamination of cell lines and their raw materials by microbiological agents is constant in cell cultures laboratories. Knowing that these contaminations are directly related to assays trustfulness, reproducibility of the results and bio product safety; national and international norms and regulations have been trying to diffuse or demand that tests to detect adventitious agents be carried out routinely in cell cultures, their stocks and raw materials.

Applying Good Practices on Cells Culture techniques and a constant training of our team, implementing a routine of assays to detect fungi, bacteria and mycoplasmas in the Laboratory of Tissue Bioengineering (LABIO of INMETRO) have been allowing us to work with more safety, as well as with our partners of scientific projects and partnership agreements.

In the last few years, LABIO has been trying to diffuse among cell cultures laboratories the practice of monitoring possible contaminations with lab assays, mainly the ones caused by mycoplasmas. Included in these activities carried out within this period are the availability of courses and subjects, Standard Operational Procedures and the implementation of mycoplasmas detection techniques, described here in 5 laboratories of the National Network of Alternative Methods – RENAMA. We still support the Rio de Janeiro Cells Bank (BCRJ) – a non-profitable company created in INMETRO – the first initiative in the country to implement interlaboratorial comparisons on the same mycoplasmas detection assays that were performed twice within this period.

As international directives establish that two independent assays to detect mycoplasmas be used, we implemented a commercial detection method by PCR. However, this method is temporarily inoperative due to its high maintenance cost. On the other hand, we have been working to develop and validate new molecular methods for this purpose, which we expect to share with the scientific community in a near future.

5 Acknowledgments

We thank the institutions which have been supported us throughout these 6 years: MCTI/ CNPq, PRONAMETRO/INMETRO, FAPERJ, and CAPES. We also thank the members of the National Network of Alternative Methods (RENAMA) that have been our partners in many projects to disseminate quality control methods of cell cultures and their raw materials.

6 Bibliographic References

[1] Enders J F, Weller T H and Robbins F C 1949 Cultivation of the Lansing Strain of Poliomyelitis Virus in Cultures of Various Human Embryonic Tissues *Science* **109** 85–7

[2] Farin H F, Jordens I, Mosa M H, Basak O, Korving J, Tauriello D V F, de Punder K, Angers S, Peters P J, Maurice M M and Clevers H 2016 Visualization of a short-range Wnt gradient in the intestinal stem-cell niche *Nature* **530** 340–3

[3] Jung D J, Shin T H, Kim M, Sung C O, Jang S J and Jeong G S 2019 A one-stop microfluidic-based lung cancer organoid culture platform for testing drug sensitivity *Lab on a Chip* **19** 2854–65

[4] Wurm F 2013 CHO Quasispecies—Implications for Manufacturing Processes *Processes* **1** 296–311

[5] Nims R W and Price P J 2017 Best practices for detecting and mitigating the risk of cell culture contaminants *In Vitro Cellular & Developmental Biology - Animal* **53** 872–9

[6] Ryan J 2008 Understanding and managing cell culture contamination. *Technical Bulletin*

[7] Taylor-Robinson D 1998 Antibiotic susceptibilities of mycoplasmas and treatment of mycoplasmal

infections *Journal of Antimicrobial Chemotherapy* **40** 622–30
[8] Folmsbee M, Howard G and McAlister M 2010 Nutritional effects of culture media on mycoplasma cell size and removal by filtration *Biologicals* **38** 214–7
[9] Cobo F, Cortés J L, Cabrera C, Nieto A and Concha Á 2007 Microbiological contamination in stem cell cultures *Cell Biology International* **31** 991–5
[10] Xiong T, Wang D, Tan B, Jiang W, Dai G, Zhao H, Li S and Li Y 2015 Contamination of Phenylobacterium in several human and murine cell cultures *Acta Microbiologica Sinica* **55(2)** 176-86
[11] Young L, Sung J, Stacey G and Masters J R 2010 Detection of Mycoplasma in cell cultures *Nature Protocols* **5** 929–34
[12] Rivera A 2009 Cell cultures contaminations by mycoplasmas *African Journal of Microbiology Research* **3** 637-40
[13] Molla Kazemiha V, Amanzadeh A, Memarnejadian A, Azari S, Shokrgozar M A, Mahdian R and Bonakdar S 2014 Sensitivity of biochemical test in comparison with other methods for the detection of mycoplasma contamination in human and animal cell lines stored in the National Cell Bank of Iran *Cytotechnology* **66** 861–73
[14] Drexler H G and Uphoff C C 2002 Mycoplasma contamination of cell cultures: Incidence, sources, effects, detection, elimination, prevention *Cytotechnology* **39** 75–90
[15] Olarerin-George A O and Hogenesch J B 2015 Assessing the prevalence of mycoplasma contamination in cell culture via a survey of NCBI's RNA-seq archive *Nucleic Acids Research* **43** 2535–42
[16] Jarvis M F and Williams M 2016 Irreproducibility in Preclinical Biomedical Research: Perceptions, Uncertainties, and Knowledge Gaps *Trends in Pharmacological Sciences* **37** 290–302
[17] Langdon S P 2004 Cell Culture Contamination: An Overview *Cancer Cell Culture* **88** 309–18
[18] Fusenig N E, Capes-Davis A, Bianchini F, Sundell S and Lichter P 2017 The need for a worldwide consensus for cell line authentication: Experience implementing a mandatory requirement at the International Journal of Cancer *PLOS Biology* **15** e2001438
[19] Dabrazhynetskaya A, Volokhov D V, David S W, Ikonomi P, Brewer A, Chang A and Chizhikov V 2011 Preparation of reference strains for validation and comparison of mycoplasma testing methods *Journal of Applied Microbiology* **111** 904–14
[20] ANSI/ATCC ASN-0002-2021 - Authentication Of Human Cell Lines: Standardization Of Short Tandem Repeat (STR) Profiling. https://www.ansi.org/
[21] Agência Nacional De Vigilância Sanitária (Brazil and Fundação Oswaldo Cruz 2010 *Farmacopeia brasileira.* (Brasília: Agência Nacional De Vigilância Sanitária)
[22] Kljavin I J 2011 Broadening Our Expectations for Viral Safety Risk Mitigation *PDA Journal of Pharmaceutical Science and Technology* **65** 645–53
[23] Mirjalili A, Parmoor E, Moradi Bidhendi S and Sarkari B 2005 Microbial contamination of cell cultures: A 2 years study *Biologicals* **33** 81–5

Convolutional Neural Network for non-invasive magnetic foreign body localization in the human body

M Rogozinski[1], C R H Barbosa[1] and E Costa Monteiro[1]

[1] Pontifical Catholic University of Rio de Janeiro, Postgraduate Programme in Metrology, Rua Marquês de São Vicente, 225, Rio de Janeiro, Brazil

E-mail: hall@puc-rio.br

Abstract. Ferromagnetic foreign bodies accidentally inserted in patients usually need to be surgically removed. Their location can be estimated by measuring the magnetic field generated by the object and solving the magnetic inverse problem to locate the source based on the magnetic field maps configuration. Considering ferromagnetic straight needles (hypodermic or sewing, for example) and a computer simulation of the magnetic flux density based on Biot-Savart's Law, this paper presents an algorithm based on Convolutional Neural Networks (CNN) to find the depth, the angles of inclination and rotation and the center of the needle inside the human body. The proposed model presents low RMSE values and type A measurement uncertainty for the depth, the inclination angle, the rotation angle, and for the displacement Δ between the midpoint of the distance connecting the magnetic field extreme values and the foreign body center, all of which adequate for use in therapeutic procedures.

1. Introduction

Foreign bodies with ferromagnetic properties needing surgical removal are frequent in medical practice. Their location is conventionally performed by employing radiography, computed tomography, and radioscopy procedures [1,2]. However, besides their drawback regarding ionizing radiation exposure, these available strategies do not provide accurate object position information, leading to long-lasting and unsuccessful surgical procedures [1]. In 2000, a non-invasive and innocuous technique based on magnetic field maps was developed, demonstrating highly accurate ferromagnetic foreign bodies localization, thus providing their rapid and successful surgical removals [1,3]. A Superconducting Quantum Interference Device (SQUID), the most sensitive magnetometer, was employed for the magnetic mapping. Despite the relevant advantages, SQUID systems' operation presents drawbacks of high-cost and cryogenic temperature requirements.

Low-cost magnetic transducers to locate ferromagnetic objects in the human body, based on the Giant Magnetoimpedance (GMI) or Giant Magnetoresistance (GMR) effects, have been recently investigated [4–7]. These are initiatives to overcome the cost disadvantage of the non-invasive and innocuous technique when employing a superconducting sensor, thus allowing it to be better suited to comply with the Biometrology Principles recommended for incorporating the device in the healthcare sector [8].

A group of researchers at Pontifical Catholic University of Rio de Janeiro (PUC-Rio) is currently testing the use of a portable device with low-cost GMR sensors to find the depth h, angle of inclination θ, angle of rotation $\sphericalangle$, and geometric center of a ferromagnetic foreign body in the human body. As many computer vision investigations applied to medical images have been performed recently using convolutional neural networks (CNN) [9,10], incipient initiatives are being implemented for applying CNN architectures for estimating the location parameters of this new technique being developed [11]. This paper addresses a more advanced step of this research by developing a CNN architecture able to solve the magnetic inverse problem and accurately provide these positioning parameters, based on a simulation of the magnetic field map generated by the foreign body, as would be detected by a magnetometer integrated into a portable device.

The rest of the paper is organized as follows. Section 2 presents the materials and methods, including the proposed CNN architecture, the dataset used in our experiments, the hyperparameter configuration, and the experimental protocol. Section 3 shows the results obtained, while Section 4 summarizes the main conclusions.

2. Materials and Methods

2.1. Ferromagnetic Foreign Body

Figure 1 illustrates a needle with the parameters used in this study. The inclination angle θ, rotation angle ϕ, and depth h of the needle are direct outputs of the network. The geometric center of the needle, represented in figure 1 by p, is obtained by calculating the center between magnetic field extreme values and adding to it a displacement value Δ, also provided by the CNN. Figure 2 shows an example image of the magnetic field generated by the ferromagnetic foreign body, which serves as the input for our CNN, with added markings to highlight some parameters. The white markings are the minimum and maximum values of the magnetic field, and the black marking is the middle point between those extremes, while the red marking is the geometric center of the needle. Our CNN returns the value Δ, which corresponds to the difference between the center of the magnetic field extreme values and the needle's geometric center.

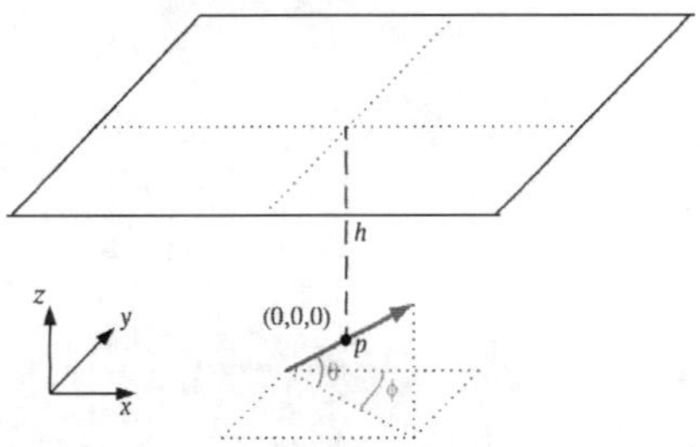

Figure 1. Visualization of a needle position (in red) with (0,0,0) as its center point p, showing the needle inclination angle θ, rotation angle ϕ, and depth h.

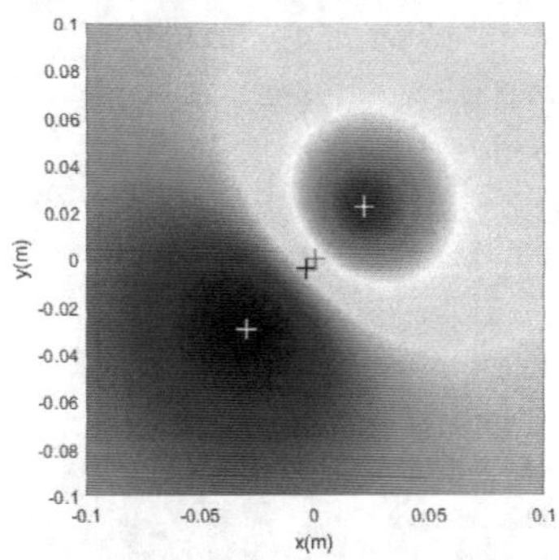

Figure 2. Colormap image representing the magnetic field with white markings on its extreme values, a black marking at the center between their positions, and a red marking at the geometric center of the ferromagnetic object that generated the map.

2.2. Dataset

The dataset was built by simulating the magnetic field data that would be read by our portable device during a complete scan. Each scan corresponds to a path taken by the device in a zig-zag pattern over a 20 cm x 20 cm area centered on a known reference, consisting of a marker affixed to the patient skin. The output values obtained from the CNN, along with the reference point marked on the patient's skin, will be used by the medical team to guide the surgical strategy.

The magnetic field data was generated based on the Biot-Savart law for a magnetic dipole, integrated over the length of the needle (50 mm). The scans were created by applying random variations to a reference path in the positions x, y and z, and in the z-axis' rotation (yaw). Rotations in the x and y axes (roll and pitch, respectively) were kept constant at zero.

Four different irregular paths were used and, for each one of them, 2 376 images were created representing the magnetic field generated by the dipole. These images encompass a variation in depth h from 50 mm to 150 mm, in 10 mm steps, a variation in inclination angle θ from -60° to 60°, in 15° steps, and a variation of rotation angle ϕ from 0° to 360°, in 15° steps. The delta Δ value was calculated from the magnetic field values for each entry according to the procedure presented in [3]. The final dataset comprises 9 504 images with corresponding h, θ, ϕ, and Δ values. Figure 3 shows in (a) an undisturbed scan path, with a perfect magnetic field map generated and, from (b) to (e), the four paths used in our dataset, with their corresponding generated magnetic field maps for the parameters h = 60 mm, θ = 15°, ϕ = 45° and Δ = 0.56 mm.

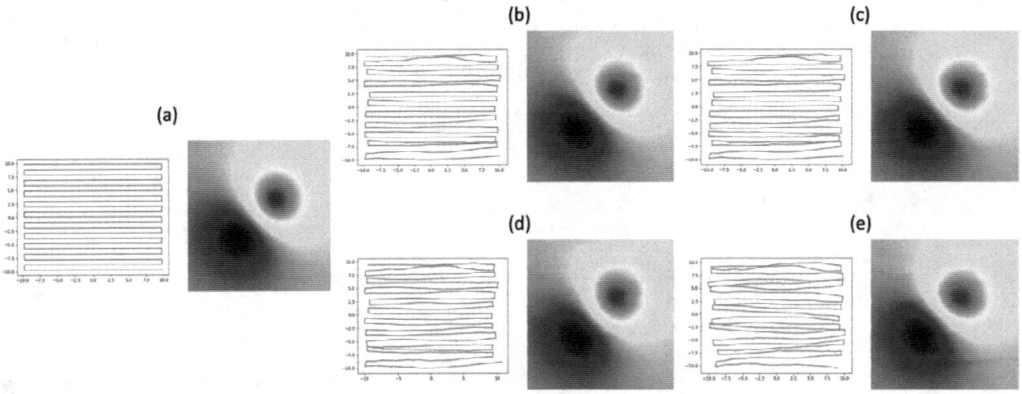

Figure 3. A perfect scan and magnetic field map (a) and the tracking variations used in our dataset (b, c, d, e) with the corresponding magnetic field map, for h = 60 mm, θ = 15°, ϕ = 45° and Δ = 0.56 mm.

Most images with imperfect scan paths or not completely centered have missing pixels near their edges. Since interpolation cannot be applied in those situations, we have created a procedure to fill in those missing pixels. The procedure checks each image pixel, from the image's center to its edges, in a spiral movement. If any pixel is missing, it fills it with the average value of the available surrounding pixels. Figure 4 shows the same image as figure 3e after a scan simulation without (figure 4a) and with (figure 4b) our procedure to fill in missing pixels.

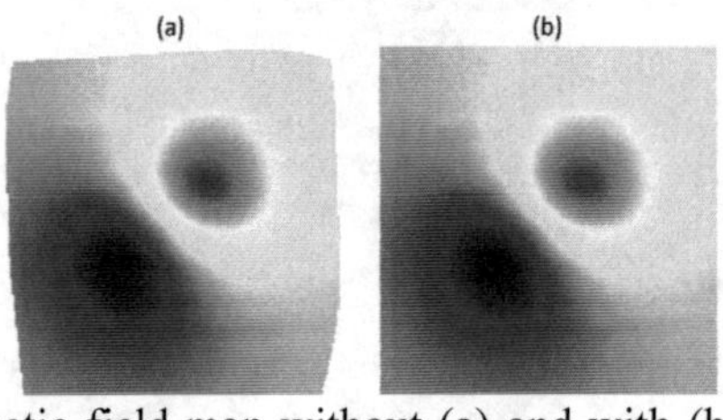

Figure 4. An image of the magnetic field map without (a) and with (b) the method to fill in missed values from the scan procedure.

2.3. Convolutional Neural Network Model

The proposed network is based on a shorter version of a VGG architecture [12] setup for regression, with branches for each output value to be found (h, θ, ϕ and Δ). The first layers of the network are shared, and the last layers are task-specific. We expected that this approach provides good generalization while keeping most of the learnable parameters specific for each task.

At its input, the network expects an image of the magnetic field values with 80 x 80 pixels with three color channels in RGB format for each pixel (red, green, and blue values). The network uses a linear activation function in the last layer with a mean squared error loss function in order to produce real numbers as the output for each task. Figure 5 shows the proposed network with the types of layers used and the number of filters for each layer.

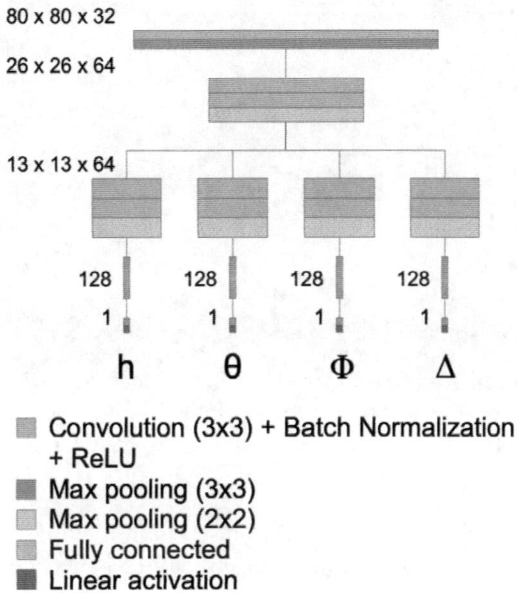

Figure 5. Proposed network showing the type of layers, the number of filters per layer, and the output for each task: the depth h, the inclination angle θ, the rotation angle ϕ, and the displacement Δ between the position of the center of the foreign body and the midpoint between the positions of the extreme magnetic field values.

2.4. Hyperparameter tuning

The model was tested with different numbers of layers and different numbers of filters per layer in search of the best results considering the size of our dataset. The final configuration, shown in figure 5, has a total of 1 534 788 learnable parameters, with 56 960 of those being shared between h, θ, ϕ and Δ. Since the rotation can also be inferred from the magnetic field values directly, we also tested an equivalent model without the rotation branch with a total of 1 165 251 learnable parameters. Both models, with and without the rotation branch, were also tested with absolute Δ values only and considering negative and positive Δ values.

2.5. Experimental protocol

The dataset was randomly separated into disjoint sets, 60 % for training, 20 % for validation, and 20 % for testing. After experimental analysis, we used the Adam optimizer with a learning rate of 0.0001, and dropout was used to help prevent overfitting. Early stopping with a patience of 10 and model checkpoint callbacks were employed to control the training time and save only the best model during training. For the code, we used the Keras framework with Tensorflow backend. Figure 6 shows the evolution of the training and validation data for all values associated with the parameters of the magnetic source location (h, θ, ϕ and Δ).

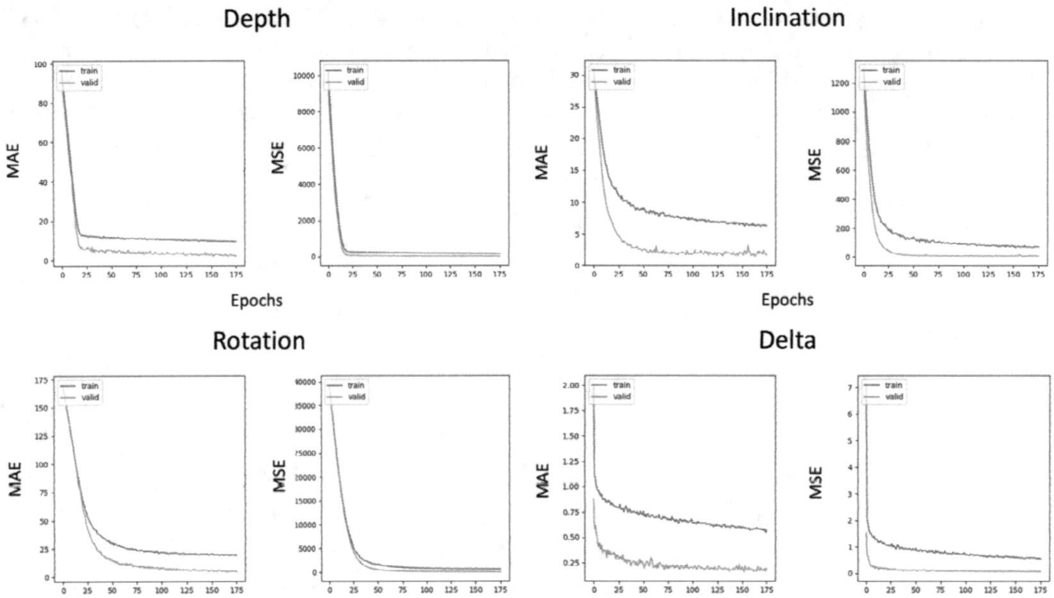

Figure 6. Evolution of Mean Absolute Error (MAE) and Mean Square Error (MSE) for the training (blue) and validation (orange) data over the epochs for the depth h, inclination θ, rotation ϕ, and delta Δ values.

3. Results

Table 1 shows the results obtained for the model presented in figure 5 for each output variable. The model was trained using only absolute Δ values because it presented better results than using negative and positive Δ values. The model with the rotation branch also provided better results than the model with only the depth, inclination, and Δ branches. The final application should consider the absolute Δ values and invert the signal when the rotation is between 90° and 270°, with negative inclination, or when the rotation is between 270° and 90°, with positive inclination.

Table 1. Root Mean Square Error (RMSE), maximum error, and expanded type A uncertainty for each output variable for the test data.

Output variable	RMSE	Max. Error	Expanded Type A uncertainty
depth h	2.8 mm	12.6 mm (@ 130 mm)	5.6 mm
inclination θ	2.2°	7.0° (@ 60°)	4.4°
rotation ϕ	5.5°	19.9° (@ 345°)	11.1°
displacement Δ	0.2 mm	1.1 mm (@ 8.6 mm)	0.4 mm

Figure 7 presents the scatter plots between predicted and true values, and figure 8 presents the histograms of the prediction errors, both for the test data. The proposed model shows an improvement in maximum error for Δ compared to [3] while providing similar results for the other variables in terms of expanded type A uncertainty as those obtained in [11].

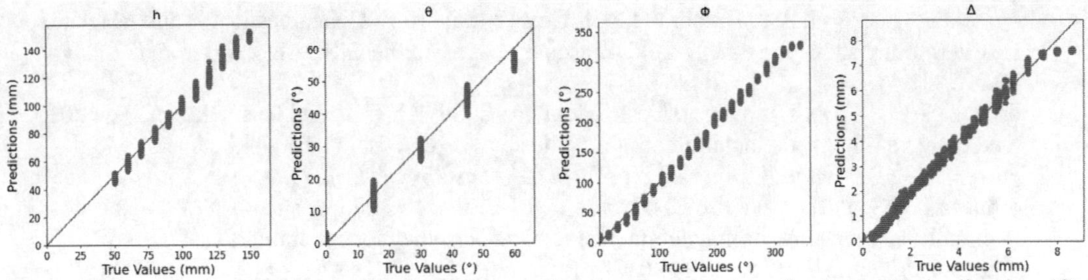

Figure 7. Scatter plot of the test data between predicted and true values for depth (h), inclination (θ), rotation (ϕ), and displacement (Δ) values.

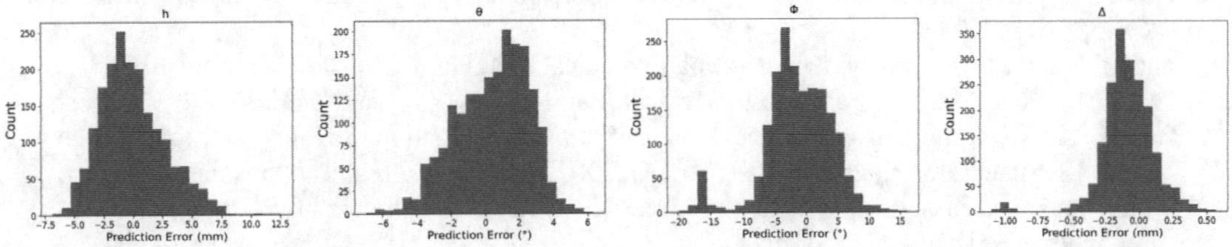

Figure 8. Histogram of the predicted error in the test data for depth (h), inclination (θ), rotation (ϕ), and displacement (Δ) values.

4. Conclusion and future work

This paper proposed a novel CNN architecture to locate foreign objects in the human body based on their magnetic field map image. Our best model achieved an RMSE value in the test data of 2.8 mm for the depth h, 2.2° for the inclination angle θ, 5.5° for the rotation angle ϕ, and 0.21 mm for the displacement Δ between the midpoint of the distance connecting the magnetic field extreme values and the foreign body center.

The obtained low RMSE and type A uncertainty values, also providing information regarding Δ values, a crucial localization parameter, with better performance than [3], point toward the promising feature of the proposed CNN to contribute to the accuracy of the non-invasive localization of ferromagnetic foreign bodies using a low-cost and portable transducer, thus providing the essential information for a safe, quick and successful surgical procedure for object removal [8]. Future steps in the study involve implementing the developed algorithm to solve the inverse problem from experimental magnetic maps.

Acknowledgments

We thank the financial support by the Brazilian funding agencies CNPq, FINEP and FAPERJ. This work was financed in part by the Coordenação de Aperfeiçoamento de Pessoal de Nível Superior – Brasil (CAPES) – Finance Code 001.

References

[1] Costa Monteiro E, Barbosa C H, Lima E A, Ribeiro P C and Boechat P R M 2000 Locating steel needles in the human body using a SQUID magnetometer

[2] Aras M H, Miloglu O, Barutcugil C, Kantarci M, Ozcan E and Harorli A 2010 Comparison of the sensitivity for detecting foreign bodies among conventional plain radiography, computed tomography and ultrasonography *Dentomaxillofac Radiol* **39** 72–8

[3] Barbosa C H, Costa Monteiro E, Lima E A, Santos S F, Cavalcanti E G and Ribeiro P C 2001 Improvement of a technique for localization of steel needles in humans using a SQUID magnetometer *IEEE Transactions on Applied Superconductivity* **11** 677–80

[4] Pompéia F, Gusmão L A P, Barbosa C R H, Costa Monteiro E, Gonçalves L A P and Machado F L A 2008 Ring shaped magnetic field transducer based on the GMI effect *Meas. Sci. Technol.* **19** 025801
[5] Fortaleza L G e S, Barbosa C R H, Costa Monteiro E, Silva E C da and Gusmão L A P de 2015 Detecting non-magnetic metallic foreign bodies by GMR sensors through the use of eddy currents *XI International Congress on Electrical Metrology (SEMETRO)* vol 1 pp 1–4
[6] Fortaleza L G S, Costa Monteiro E, Barbosa C R H, Silva E C and Gusmão L A P 2018 Biomedical comparison of magnetometers for non-ferromagnetic metallic foreign body detection *J. Phys.: Conf. Ser.* **1044** 012013
[7] Oliveira B R, Louzada D R, Costa Monteiro E and Barbosa C R H 2020 Automatic system for locating magnetic foreign bodies using GMI magnetometer *Journal of Physics*
[8] Costa Monteiro E and Leon L F 2015 Metrological Reliability of Medical Devices *J. Phys.: Conf. Ser.* **588** 012032
[9] Suzuki K 2017 Overview of deep learning in medical imaging *Radiol Phys Technol* **10** 257–73
[10] Giger M L 2018 Machine Learning in Medical Imaging *J Am Coll Radiol* **15** 512–20
[11] Rogozinski M, Barbosa C and Costa Monteiro E 2021 Localization of magnetic foreign bodies using CNN and GMI magnetometer IMEKO XXIII World Congress (Yokohama, Japan)
[12] Simonyan K and Zisserman A 2015 Very Deep Convolutional Networks for Large-Scale Image Recognition *arXiv:1409.1556 [cs]*

pH measurements hydrous ethanol fuel

N C Atilio[1], **F L Fertonani**[2], **E C de Oliveira**[1,3]

[1] Programa de Pós-Graduação em Metrologia, Pontifícia Universidade Católica do Rio de Janeiro, 22453-900 Rio de Janeiro – RJ, Brasil
[2] Universidade Estadual Paulista, UNESP, Rua Cristovão Colombo, 2265, Jardim Nazareth, CEP 15054- 000 São José do Rio Preto – SP, Brasil.
[3] Logistics, Operational Planning and Control, Measurement and Product Inventory Management, PETROBRAS S.A., Rio de Janeiro 20231-030, Brazil.
E-mail: nacambiaghi_@outlook.com

Abstract: The pH is one of the quality parameters of hydrous ethanol fuel (HEF), being established by Brazilian standard ABNT NBR 10891 and by American standard ASTM D 6423. The present work aims to compare the measurement systems proposed by the standards for measuring pH in hydrous ethanol fuel and compare them with the pH range found from pH indicators, considered as a standard measurement. The results reveal that the pH values obtained by the different systems used are statistically different from each other and different from the range found by the indicators.

1. Introduction

With the expansion in economic development and scientific knowledge, concern for the environment arose due to degradation and pollution caused by human beings, generating demand for renewable energy alternatives. Due to initiatives to curb global warming, several countries have drafted mandates to increase the consumption of biofuels, which are projected to increase every year [1].

Currently, Brazil is the second country that produces the most ethanol in the world and its main raw material is sugar cane, which has a higher yield than the others because of its higher concentration of sucrose. Brazil is surpassed only by the United States, whose main raw material is corn; these two countries represent approximately 80 % of global production [1][2].

One of the parameters used to assess the quality of hydrous ethanol fuel (HEF) is the potential of the hydrogen ion, called pH. This quality parameter has been applied to indicate the risk of corrosion in vehicle engines, formation of deposits on metal parts, and clogging of filters, injection nozzles, and fuel pump, so it is of singular importance that the pH is measured from the effective way [3][4].

1.1 pH

The pH measurement is a potentiometric technique, being one of the most used techniques in analytical chemistry because it is simple and low cost compared to others. The technique is based on hydrogen ion activity, whose concepts apply to aqueous solutions [5-7].

A given medium to be considered aqueous is necessary for it to have at least 30 % water in its composition. This concept does not apply to ethanol mixtures in water; as hydrous ethanol fuel (HEF) has around 7 % water (m/m). Thus, pH measurements that are conducted on ethanol (HEF) samples can present several problems, such as: (i) instability of readings; (ii) need for long response times; and (iii) results depending on the type of electrode used. These problems are probably due to the low conductivity of the solvent, dehydration of the glass membrane of the electrode, the measurement time and the development and variation of the liquid junction potential (Ejl) [8].

1.2 Hydrous ethanol fuel pH

Brazilian standard, ABNT NBR 10891, and American standard, ASTM D 6423, describe the potentiometric method for determining the pH in hydrous ethanol fuel and the standard test method for determining the pHe of mixtures of ethanol fuel and denatured ethanol fuel, respectively. The American

standard is applied only to anhydrous ethanol fuel, which imposes the use of the Orion electrode, Ross Sure-Flow, and it uses an internal solution of potassium chloride in an aqueous medium [10-11].
The use of the ABNT NBR 10891, Brazilian standard, is indicated for hydrous ethanol fuel, being the focus of the research. The pH of hydrated ethanol is measured by means of a combined pH glass electrode having the external electrode (Ag/AgCl electrode external to the glass bulb) filled with a solution of lithium chloride in an ethanol medium; however, the internal electrode solution (Ag/AgCl electrode inside the glass bulb) is still aqueous. This difference in the composition of the solutions causes a potential difference to be generated, between the internal and external reference electrodes, different from zero, which can cause distortions in the results of the pH measurement. The difference in solvent composition is also present in the calibration process of the measurement system used to take the pH readings, since the certified reference material (CRM) used is prepared in an aqueous medium. For the calibration to be reliable, this study suggests that the pH calibration buffer solutions be produced with the same solvent as the sample [4][5][10]. Therefore, this work aims to comparatively study the measurement of pH using different electrodes concerning pH indicators.

Studies were carried out by the National Institute of Metrology, Quality, and Technology (INMETRO), with samples of anhydrous and hydrous ethanol fuel, comparing the different pH electrodes from different manufacturers, including those indicated by the standards, to identify if there is any significant difference between the obtained results. This study showed that there is a significant difference between electrodes with KCl when compared to the electrode with LiCl [4].

1.3 pH indicators
An acid-base indicator is a substance that changes color with sudden variations in hydrogen concentration. It is usually a weak organic acid or a weak organic base in a very dilute aqueous solution. The undissociated indicator has a different color from the dissociated product. In the case of an acidic indicator (*HInd*), dissociation occurs according to the equilibrium:

$$HInd \leftrightharpoons H^{+} + Ind^{-} \qquad (1)$$

The color of the indicator anion (Ind^-) is different from that of the acid indicator. If the solution to which the indicator is added is acidic, that is, contains many hydrogen ions (H^+), the balance of the equation will shift to the left, making the color of the acid form of the undissociated indicator species visible. If, however, the solution is made alkaline, the balance shifts to the right (basic species), changing the color of the indicator. Each indicator has a very well-defined pH range (ΔpH), having a specific color for the acidic solution and another for the basic color of its respective range.

From the combination of more than one indicator, in separate experiments, the pH of an aqueous or alcoholic solution can be determined; as it is possible, to find a narrow and reliable range of pH variation and compare it with the pH values obtained from different electrode types for such measurement [11].

2. Materials
The experimental part was carried out in a laboratory of a partner company, SPR soluções metrológicas, located in São José do Rio Preto – SP, Brazil.

For this, the following reagents were used: hydrous fuel ethanol, obtained from a donation from a partner plant; deionized water; the pH indicators: bromocresol green indicator, bromothymol blue indicator, methyl red indicator all from the Synth brand. The following instruments were also used: pH meter – Analyser 350M, glass electrode Sensoglass SC09, glass electrode Sensotec 2.52, spectrophotometer HITACHI U-2910 and quartz cuvettes, d = 0.998 cm

3. Materials and Methods
The tests to be used depend on the behavior of the sample data, therefore, a Shapiro-Wilk statistical test was performed to assess whether they had a normal distribution. After that, the other tests

were performed: Grubbs test [12], one-way analysis of variance (ANOVA) [13], F-test: Two samples for variance and t-test: two samples assuming different variances [14].

For this work, three pH indicators were used, as specified in table 1, adapted from [11].

Table 1. Specifications of the indicators used.

Indicator	Acid form	Basic form	Δ pH
Methyl red (MR)	Red	Yellow	4.2 – 6.3
Bromocresol green (BCG)	Yellow	Blue	3.8 – 5.4
Bromothymol blue (BTB)	Yellow	Blue	6.0 – 7.6

To prepare the bromothymol blue solution, dissolve 0.1 g of bromothymol blue in 20 mL of hot anhydrous ethanol and after that, dilute with water to 100 mL. For methyl red, dissolve 0.2 g of methyl red in 60 mL of anhydrous ethanol and dilute to 100 mL with water. And finally, 0.1 g of bromocresol green was dissolved in 20 ml of anhydrous ethanol and diluted with water to 100 ml.

Spectrometry comprises a group of analytical techniques based on the properties of atoms and molecules to absorb or emit electromagnetic energy in a particular region of the electromagnetic spectrum. Spectrophotometry does it in the radiation region: UV (190 – 340) nm and visible light (350 – 700) nm. This technique uses the property of substances, organic or inorganic, in solutions to absorb or transmit light to quantify reactions. In practice, the amount of absorbed or transmitted light is proportional to the concentration of the substance in the solution. Like a digital print, knowing exactly the color absorbed allows us to identify and quantify different materials. The absorption spectrum is a scan of absorbance versus wavelength. When performing the analysis of a spectrum, it is possible to identify four types of effects, which are:

Bathochromic shift: When the absorption peak shifts to a longer wavelength;

Hypsochromic shift: When the absorption peak shifts to a shorter wavelength;

Hyperchromic effect: Increased absorption peak intensity;

Hypochromic effect: Decrease in absorption peak intensity [15].

The pH measurement of a hydrated ethanol fuel sample was performed using two different glass electrodes, the values obtained were compared with the values selected spectrophotometrically from the pH indicators.

3.1. pH measurement by system proposed by the standards

For the measurements, two combined glass electrodes were used, one with an internal solution of lithium chloride 3 mol L^{-1} in ethanol and the second with an internal solution of potassium chloride (KCl) 3 mol L^{-1} in water. The procedures adopted for the measurements with the ethanol electrode were those described by the ABNT NBR 10891 standard, for which the measurement time used is 2 minutes and the calibration is performed on certified reference material of pH buffer solutions with nominal values of 4.01 and 7.0 at 25°C.

For the glass electrode with an external electrode containing an aqueous solution of KCl, the procedures adopted were those described by the ASTM D 6423 standard, with a measurement time of only 30 seconds and the calibration being performed in the same way as the previous electrode. The electrode model proposed by this test method is not the same as the one used; however, the external electrode solution has the same composition, KCl in water. Measurements were performed on samples of hydrous ethanol fuel over ten weeks.

3.2. UV-vis

Acidified and basified solutions were prepared in aqueous and ethanolic medium (EHC) and to each solution was added an equal amount of indicator, in volume and concentration. After the addition of the indicator, the spectra were obtained, in the visible spectral region, with a scanning interval from 350 to 700 nm, step size 0.5 nm, quartz cuvette with optical path of 10 mm, SBW of 1.5 nm, scan speed equal to 400 nm min-1, mean response time for all solutions of the 3 indicators, in water, acidified water and basified water and for ethanol, acidified ethanol and basified ethanol. The acidified and basified solutions were obtained by adding a solution of HCl and NaOH (0.1 mol L^{-1}), respectively, in equal volumes.

4. Results and discussion

4.1. pH results obtained by the measurement system proposed by the standards.

The results obtained from the different electrodes are presented in tables 2 and 3.

Table 2. pH values obtained from the glass electrode filled with water/KCl 3 mol L^{-1}

	n1	**n2**	**n3**
Day 1	6.96	6.89	6.91
Day 2	6.87	6.80	6.86
Day 3	6.93	6.89	6.92
Day 4	6.86	6.81	6.84
Day 5	6.96	6.94	6.98
Day 6	6.98	6.96	6.92
Day 7	6.92	6.96	6.98
Day 8	6.89	6.92	6.96
Day 9	6.98	6.91	6.94
Day 10	7.03	7.05	6.98

Table 3. pH values obtained from the glass electrode filled with EtOH/LiCl 3 mol L^{-1}

	n1	**n2**	**n3**
Day 1	6.38	6.38	6.38
Day 2	5.97	5.97	5.97
Day 3	5.98	5.96	5.96
Day 4	6.20	6.17	6.21
Day 5	6.96	6.96	6.96
Day 6	6.47	6.56	6.49
Day 7	6.50	6.50	6.51
Day 8	6.55	6.65	6.55
Day 9	6.87	6.78	6.83
Day 10	6.72	6.70	6.70

4.1.1. Normality test.

The Shapiro-Wilk test was applied to both samples, the results of the W Test, together with the sample mean value and standard deviation of each distribution, are shown in table 4.

Table 4. Shapiro-Wilk test.

	Electrode water/KCl	Electrode EtOH/LiCl
W calculated	0.971	0.921
W critical, α = 0.05	0.927	0.927
Sample mean	6.93	6.46
Standard deviation	0.06	0.33

Analyzing table 4, it is possible to observe that the W value calculated for the distribution of data obtained using the electrode prepared with H_2O/KCl 3 mol L^{-1}, corresponding to 0.971, is bigger when compared to the critical W value, 0.927; this demonstrates that the distribution is Gaussian. The same does not apply to the distribution obtained by the electrode prepared with EtOH/LiCl 3 mol L^{-1}. The calculated W value, corresponding to 0.921, is smaller than the critical W value, showing that the distribution is not Gaussian. However, the calculated W value is very close to the critical W value; and as tests for normal distribution are more rigorous, it was decided to use parametric tests.

Comparing the values obtained in tables 2 and 3, it is possible to observe that the difference between the means is 0.47. The standard deviation obtained for pH measurements from the electrode using an EtOH/ LiCl 3 mol L^{-1} solution is greater, suggesting that this electrode generates a greater variation for the measurement results.

4.1.2. Grubbs test

Calculations were performed for the two distributions, pH values obtained from the two different electrodes, and the results are shown in table 5. From table 5, it is possible to state that there are no extreme values (outliers) in any of the distributions since for all suspect values, the calculated G found is less than the value of critical G.

Table 5. Grubbs test results.

	Electrode water/KCl		Electrode EtOH/LiCl	
Suspicious Values	6.80	7.05	5.96	6.96
Calculated G	2.15	2.09	2.15	2.09
Critical G (α = 0.05)	2.17		2.17	

4.1.3. ANOVA

Starting from the results in Tables 2 and 3, different tests were performed. The first being the statistical test of analysis of variance (ANOVA), to verify the stability of the HEF sample over the study time.

Table 6. Analysis of the variance of the results obtained by the glass electrode containing an external electrode containing water/KCl 3 mol L^{-1}.

Source of variation	Sum of squares	Degrees of freedom	Mean squares	$F_{calculated}$	$F_{critical}$
Between groups	0.0032	2	0.0016	0.44	3.35
Within groups	0.0968	27	0.0036		

Table 7. Analysis of the variance of the results obtained by the glass electrode containing an external electrode containing EtOH/LiCl 3 mol L^{-1}.

Source of variation	Sum of squares	Degrees of freedom	Mean squares	$F_{calculated}$	$F_{critical}$

Between groups	0.0002	2	0.0001	0.001	3.35
Within groups	3.0931	27	0.1146		

From the results of the analysis of variance for the data obtained by the glass electrode with an external electrode filled with an aqueous solution, the following values were obtained: $F_{calculated} = 0.44$, which is less than the value of $F_{critical} = 3.35$, therefore the sample is considered statistically stable. Moreover, although the values obtained by the glass electrode containing the external electrode filled with ethanolic solution, EtOH/LiCl 3 mol L^{-1}, show greater variation, the solution still presents stability since the value of $F_{calculated} = 0.001$ is smaller than the value of $F_{critical} = 3.35$.

4.1.4. F-Test: Two samples for variance and t-Test.

The F test was performed to check whether the variances of the samples obtained from the different electrodes were equal, the result is shown in table 8. From the table, the difference between the variance values found is evident

Table 8. F-test.

F-Test: two samples for variances		
	EtOH/LiCl	Water/KCl
Mean	6.46	6.93
Variance	0.11	0.0034
Observations	30	30
Degrees of freedom	29	29
F	30.91	
P(F<=f) one-tail	$4.11E^{-15}$	
F Critical one-tail	1.86	

From the F-Test it is noticeable that the samples have different variances. Therefore, the t-Test was applied assuming different variances to verify if the distributions are statistically equal or unequal. The result is shown in table 9.

Table 9. t-test: two samples assuming unequal variances.

	EtOH/LiCl	**Water/KCl**
Mean	6.46	6.93
Variance	0.10	0.0034
Observations	30	30
Hypothesized mean difference	0	
Degrees of freedom	31	
$t_{calculated}$	-7.70	
P(T<=t) one-tail	$5.37E^{-09}$	
t critical one-tail	1.69	
t critical two-tail	2.04	

From the values shown in table 9, it is possible to observe that the modular value of calculated t corresponding to 7.70 (absolute value) is higher when compared to the two-tail t critical value of 2.04, thus showing that the values obtained by the electrodes are statistically different.

4.2. pH indicators using the spectrophotometer.

Figure 1 shows the spectrum obtained by the bromothymol blue (BTB) indicator for the solutions described previously in experimental part. It is possible to observe that the BTB indicator in basified water, H_2O (OH^-), and basified ethanol, EtOH (OH^-), solutions present a maximum absorbance peak at 617 nm and 628 nm, respectively, showing a hypsochromic displacement of water about ethanol solutions. As for the acidified water, H_2O (H^+), and acidified ethanol, EtOH (H^+), solutions, the maximum absorbance peaks are at 435 nm and 423 nm, respectively; in this condition, water has a bathochromic shift in relation to ethanol solutions. There is a small difference between the spectra when comparing the aqueous and ethanolic medium with the use of the indicator. However, such effects only reinforce the colors, not affecting the turning range of the indicators, showing that it is possible to use the indicators in an ethanol medium; the same considerations apply to the other indicators. It is possible to analyze that the spectrum obtained from the hydrous ethanol fuel sample, (HEF), red curve of BTB indicator with a maximum absorbance peak at 424 nm, has a very similar behavior with acidified solutions. As the acid range of this indicator is 6.0, the spectrum obtained indicates that the pH of the solution is less than or equal to 6.0.

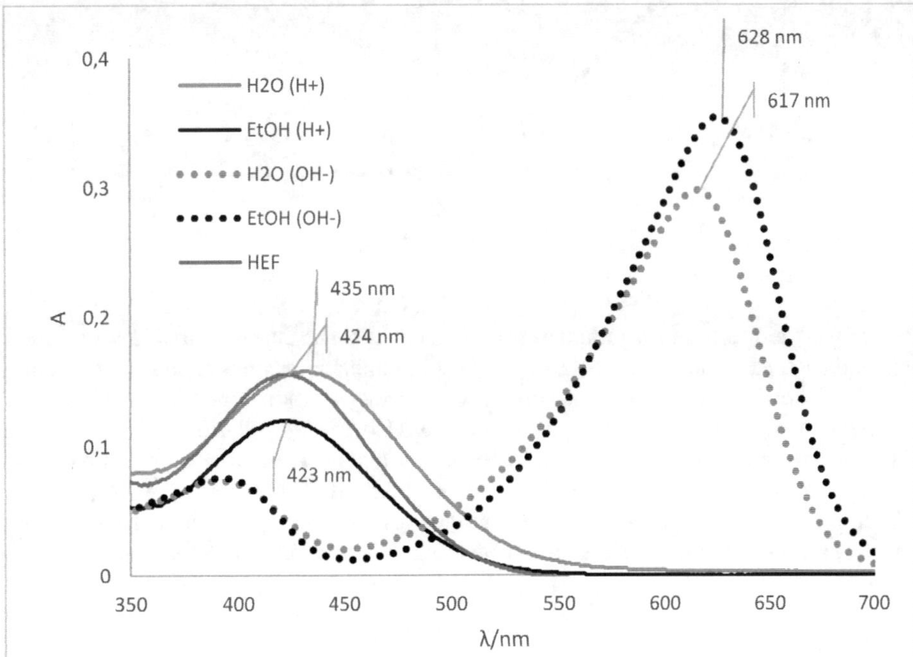

Figure 1. Spectrum obtained from the bromothymol blue (BTB) indicator.

Figure 2 shows the spectrum for methyl red (MR). The curves obtained for basified water, H_2O (OH^-), and basified ethanol, EtOH (OH^-), MR solutions have maximum peaks at 430 nm and 404 nm, respectively; showing a bathochromic shift of the aqueous solution in relation to the ethanolic one. As for the acidic medium, the maximum absorbance peak for MR in acidified water, H_2O (H^+), is 522 nm and for acidified ethanol, EtOH (H^+), 499 nm, repeating the bathochromic shift. Analyzing the spectrum found for the studied HEF sample, it is possible to see that two peaks were found, 492 and 411 nm, which are very close to the peaks found for the basified and acidified medium, showing that there is the presence of two-colored species, red and yellow, showing that they are in similar concentrations.

As the pH range of this indicator is 4.2 and 6.3, it is suggested that the pH of the sample is in the middle of the range, close to 5.3. Simultaneously, it observes the colors developed by the dyes under the same experimental conditions, but in test tubes, for visual comparison and confrontation with the spectra. It should be noted that the color obtained for methyl red showed an orange color, due to the presence of red and yellow simultaneously, indicating that the pH was in the middle of the range.

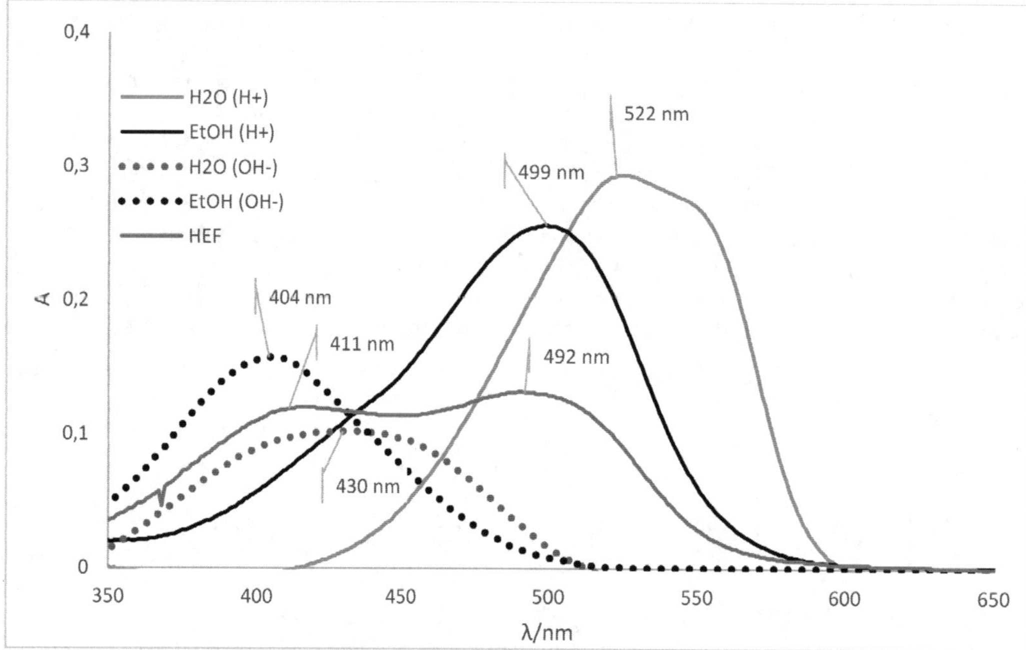

Figure 2. Spectrum obtained from the methyl red (MR) indicator.

Figure 3 shows the spectrum for bromocresol green (BCG). The curves obtained for basified water, H_2O (OH^-), and basified ethanol, EtOH (OH^-), BCG solution have maximum peaks at 616 nm and 622 nm, respectively; showing a hypsochromic shift of the aqueous solution in relation to the ethanolic one. As for the acidic medium, the maximum absorbance peak observed for the BCG acidified water, H_2O (H^+), is 445 nm and for acidified ethanol, EtOH (H^+), solutions, 426 nm, now showing a bathochromic shift. The spectrum obtained for the HEF sample in presence of BCG shows a maximum absorbance peak at 623 nm, a wavelength very close to that of the basified solutions, so it is possible to infer that the pH value is very close to the basic value of the range of the indicator used, the value specified in table 1 as 5.4

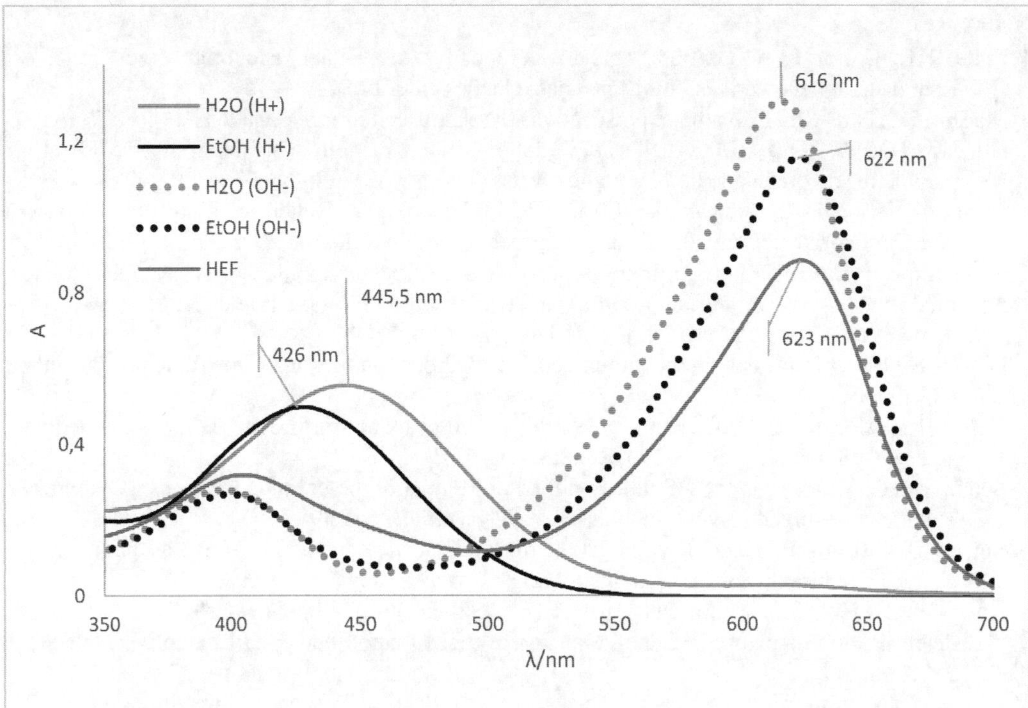

Figure 3. Spectrum obtained from the bromocresol green (BCG) indicator.

Thus, in agreement with the spectra presented, the pH of the HEF was determined to be present in the range, as follows: $5.4 \leq pH_{HEF} \leq 6.0$; that is, above the alkaline range of bromocresol green, but below the acidic range of bromothymol blue.

It is, therefore, interesting to compare the pH range obtained experimentally with the dyes with the values obtained by the two electrodes used, which indicated average pH values: pH 6.46 with a standard deviation of 0.33 - EtOH/LiCl and pH 6.93 with a standard deviation of 0.06 - water/KCl, showing that the values found by the electrodes are outside the range found by the indicators.

A comparison using the mean pH ($5.4 \leq pH_{HEF} \leq 6.0$) for the range leads to a difference of: 1.23 for the external electrode filled with water/KCl 3 mol L^{-1} and 0.66 for the external electrode filled with EtOH/LiCl 3 mol L^{-1}.

5. Conclusions

This preliminary study revealed that the pH values obtained using the different electrodes are statistically different; however, the samples remain stable over time. It is also possible to suggest that such values are different from the pH range obtained from the studies using the indicators, standards chosen for the study, allowing suggesting that the electrodes used in this work do not provide effectively reliable values for the realization of the pH measurement in hydrous ethanol fuel.

This preliminary work was essential to start the study of a new glass electrode architecture for measuring pH in hydrous ethanol fuel.

As a future work, one suggests a review article on the state of the art of pH measurement in hydrated ethanol fuel critical to current guidelines.

Journal of Physics: Conference Series **2606** (2023) 012019 doi:10.1088/1742-6596/2606/1/012019

6. References

[1] Leite R C, Cortez L A B. 2008 O etanol combustível no Brasil. Biocombustíveis no Brasil: Realidades e perspectivas, ministério das relações exteriores, p 60-75

[2] Macedo I C. 2007 Situação atual e perspectivas do etanol. Estudos avançados **21**, pp 157-165

[3] Gonçalves M A et al. 2013 Avaliação de laboratórios brasileiros na determinação de alguns parâmetros de qualidade de biocombustíveis. *Química Nova* **36** pp. 393-399

[4] Instituto Nacional de Metrologia, Qualidade e Tecnologia. Relatório final da comparação interlaboratorial para avaliação do desempenho de eletrodos na medição de pHe em etanol combustível. Rio de Janeiro: Inmetro, 2010 de Metrologia, Qualidade e Tecnologia. Relatório final da comparação interlaboratorial para avaliação do desempenho de eletrodos na medição de pHe em etanol combustível. Rio de Janeiro: Inmetro p 44

[5] Galster H. 1991 pH measurement: fundamentals, methods, applications, instrumental. Weinhein: VCH, p 356

[6] Gama M S, Afonso J C. 2007 De Svante Arrhenius ao peagâmetro digital: 100 anos de medida de acidez. Química Nova 30 n. 1, p. 231-239

[7] Senanayke G, Muir D M. 1987 Studies on the liquid junction potential corrections of electrolytes at aqueous + mixed solvent boundaries. *J. Electroanal.* pp 149-162

[8] Spitzer P, Fisicaro P, Seitz S, et al. 2009 pH and electrolytic conductivity as parameters to characterize bioethanol. Accred Qual Assur 14, p 671

[9] American society for testing and materials. ASTM D6423-20. Standard Test Method for determination of pHe of Ethanol, Denatured Fuel Ethanol, and Fuel Ethanol. United States, 2020. 5 p.

[10] Associação brasileira de normas técnicas. ABNT NBR 10891: Etanol hidratado combustível – Determinação do pH – Método potenciométrico. Brasil: ABNT; 2018. 10 p

[11] Vogel A I. 1981 Química Analítica Qualitativa. 5th ed Mestre Jou. pp 65–67

[12] de Oliveira, E. C., de Faro Orlando, A., dos Santos Ferreira, A. L., & de Oliveira Chaves, C. E. (2016). Comparison of different approaches for detection and treatment of outliers in meter proving factors determination. Flow Measurement and Instrumentation, 48, 29-35. http://dx.doi.org/10.1016/j.flowmeasinst.2016.02.002

[13] Miller J C and Miller J N. 1994 *Statistics for analytical chemistry* 3rd ed Ellis Horwood Series in Analytical Chemistry, pp 39-69 [14] de Almeida, F. C., Biazon, C. L., & de Oliveira, E. C. (2017). Uncertainty evaluation and validation of the test specimen size reduction in the determination of water content in crude oils by coulometric karl fischer titration. Petroleum Science and Technology, 35(11), 1135-1140. http://dx.doi.org/10.1080/10916466.2017.1310885

[15] Skoog A D, West M D, Holler J F, et al. 2015 *Fundamentos de química analítica* 9th ed Cengage Learning. pp 702-706

Characterization of an OCT probe for use in the diagnosis of precancerous lesions of the cervix

A H Cordes[2], I B Couceiro[1], A D Alvarenga[1], I Malinovski[1], C V de Andrade[3], F B Russomano[3] and J P von der Weid[2]

[1] Optical Metrology Division, Inmetro, Av. N. Sra. das Graças, 50, Xerém, Duque de Caxias, RJ – Brazil
[2] Center for Telecommunications Studies, Pontifical Catholic University of Rio de Janeiro, Rua Marquês de São Vicente, 225, Gávea, Rio de Janeiro, RJ, Brazil
[3] Instituto Nacional de Saúde da Mulher, da Criança e do Adolescente Fernandes Figueira, Fundação Oswaldo Cruz (IFF/Fiocruz), Av. Rui Barbosa, 716, Flamengo, RJ, Brazil

andy@opto.cetuc.puc-rio.br

Abstract: The development of an auxiliary diagnostic tool (OCT probe) for application in the assessment of cervical cancer is being carried out and may have an impact on the diagnostic process and, consequently, on the reduction of the incidence of female mortality from cervical cancer, by allowing a more accurate assessment of the extent of the lesion and by enabling the expansion of this assessment within the endocervical canal. In addition to the scientific result, the use of available equipment will boost the quality of clinical care currently available for diagnosis. Combining technical development with a precise methodology and clinical application, the probe developed will have a scientific impact of socio-economic relevance for society. As it is an optical measurement, it is necessary to evaluate the optical characteristics of the developed probes and the choice of one of them was made taking into account the measurements carried out and the application of metrological concepts.

1. Introduction

Optical Coherence Tomography (OCT) is a technique based on optical principles of light interference, allowing the generation of high resolution cross-sectional images of biological tissue microstructures with a spatial resolution of 10 μm or less [1], which is a non-invasive method very similar to ultrasound and, thus, without the need for excision and processing of the tissue to be submitted to the technique [2].

Optical coherence tomography (OCT) is a non-invasive high-resolution imaging technique that allows real time micro architectural characterization in vivo, that uses infrared light to create high resolution images, based on interferometry. It was first used to visualize human tissue in 1991 [3]. Since its development, Ophthalmic and intravascular imaging have emerged as the two largest clinical OCT applications [4], it has been used to image nontransparent tissue, such as skin, gastrointestinal mucosa and cervical epithelium [5]. There are studies showing that OCT can detect skin cancers, oral, laryngeal, esophageal cancer, genital and bladder cancer. OCT is an innovative technique which enables an optical biopsy of epithelial lesions [6].

In the specific case of cervical cancer, it is necessary to diagnose the lesions early so that treatment or surgery can be prescribed, if necessary. The OCT probe will act precisely in the evaluation of the cervical canal, helping the specialist to check if there is a lesion and what treatment will be given to the patient. The probe is being developed and an important part of this development is its initial characterization, the subject of this article.

2. Justification

The overall incidence of genital cancer is around 860 million cases/year. In Brazil, the number of new cases of cervical cancer expected for each year of the 2020-2022 triennium will be 16,590, with an estimated risk of 15.43 cases per 100,000 women. On average, it is the fourth most frequent cancer among women, behind only by skin, breast and colon cancers. Failure in the prevention and early detection of cervical cancer results in the diagnosis of cancer in advanced stages, which results in an average mortality of 50% of cases, usually women between the fourth and fifth decade of life. Classically, prevention is done by screening their precursor lesions, called grade II and III cervical intraepithelial neoplasia (CIN) and adenocarcinoma in situ. This screening also allows the detection of women already affected by cancer, but still in its initial, asymptomatic stages, that is, when the prognosis for successful treatment is still very good.

The probe being developed is to fill in the gap for younger women in the child bearing age and who do not present visible precursor lesions outside the cervical canal. Without a way to judge the level of the lesion hidden in the cervical canal the prudent choice in the face of the risk of the presence of a high grade lesion, is to perform an excisional procedure, in these cases, a type 3 excision, previously called as a cervical conization or cervical cone biopsy, taking out at least 2 cm of the tissue surrounding the cervical canal and sending it for histology diagnosis. The results of some of the histology evaluations would be negative for high grade neoplasia and the excision would have been unnecessary. After excional procedures, there is an increased risk for premature labour. Our probe is to give the gynecologist a way to diagnose precancerous lesions within the cervical canal and reduce the need for unnecessary excisions.

3. Discussion and Results

The pre-cancer lesions occur within the epithelial layer of tissue. Sometimes the lesion is located inside the endocervical canal, where it cannot be seen during colposcopy. In these cases, as the probe will be in contact with the tissue, the optical focal point of the probe needs to be within this epithelial layer which is on the order of 500 μm thick. With this in mind the focal point was placed 250 μm outside the glass covering of the probe.

The optical beam of the probe is Gaussian and is characterized by three parameters, w_0, half the beam diameter at focus, z_0, the location of the focus and z_R, the Rayleigh range which is half the depth of field over which an object can be considered to be in focus [7]. In order to characterize the probe, it was necessary to develop a manner to measure the beam width very near the probe as even without the glass covering the focus is less than 2 mm from the output of the optics and it was desired to have beam width measurements as near as 250 μm from the output in order to have sufficient data on both sides of the focus to get a good fit for the model.

With that in mind, the knife edge method [8, 9] was chosen to measure the beam width at known positions along the beam. A razor blade was passed through the beam using a Thorlabs motorized stage while measuring the optical power passed with an Agilent optical power meter, as shown in figure 1. The optical power, P, as a function of the position of the razor blade as it passes through the optical beam is an error function (equation (1)) and the beam width can be pulled from the best fit.

$$\mathrm{P} = \mathrm{P}_0 + \frac{\mathrm{P}_{\max}}{2}\left(1 - \mathrm{erf}\left(\frac{\sqrt{2}\,(\mathrm{x}-\mathrm{x}_0)}{\mathrm{w}}\right)\right) \tag{1}$$

where P_0 is the background power (μW), P_{max} is the maximum power (μW), x_0 is the midposition (μm) of shift between the P_0 and P_{max}, w is the beam radius (μm) and erf is the standard error function.

A PI motorized stage was used to control the distance from the razor blade to the probe and the beam width measurement was repeated at various distances from the probe, with more measurements taken near the focus. The distance from the probe to the razor blade was measured by taking OCT measurements of the razor blade at a repeatable X position, found by moving the blade in X until a determined amount of the beam was blocked. The beam characteristics of w_0, z_0 and z_R were derived from the best fit of the beam widths vs distances to the Gaussian beam equations (equations (2) and (3)).

Ten probes of Type 1 and ten probes of Type 2 were measured. Two of the Type 1 and two of the Type 2 probes were selected for ten more measurements each. The results were analyzed statistically to evaluate the quality and repeatability of the construction of the probes (relative to the design criteria) and the quality, reproducibility and repeatability of the measurement itself.

Figure 1. Hardware for the knife-edge measurement of beamwidth. The probe is held stationary from the bottom left while the razor blade is moved in X in front of it by the Thorlabs stage. The power is measured by the Agilent power meter. The distance Z from the probe to the razor blade is controlled by the PI stage (blue) but the distance is measured by OCT using the probe itself.

The beam width w is extracted from the optical powers measured as the razor blade passes in front of the probe using equation (1) and a Levenberg-Marquardt non-linear curve fit. The powers are measured at in 1 μm spacing in order to have sufficient data across the expected 10 μm focus and beginning far enough away from the beam to find a baseline long from the effects of diffraction as shown in figures 3b and 4b.

The PI stage then moves the razor blade away from the probe and the beam width is measured again. To measure the z position, 15 OCT measurements of the razor blade´s position are taken before the beam width measurement begins. The z positions are extracted from the resulting Z-scan (figure 2) by using peak detection on the individual A-scans (figures 3a and 4a) and averaging over the 15 measurements. To guarantee that the OCT measurement is taken consistently at all distances from the probe, the Thorlabs stage is used to position the razor blade just inside the beam where the measured power is equal to a pre-determined value. More beam width measurements are taken near the focus where the beam shape is changing rapidly and fewer further away where the beam width changes in a linear manner. The beam parameters z_0 (waist location, in μm), w_0 (waist radius, in μm) and z_R (Rayleigh range, in μm) are read from the parameters of the Levenberg-Marquart non-linear best fit for the measured beam widths and z positions to the standard gaussian beam equations shown in equations (2) and (3).

$$w(z) = w_0\left(1 + \left(\frac{z-z_0}{z_R}\right)^2\right)^{1/2} \tag{2}$$

$$z_R = \frac{\pi w_0^2}{\lambda} \tag{3}$$

Note that the zero position in the OCT measurement is the optical reference which is the entrance to the prism in Type 1 probes and the exit of the prism in Type 2 probes (figure 6).

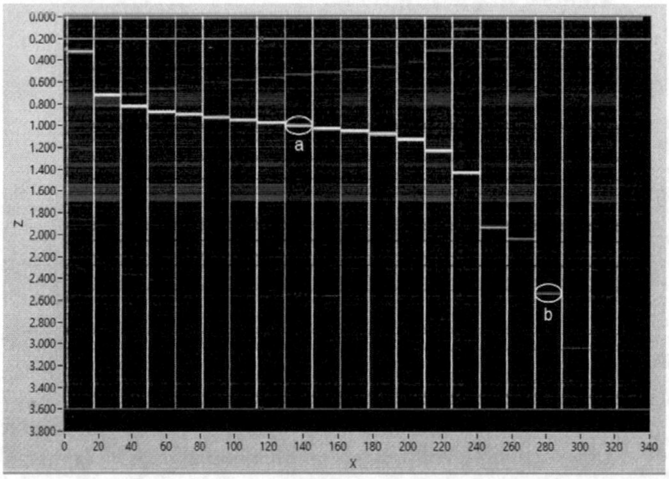

Figure 2. A Z-scan of the razor´s position at each beam waist measurement for a Type 2 probe. The Z-scan consists of 15 A-scans taken at each razor position. The location of the razor blade was determined in each of the 15 A-scans by peak detection and averaged together for a final result. (a) and (b) mark the representative measurements which are used to explain the process further on. (a) is near the focus while (b) is far from focus.

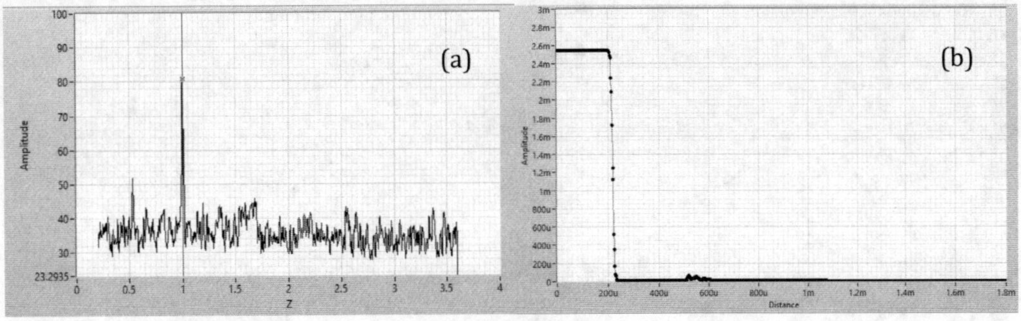

Figure 3. A beam width measurement pair for the measurement position marked (a) in figure 2. On the left in figure 3a is one of the OCT A-scans of the razor in the beam path. The Z position is found through a peak find, in this case 1.00195 mm from the probe's reference. Fifteen such measurements are averaged for a final result. On the right in figure 3b are the power measurements as the razor moves through the beam. As the beam was near focus the transition from high to low in the beam width measurement happens over just a short distance. In this case, the best fit to equation (1) returned a width of 10.1796 μm.

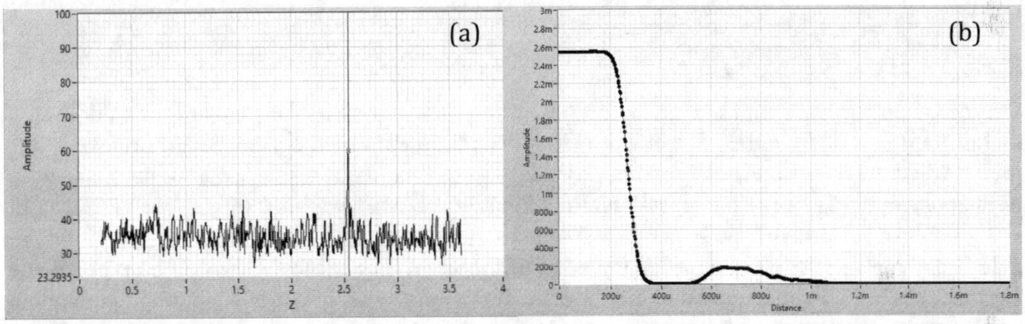

Figure 4. A beam width measurement pair for the measurement position marked (b) in figure 2. On the left in figure 4a is one of the OCT A-scans of the razor in the beam path. The Z position is found through a peak find, in this case 2.53937 mm from the probe's reference. Fifteen such measurements are averaged for a final result. On the right in figure 4b are the power measurements as the razor moves through the beam. As the beam was far from focus, the transition from high to low in the beam width measurement happens over a much longer distance. In this case, the best fit to equation (1) returned a width of 62.8807 μm.

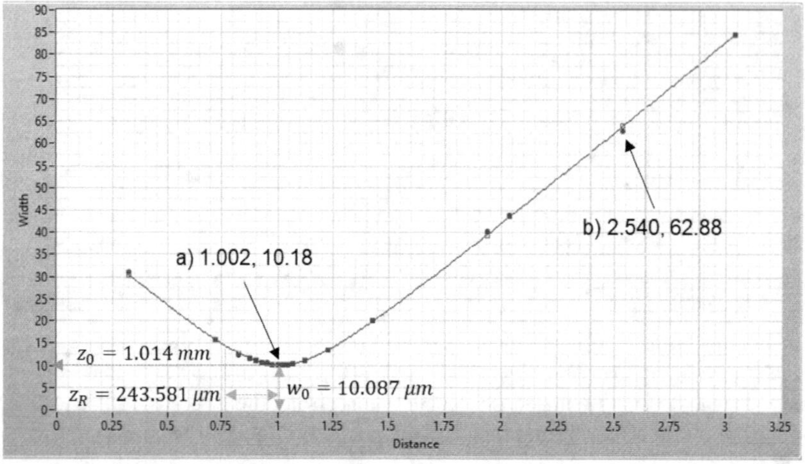

Figure 5. Beam width vs Z-location Type 2 Probe 07 (Gillette25 Measurement 2). Blue squares are the measured values; red squares are from the best fit of the measured values to equations (2) and (3). The blue curve passes through the best fit values. From the best fit the beam parameters are directly read. In this particular case $w_0 = 10.087\ \mu m$, $z_0 = 1.014\ mm$ and $z_R = 243.581\ \mu m$.

Target values for w_0, z_0 and z_R of the probes were chosen to place the focus in the middle of the epithelial tissue outside the transparent cap or cover with a depth of focus sufficient to cover typical epithelial thickness (0.5 mm). Component values were obtained using Zmax and sent to the manufacturer for production. The target values for Type 1 probes are: w_0 = 10.00 μm, z_0 = 2533.65 μm, z_R = 239.82 μm. For probe Type 2, the values are: w_0 = 10.00 μm, z_0 = 1030.00 μm, z_R = 239.82 μm. The difference in z_0 is due to the reference being in different places for the two probe types. Type 1 probes have their 33% reflection coating reference at the beginning of the 90-degree prism while the Type 2 probes have theirs at the exit (figure 6). The physical difference in path length is 1000 μm through BK7 which optically at 1310 nm is 1503.65 mm.

Figure 6. Type 1 probes (top) have a 33% reflective coating at the beginning of the right-angle prism at the output of the probe (shown in yellow). Type 2 probes (bottom) have a 33% reflective coating at the exit instead (shown in yellow).

We took measurements on all 10 Type 1 probes and all 10 Type 2 probes as shown in tables 1 and 2. The values shown for four probes, Type 1 Probes 2 and 6 and Type 2 Probes 7 and 10 are the averages of ten measurements each rather than a single measurement. Ideally, all values would be the average of ten measurements but there were time constraints and each measurement takes half a day. The standardized values are the number of standard deviations above or below the mean for all the probes of a given type and are calculated as the difference between value and average, divided by the standard deviation. Two Type 2 probes were discarded, Probes 04 and 05, based on their negative z_0 measurements, probably due to a problem with the Grin lens or the 33% reflective coating.

Table 1. Type 1 probe measurements for w_0, z_0 and z_R.

Probe	w_0 (μm)	z_0 (mm)	z_R (μm)	w_0 Standardized	z_0 Standardized	z_R Standardized
Type1-01	10.37	2.532	231.85	2.00	-0.09	-0.95
Type1-02	9.88	2.528	233.55	-0.36	-0.43	-0.70
Type1-03	9.69	2.530	230.35	-1.29	-0.28	-1.17
Type1-04	9.75	2.517	234.08	-1.01	-1.36	-0.63
Type1-05	10.06	2.526	238.32	0.48	-0.61	-0.01
Type1-06	10.11	2.550	244.57	0.71	1.36	0.90
Type1-07	10.08	2.551	251.54	0.58	1.45	1.91
Type1-08	10.08	2.529	242.55	0.57	-0.33	0.60
Type1-09	9.80	2.522	233.44	-0.76	-0.96	-0.72
Type1-10	10.11	2.548	232.27	0.72	1.24	-0.89

Table 2. Type 2 probe measurements for w_0, z_0 and z_R.

Probe	w_0 (μm)	z_0 (mm)	z_R (μm)	w_0 Standardized	z_0 Standardized	z_R Standardized
Type2-01	10.73	1.007	250.22	0.95	0.24	-0.05
Type2-02	10.21	0.986	238.95	-0.54	-1.16	-0.99
Type2-03	9.86	1.000	245.80	-1.56	-0.23	-0.42
Type2-04	9.28	-0.053	123.35	NA	NA	NA
Type2-05	34.99	-0.043	368.73	NA	NA	NA
Type2-06	10.68	1.002	257.22	0.80	-0.08	0.53
Type2-07	10.10	1.009	245.38	-0.88	0.40	-0.45
Type2-08	10.83	1.027	277.50	1.25	1.62	2.21
Type2-09	10.25	0.981	242.81	-0.44	-1.50	-0.67
Type2-10	10.55	1.014	248.81	0.43	0.71	-0.17

The mean and standard deviation related to each probe are shown in tables 3 and 4. The negative z_0 values for Type 2 Probe 4 and Probe 5 indicates a gross manufacturing problem. Their values are not reflected in the statistics below.

Table 3. Statistical evaluation of the Type 1 probes.

Type 1	$\mathbf{w_0}$ (μm)	$\mathbf{z_0}$ (mm)	$\mathbf{z_R}$ (μm)
Average	9.99	2.533	237.25
StDev	0.21	0.012	6.89
StDev % of Avg	2.1%	0.5%	2.9%

Table 4. Statistical evaluation of the Type 2 probes.

Type 2	w_0 (µm)	z_0 (mm)	z_R (µm)
Average	10.40	1.003	250.84
StDev	0.35	0.015	12.05
StDev % of Avg	3.3%	1.5%	4.8%

A global evaluation of the Type 1 and Type 2 probes for the values of w_0 and z_R can be done. For z_0 a global evaulation is not possible because the baseline is different for each probe type and therefore cannot be compared directly. The global evaluation results of w_0 and z_R are shown in table 5.

Table 5. Statistical evaluation of all probes without Type 2 Probe 04 and Probe 05. The target z_0 is different for each probe type and so can′t be compared directly.

Global	w_0 (µm)	z_R (µm)
Average	10.17	243.29
StDev	0.34	11.54
StDev % of Avg	3.4%	4.7%

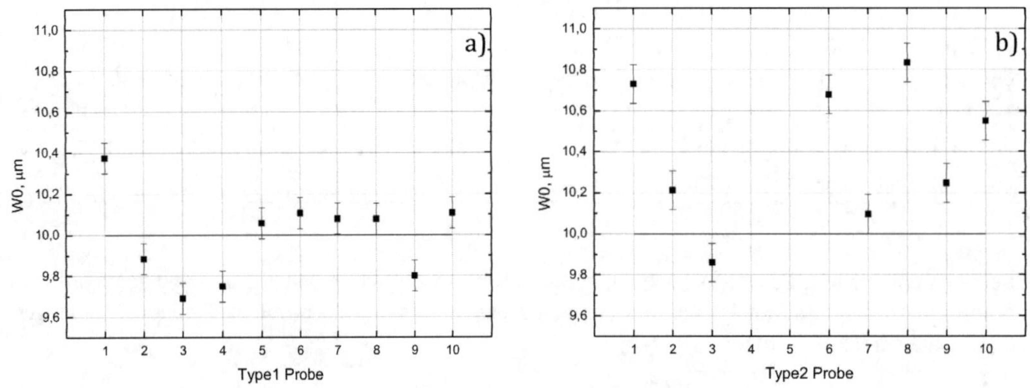

Figure 7. w_0 measurements for the (a) Type 1 and (b) Type 2 probes. The error bars are 0.08 µm for the Type 1 probes and 0.09 µm for the Type 2 probes. The solid horizontal lines are the target values, 10 µm for both types.

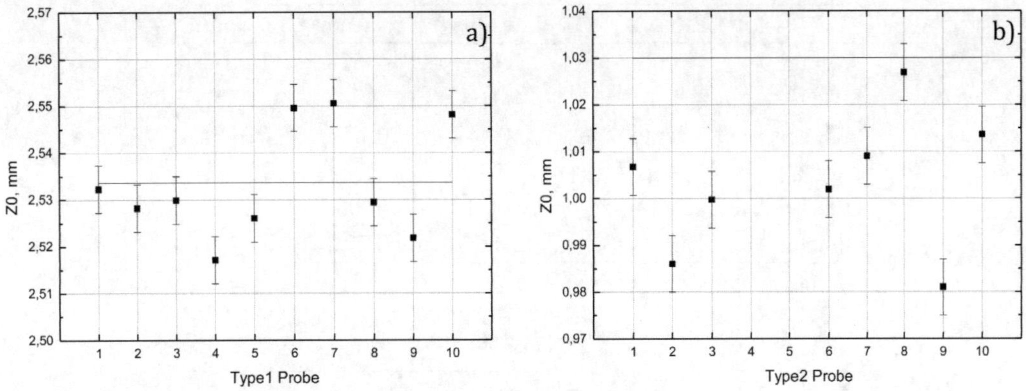

Figure 8. z_0 measurements for the (a) Type 1 and (b) Type 2 probes. The error bars for the Type 1 probes are 5.1 µm and 6.1 µm for the Type 2 probes. The solid horizontal line are the target values, 2.53 mm for the Type 1 probes and 1.03 mm for the Type 2 probes.

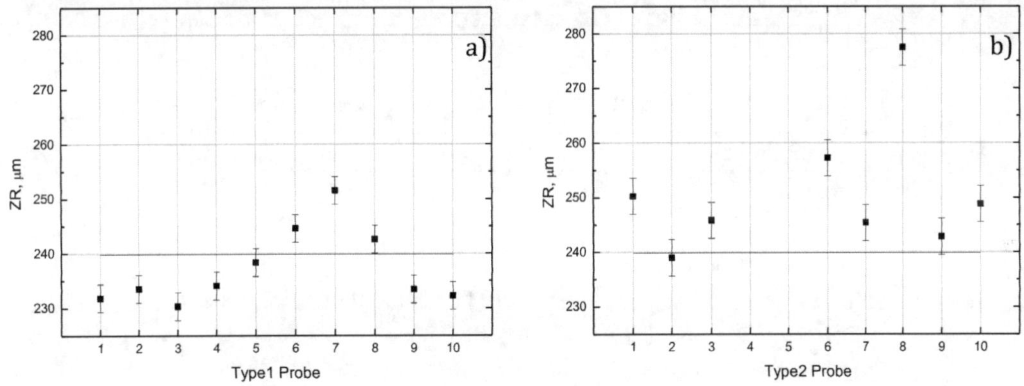

Figure 9. z_R measurements for the (a) Type 1 and (b) Type 2 probes. The error bars for the Type 1 probes are 2.5 µm and on the Type 2 probes 3.32 µm. The solid horizontal lines are the targets, 239.82 µm in both cases.

Finally, we chose four probes, two of Type 1 and two of Type 2, which were the closest to their respective target values. While all the probes other than the obviously bad Type2 04 and 05 had acceptable performance, we will want to use the best possible probes in order to minimize whatever effects deviations from the target values might have on downstream results.

These four probes were each tested ten times and the statistics tabulated in tables 6, 7, 8 and 9. As the devices are 100% passive, the variations in the values measured are due to the measurement itself and give the uncertainty in the measurement. These uncertainties are shown in the measurements of all the probes.

Table 6. Statistics of 10 measurements of Type 1 Probe 2.

Type 1 Probe 2	w_0 (µm)	z_0 (mm)	z_R (µm)
Average	9.88	2.528	233.55
StDev	0.055	0.003	1.35
StDev as % of Average	0.56%	0.13%	0.58%
Target	10.00	2.534	239.82
Offset = Average - Target	-0.12	-0.006	-6.27
Offset as % of Target	1.2%	0.22%	2.6%

Table 7. Statistics of 10 measurements of Type 1 Probe 6.

Type 1 Probe 6	w_0 (µm)	z_0 (mm)	z_R (µm)
Average	10.11	2.55	244.6
StDev	0.08	0.005	2.54
StDcv as % of Avcragc	0.75%	0.20%	1.04%
Target	10.00	2.53	239.82
Offset = Average - Target	0.11	15.9	4.75
Offset as % of Target	1.1%	0.63%	1.98%

Table 8. Statistics of 10 measurements of Type 2 Probe 7.

Type 2 Probe 7	w_0 (µm)	z_0 (mm)	z_R (µm)
Average	10.10	1.009	245.38
StDev	0.09	6.06	3.32
StDev as % of Average	0.94%	0.60%	1.36%
Target	10.00	1.030	239.82
Offset = Average - Target	-0.10	21.00	-5.56
Offset as % of Target	0.1%	2.04%	2.32%

Table 9. Statistics of 10 measurements of Type 2 Probe 10.

Type 2 Probe 10	w_0 (µm)	z_0 (mm)	z_R (µm)
Average	10.55	1.014	248.81
StDev	0.08	2.92	1.92
StDev as % of Average	0.71%	0.29%	0.77%
Target	10.00	1.030	239.82
Offset = Average - Target	0.55	-16.44	8.99
Offset as % of Target	5.50%	1.60%	3.75%

4. Conclusions

A system to measure the beam characteristics of an optical probe fabricated to make OCT measurements of precancerous lesions of the cervix was constructed. The knife edge method was chosen in order to measure the beam width very near the output of the optical probe. Measurements as near as 250 μm from the 90-degree prism were taken. It was used to measure twenty probes, ten each of two different types, Type 1 and Type 2, each of which had a different focal distance. Two of the Type 2 probes were discovered to be defective. Ten measurements were performed on each of two representatives each of the two different probe types to evaluate the performance of the measurement system and to establish error margins for the measured values. The probes had been built according to specific values for the parameters w_0, z_0 and z_R. The measurements showed compliance with the technical specifications established in their construction. The results obtained showed that 18 of the 20 probes are able to be used for the diagnosis of cancer precursor lesions in real-time measurements.

Acknowledgments
We would like to thank FAPERJ, CAPES and CNPq for their support.

References

[1] Popescu D P, Choo-Smith L-P, Flueraru C, Mao Y, Chang S, Disano J, Sherif S and Sowa M G 2011 Optical coherence tomography: fundamental principles, instrumental designs and biomedical applications *Biophys. Rev.* **3** 155–69

[2] Al-Mujaini A, Wali U K and Azeem S 2013 Optical coherence tomography: clinical applications in medical practice *Oman Med. J.* **28** 86–91

[3] Huang D *et al* 1991 Optical coherence tomography *Science* **254** 1178–81

[4] Fujimoto J and Swanson E 2016 The development, commercialization, and impact of optical coherence tomography *Invest. Ophthalmol. Vis. Sci.* **57** OCT1–13

[5] Escobar P F, Belinson J L, White A, Shakhova N M, Feldchtein F I, Kareta M V and Gladkova N D 2004 Diagnostic efficacy of optical coherence tomography in the management of preinvasive and invasive cancer of uterine cervix and vulva *Int. J. Gynecol. Cancer* **14** 470–4

[6] Wessels R, De Bruin D M, Faber D J, Van Leeuwen T G, Van Beurden M and Ruers T J M 2014 Optical biopsy of epithelial cancers by optical coherence tomography (OCT) *Lasers Med. Sci.***29** 1297–05

[7] Cordes A H, Couceiro I B, Alvarenga A D, Malinovski I, Dominguez C T, de Andrade C V, Russomano F B and von der Weid J P 2021 Practical considerations for OCT applications *J. Phys. Conf. Ser.* **1826** 012064

[8] Guedes M B and Alvarenga A P D 2021 Measurement of the beam diameter at the polarimeter for quartz control plate calibration *J. Phys. Conf. Ser.* **1826** 012062

[9] Siegman A E 1993 Defining, measuring, and optimizing laser beam quality *Proc. SPIE 1868, Laser Resonators and Coherent Optics: Modeling, Technology, and Applications (Los Angeles) vol 1868*

Digital enhancement of OCT images toward better diagnostic of precursor cancer lesions in the uterine colon

I Malinovski[1], I B Couceiro[1], A D Alvarenga[1], A H Cordes[2], C V de Andrade[3], F B Russomano[3] and J P von der Weid[2]

[1] Optical Metrology Division, Inmetro, Av. N. Sra. das Graças, 50, Xerém, Duque de Caxias, RJ – Brazil
[2] Center for Telecommunications Studies, Pontifical Catholic University of Rio de Janeiro, Rua Marquês de São Vicente, 225, Gávea, Rio de Janeiro, RJ, Brazil
[3] Instituto Nacional de Saúde da Mulher, da Criança e do Adolescente Fernandes Figueira, Fundação Oswaldo Cruz (IFF/Fiocruz), Av. Rui Barbosa, 716, Flamengo, RJ, Brazil

E-mail: imalinovski@colaborador.inmetro.gov.br

Abstract. We present a combination of image processing operations, which improve the quality of OCT images. The sequence is: (1) Gaussian blur, (2) Enhance Local Contrast (CLAHE), (3) FFT based band-pass filter, (4) Background Subtraction, (5) adjustment of the contrast of the image with light saturation (Enhance Contrast) and finally (6) reassignment of the palette such as to enhance the bright part of the images. The multi-step treatment reduces pixel-to-pixel noise, improves the homogeneity of the contrast, supresses vertical strips, reduces cosmetic distractions, equalizes contrast across a collection of images, and uses the pallet to emphasise areas of greater intensity. The enhanced images show a significant improvement in their quality, with details that do not appear in the original OCT image. The processing of these images and the comparison with untreated OCT images are shown throughout the paper.

1. Introduction

Image enhancement is the process of digitally manipulating a stored image or live stream of images using specific software. The tools used for image enhancement include many different kinds of software such as filters, image editors and other tools for changing various properties of an entire image or parts of it.

Among other types of images especially important is medical image enhancement [1, 2]. Medical image plays important role in clinical diagnostic applications. The importance of this field has inspired creation of the various software projects aimed to help scientists and clinicians to analyze medical images more effectively. The most successful and famous software package is ImageJ. It was developed in cooperation between the National Institute of Health (NIH, USA) and the Laboratory for Optical and Computational Instrumentation (LOCI, University of Wisconsin, USA).

ImageJ is a Java-based multiplatform image processing software developed as an open project in the public domain [3, 4]. Since its first version it has got a lot of popularity and was hugely increased by multiple collaborators that developed plug-in filters as add-on's.

Optical coherence tomography (OCT) is relatively new tool in the area of cervical cancer diagnostics, but it is quite attractive due to its advantage that it allows rapid and non-invasive evaluation of the cervix [5, 6].

It was previously demonstrated that it is fair diagnostic modality [5]. The problem of the OCT diagnostic is that the brightness (intensity of backscatter) is the main criteria of the analysis. In multiple cases the difference in brightness is not well pronounced and might be confusing. Further, intrinsic noise is always present in OCT imaging, which also complicates the usage.

2. Results and discussion

The OCT system used was a commercial spectral domain system from Thorlabs (TELESTO II) with 170 nm of optical bandwidth centered around 1300 nm. The instrument was calibrated using an intermediate height calibration standard. This secondary standard was traceable to SI via an absolute Linnik type interferometer [6]. Measurements were done in 3D mode with multiple B scans covering the volume of the interest (VOI). The same VOI was independently screened with regular histology using an Axio Z1 (Leica, Germany) microscope. Special care was taken to achieve close fit between histology and OCT data.

We started from the idea that modern computing and image processing should be able to help in such a diagnostic in both the cases of static images and live mode. Indeed, we have as the typical case tiny differences in brightness being used as the principal criterion for classification decisions. So, it is considered self-evident that enhancement of those differences must help in making a correct diagnostic. We will concentrate on results within this public available ImageJ (and variants such as Fiji) to make the results repeatable to anyone.

We have found that no single filter gives significant visual improvement of the raw OCT output for the lesion detection. This observation reflects the fact that there are several defects in OCT imaging that are caused by the hardware. Those defects might be invisible in the raw image, but became significant whenever one tries to enhance the image. A combination of filters was found to produce the results with the desired enhancement of the cancer-affected areas. Multiple steps are necessary to eliminate undesired defects one by one as explained in each step of the proposed processing.

We will list all the steps mentioning the purpose of each. The optimal parameters of the filters used are indicated in the table 1. The reasoning behind the choice of the optimal parameters of the filters are discussed in the text. Note that we suggest making all the processing in 16-bit resolution mode in order to preserve tiny details that are lost in 8-bit mode.

1. Gaussian Blur filter. First, we apply Gaussian blur because images are subject to pixel-to-pixel noise as a result of the OCT measuring technique. The parameter of radius (called "sigma" in ImageJ) will depend on hardware. In our case, we found that an optimal radius is somewhere between 1.5 and 2.0 pixels. With a higher radius, borders of the areas of the interest become less clear, while a lower radius value does not remove the pixel-to-pixel noise. Figure 1 shows the effect of the optimal radius for the Gaussian Blur. The untreated image is in (a) and the treated image is in (b). A plot of the pixel intensities along the blue lines in each image are shown to the right of each image. Here the desired reduction in jitter while retaining details can most easily be seen; the plot in (a) is much more ragged than that of (b) but (b) still retains enough fluctuations in intensity to give details.

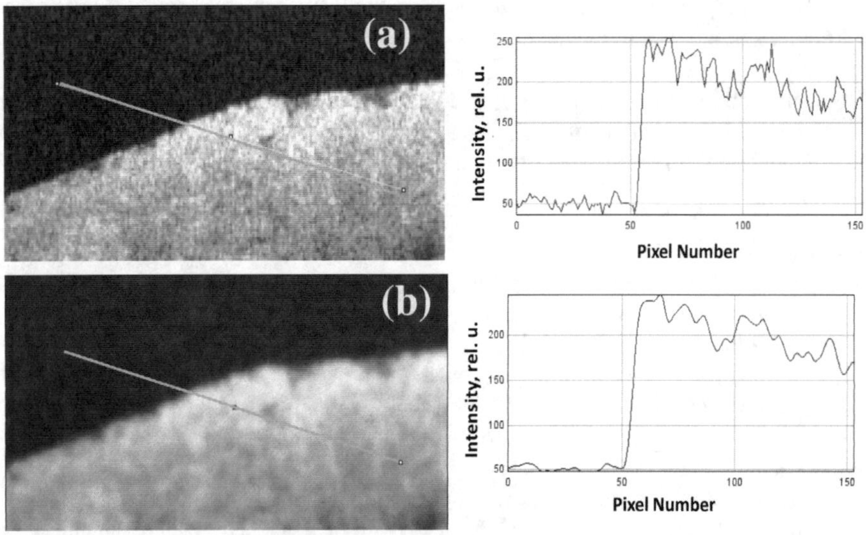

Figure 1. Desired effect of Gaussian Blur filter. Images (left) and corresponding intensity plots from underneath the cyan line (right). Image (a) is original OCT non-filtered portion (zoomed) and (b) is the image portion after filter.

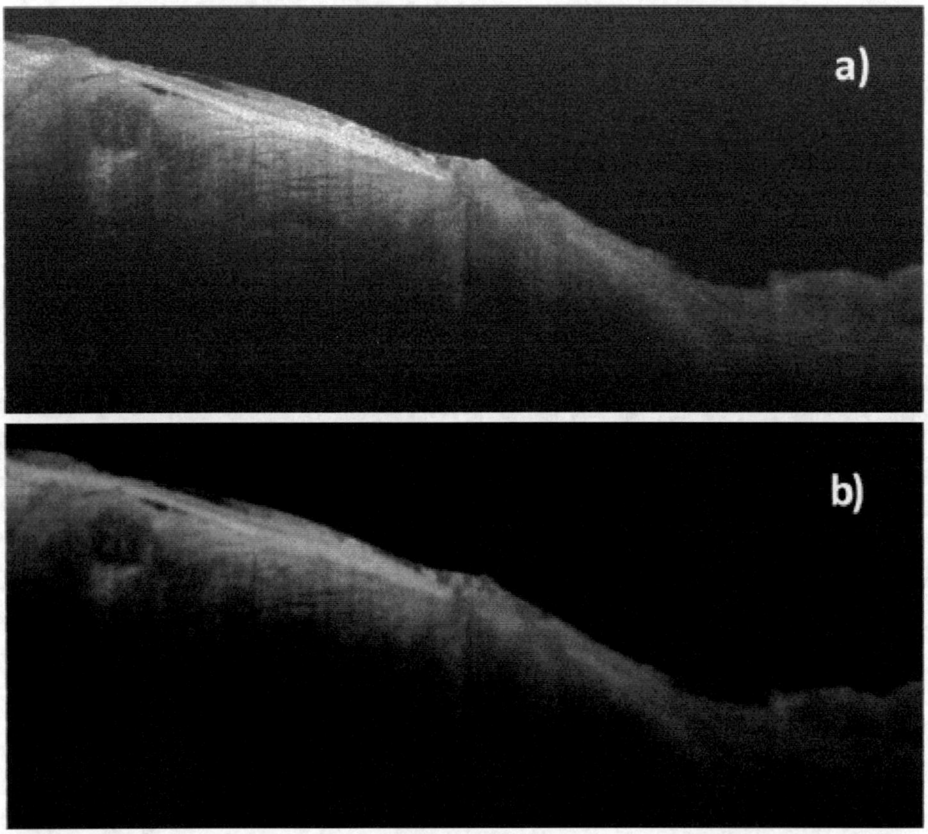

Figure 2. Raw OCT image (a) and filtered (b) image after smoothing with Gaussian blur and removing background.

2. Enhance Local Contrast (CLAHE) filter. Another defect of the measuring technique is a variation in the contrast of the image because OCT works with a different focus at the top, middle and end of the measuring area in the vertical direction. This difference in focus affects the contrast of the image. We found that the CLAHE filter helps a lot to improve the homogeneity in the contrast.

This filter applies a contrast limited adaptive histogram equalization (CLAHE) method that is known to be useful for the enhancement of local features. We will demonstrate effect of this filter on an example starting with original OCT image and continuing with blurred and CLAHE filtered image. (See figures 2 and 3 below). This example is useful as there is an area of interest which is at an angle, declining from left to right from the left side of the image. The area of interest moves into focus at the middle of the image and so appears brighter, with more contrast.

Now we change the palette to make the area of interest more visible. This emphasizes the appearance of the effect of the CLAHE filter.

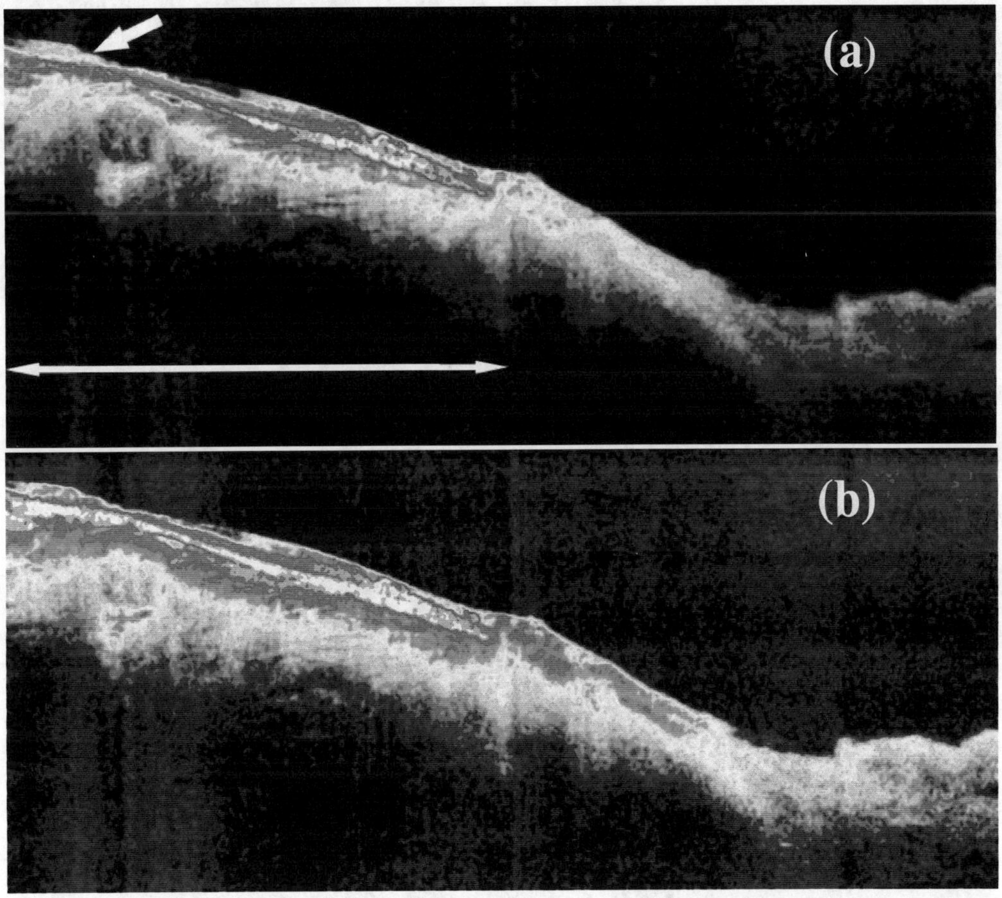

Figure 3. Effect of CLAHE filter. The images are colored with 16-color palette for better perception of the effects of the CLAHE filter. Case (a) is the same as figure 2 (b), just with the different pallet, and (b) is the CLAHE filtered image. The double-ended arrow in lower part of the image (a) indicates the extent of the area of interest, in this case a CIN affected area as identified from histology. We would expect that the affected areas would all have the same intensity (contrast) and hence the same color, in this pallet white and pink corresponding to the maximum intensity. The short white arrow in upper part of (a) shows the portion of the area of interest which is not the expected intensity due to being higher in the image and further from focus. In (b) the same region is all the same color due to the effect of the CLAHE filter.

The parameters used were a blocksize of 549, histogram of 256, a maximum slope of 1.5 and mask = None. The values were found through trial and error, fixing values when the improvement matched the expected result. The most important value was the slope of 1.5. The most important parameter of CLAHE is the maximum slope that decides level of enhancement. Balancing between output noise and desired contrast improvement one can find the optimum slope for the specific equipment. In our case it was 1.5. Rest of parameter one can keep as suggested by imageJ. For us was important improvement of homogeneity in order to compensate out of focus areas

3. FFT based band-pass filter. Note that this filter is used mostly to suppress the vertical strips which are typically present in OCT imaging. The parameters we use in this filter are: large = 600, small = 1.1, suppress = Vertical, tolerance = 1, autoscale off. Here we use a "small" radius of 1.1 for smoothing because we already have smoothed the image in the Gaussian blur step. We do not need to filter large features, so we chose 600 for the "large" parameter, which is over half of the image width. The "suppress" is vertical since it is the main desired effect. Finally, "tolerance" is set to 1 but can be anything between 1 and 5 degrees without much difference.

4. Subtract Background filter. We find this filter useful especially when an inhomogeneous background is still visible after the FFT filter. The CLAHE filter typically brings some gray background which can be distracting. The effect of the subtraction is mostly cosmetic, reducing visual noise as illustrated in Figure 4, the area of interest, the white colored area, is not affected but broad swaths of the image away from this region are suppressed. The best results were obtained with option: rolling = 800.

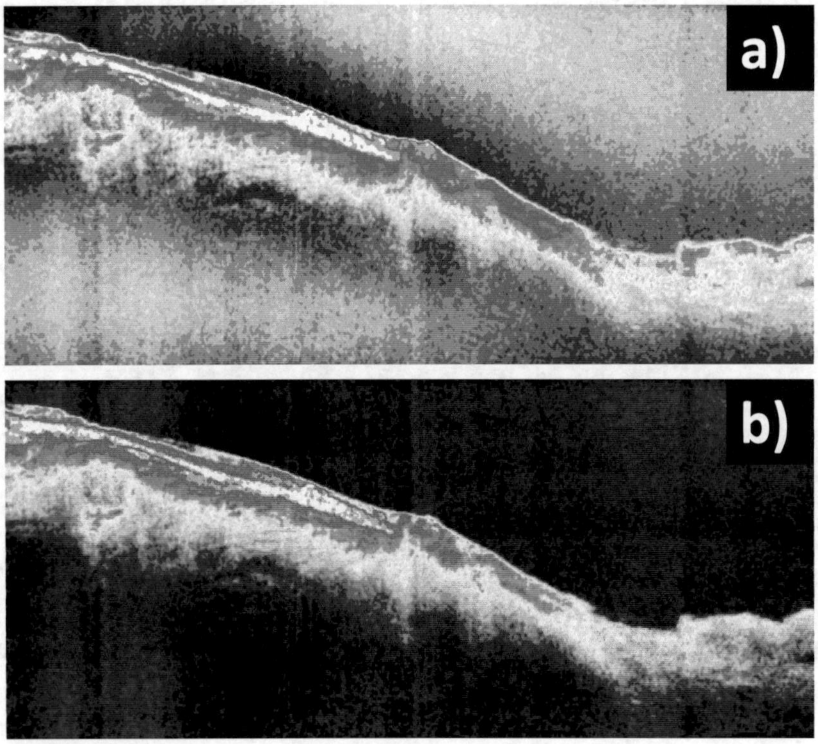

Figure 4. Effect of the background subtraction. Image (a) is the OCT data before background subtraction and (b) is the result after subtraction.

5. Enhance Contrast filter. Now we adjust the contrast of the image with light saturation. This could be considered an optional step in the case of a single image because normally contrast is auto-adjusted by the OCT software. However, we found it useful to equalize all the images in a collection of evaluated

Journal of Physics: Conference Series **2606** (2023) 012021 doi:10.1088/1742-6596/2606/1/012021

cases. Increasing contrast with a little saturation is desired because: 1) both FFT filter and background removal effectively reduces the contrast, 2) it is logical to increase the contrast because what we are looking for are the areas of maximum intensity with expected small difference from neighboring areas. The parameters are chosen by subjective visual evaluation with criteria of best perception without losing much of the important details. The best parameters in our case were: saturated = 0.3, normalize & equalize histogram None.

6. Reassign the palette. Finally, we reassign the palette such as to enhance the bright part of the images. In this final step, artificial coloring was tried in the manner of Figure 4, but it was found that most in our group considered these images confusing. Therefore, a black and white palette modified to increase perception without using artificial coloring was applied. The modification goes from black to white for intensities 0 to the middle intensity and returns from white to black from the middle intensity to the maximum. This operation makes the maximum signal intensity black, same as the background, which can be confusing in itself at first, however it provides two advantages. First, the full dynamic range of intensity is used over half the intensity range, gaining one bit of dynamic range, and second high intensity areas are automatically bordered by a brighter area. This built in "circling" attracts the attention of the viewer to the higher intensity areas most likely to be of interest more effectively than if the higher intensity area had been left white as in a traditional palette. The palette used is shown in figure 5.

Table 1. List of the filters used in study with exact parameters applied.

#	Filter name	Parameters
1	Gaussian blur	radius 1.5 pixels
2	CLAHE	blocksize 549, histogram 256, maximum slope=1.5, mask = None.
3	FFT band-pass	large = 600, small = 1.1, suppress = Vertical, tolerance = 1, autoscale off
4	Enhance contrast	saturated = 0.3, normalize & equalize histogram None
5	Enhance palette	user dependent

The steps above (all but number 4) were applied to all raw images for evaluation of the enhancement related improvement in sensitivity and specificity.

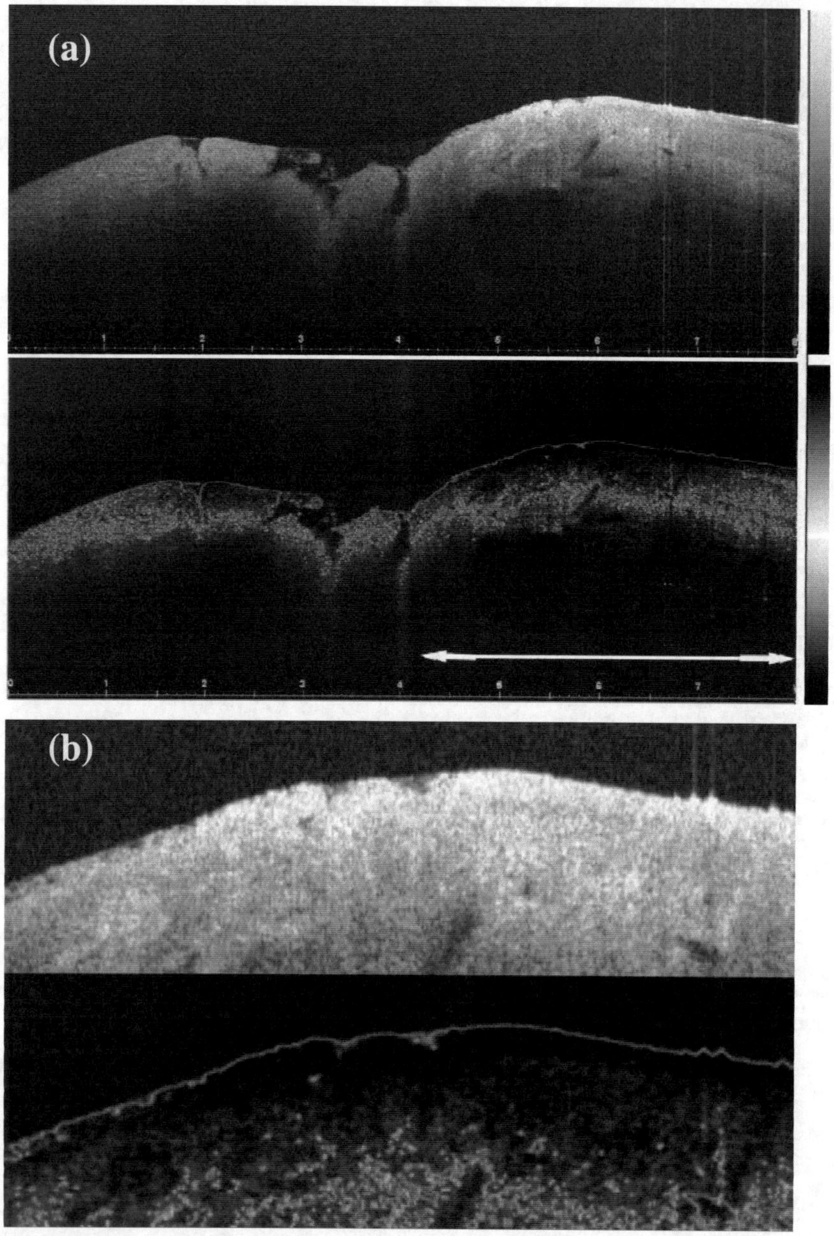

Figure 5. Original OCT (upper) and filtered (lower) images for comparison (sample (a)). Zoomed are the areas affected by CIN as confirmed with histology (sample (b)). The original and modified palettes are shown on the right of the sample (a).

3. Conclusions

From our study, we have concluded that multiple filters, in the correct combination, provide a significant improvement in the quality of OCT images, which aids in the identification of lesions in the samples evaluated. We consider that further development in both the OCT hardware and image processing software can potentially result in new and more reliable diagnostic capabilities, that is highly desirable in many areas for example early cervical cancer detection.

References

[1] Bankman I 2008 *Handbook of Medical Image Processing and Analysis* (New York: Elsevier)
[2] Dougherty G 2011 *Medical Image Processing: Techniques and Applications* (New York: Springer)
[3] Schneider C A, Rasband W S and Eliceiri K W 2012 NIH Image to ImageJ: 25 years of image analysis *Nat Methods* **9** 671–675
[4] Collins T J 2018 ImageJ for microscopy *BioTechniques* **43** 25–30
[5] Gallwas J, Turk L, Friese K and Dannecker C 2010 Optical coherence tomography as a non-invasive imaging technique for preinvasive and invasive neoplasia of the uterine cervix *Ultrasound Obstet. Gynecol* **36** 624–629
[6] Gallwas J K S, Turk L, Stepp H, Mueller S, Ochsenkuehn S, Friese K and Dannecker C 2011 Optical coherence tomography for the diagnosis of cervical intraepithelial neoplasia *Lasers Surg Med* **43** 206–212

Journal of Physics: Conference Series **2606** (2023) 012022 doi:10.1088/1742-6596/2606/1/012022

Comparison between methods of electric current generation applied to intermediary checking for electrometers calibration

L Machado[1] and Barbosa, R.A.[1]

[1] Instituto de Radioproteção e Dosimetria

E-mail: machado.liana@gmail.com

Abstract. The Intermediary Checking of the calibration the electrometers, to ensure that the calibration factor of a given instrument is valid for a longer time, has been developed at LNMRI-IRD since 2019. It works with three of the methods for generating electric current of small values (pA): capacitance, resistance (Ohm) and ionization current (radiation sources). In this article, we will evaluate the performance of electrometers using the ionization current method and we will make a comparison between the results obtained by this method with the results that uses the Ohm method. From this comparison, we could see that the method using radiation sources had a better performance of the electrometers, however, in terms of cost-benefit, the Ohm method better serves the intermediary checking of the calibration the electrometers.

1. Introdution

The National Metrology Laboratory of Ionizing Radiation (LNMRI) is located at the Institute of Radiation Protection and Dosimetry (IRD) and has responsibilities as a National Laboratory by designation of the National Institute of Metrology, Standardization and Industrial Quality (Inmetro). In the R&D context, LNMRI/ IRD develops primary standardization techniques in the areas of Radiodiagnosis, Radioprotection and Radiotherapy. In addition, it provides metrological traceability to the measurements of radiation doses received by patients undergoing radiological examinations and cancer treatments. It also contributes to the correct dose monitoring of workers in the nuclear industry and radioactive facilities, which are called OEP (Occupationally Exposed Person). For this reason, it is so important to guarantee the accuracy of measurements in these areas of application of ionizing radiation.

Dosimeters are instruments built by two devices: an ion chamber and an electrometer. The first device converts the received ionizing radiation dose to electrical current (ionization current), and the second device measures the electrical charge from the chamber. LNMRI/ IRD periodically sends its standards to the International Reference Laboratories, PTB-Germany and BIPM-France. However, after the automatization of procedures, the International Laboratories started calibrating only the ion chambers. The calibration of electrometers, associated with the National Standards, became the responsibility of the LNMRI/ IRD. [1]

There aren't laboratories in the country that calibrate electrometers at small values of electric current: 10^{-12} A. The calibration of electrometers abroad, as is currently done, represents a high cost, in addition to presenting administrative, operational and customs difficulties.

As already presented in [2] and [3], the LNMRI/ IRD started the implementation of the intermediary checking of calibration [4], to guarantee that the calibration factor of a given instrument has validity for a longer time and, therefore, such an instrument could continue to be used in the laboratory routines of the LNMRI/ IRD.

The system applied to the intermediary checking was intended to monitor the drift of a measurement standard in the period between two calibrations, that is, to assess whether the characteristics of said standard have not changed significantly since the last calibration and, consequently, if the calibration certificates values remain valid. Intermediary checking is an essential procedure to ensure metrological consistency [4].

At the article [5], is analyzed the interference of the application or not of the REL function in the performance of the electrometers, during the Intermediate Checking. It shows that this function, widely used during the measurement, also affected the performance of the electrometers submitted to the test.

Since 2000, three of the existing methods for generating small-value electric current (pA), are applied at the LNMRI/ IRD: capacitance, resistance (Ohm) and ionization current (radiation sources). Through the previous articles, [2], [3] and [5] we applied the second method in the performance evaluation of electrometers during "Intermediary Checking" tests. At this phase of the project, the use of the third method will begin. For this purpose, we will be making use of the electric current quantity from dosimetry procedures for gamma and X-rays radiation sources [7].

2. Methodology

2.1. Current source

The ionization current source, used to generate the reference current for the test described here, was a Standard Imaging well-type chamber, model HDR 1000 Plus. The radiation source was a ^{226}Ra test source. The same tests described in [5] were repeated. The concept was to compare the performance presented by a group of electrometers, when connected to an ionization current source against the performance presented when connected to a current source (Ohm).

2.2. Electrometer

To measure the electric current generated by the source, four KEITHLEY electrometers, model 6517 [6] were used.

2.3. Measurements

The form and sequence of measurements were the same as those presented in the article [5]. This procedure was necessary to reach the objective of the present article, which is the comparison between the performances of the electrometers against a current source (ionization) and current source (Ohm) methods. The measurement groups involved 6 (six) series of 100 (one hundred) measurements; from which three groups of 10 (ten) measurements were withdrawn and analyzed in each series.

3. Results

This section shows the performance results of four LNMRI/ IRD electrometers that were used within the intermediary checking of the calibration study. Table 1 presents the performance of these electrometers when subjected to the current generated by the ionization current method. Table 2 presents the results of comparing the performance of the same group of electrometers when subjected to the current generated by the Ohm method and the ionization current method. For this purpose, we used data obtained in the previous articles [2], [3] and [5].

Table 1. Performance of electrometers by the ionization current method

Standard Deviation	*Measurements*			
	#885222	#1211856	#1215369	#4043232
0,03 – 0,10	*17*	*14*		*18*
0,11 – 0,15	*1*	*3*		
0,16 – 0,20		*1*		
0,21 – 1,50				
1,51 – 2,50			*1*	
2,51 – 3,00			*7*	
3,01 – 3,50			*2*	
Total	*18*	*18*	*18*	*18*

Table 2. Comparison between the two methods

Standard Deviation	*Measurements*							
	#885222		*#1211856*		*#1215369*		*#4043232*	
	^{226}Ra	*Ohm*	*^{226}Ra*	*Ohm*	*^{226}Ra*	*Ohm*	*^{226}Ra*	*Ohm*
0,03 – 1,0	*18*	*14*	*18*	*8*		*15*	*18*	
1,1 – 1,4		*3*		*6*		*1*		*4*
1,5 – 2,0		*0*		*4*	*1*	*2*		*1*
2,1 – 2,4		*1*			*8*			*1*
2,5 – 3,5					*9*			
Total	*18*	*18*	*18*	*18*	*18*	*18*	*18*	*18*

4. Conclusion

Observing the results of table 1, which used the ionization current method, the electrometer #1215369 presented a performance quite inferior to the remaining electrometers. While all other electrometers presented an average standard deviation value lower than, or equal to 0.2%, the mentioned electrometer presented an average standard deviation near to 3.5%.

Still on the subject of the ionization current method, three of the four electrometers used in the analysis have presented performances below 0.2% of average standard deviation. However, through the Ohm Method, these same three electrometers presented performances that ranged from 0.2% to 2.4% of average standard deviation.

However, to obtain measurements necessary to evaluate the performance of electrometers, the Ionization Current method needs a period four times longer, when compared to the Ohm method.

Observing table 2, we see that about 70% of the performance of electrometers using the Ohm method is bellow or equal to 1%; which can be considered a good performance for the intermediary checking of the calibration; with results being obtained in a much shorter period.

Finally, when it comes to the inconvenience of using radioactive sources, applied to the ionization current method, we can mention:

- High financial cost compared to resistance (Ohm);
- Serious radiation protection problems;
- Infeasibility of operation by personnel who are not OEI (occupational exposed individual);
- Demand for a protection system against theft or robbery, due to national security issues, among others.

Thus, it can be concluded that, in relation to cost-benefit, the Ohm method would better serve the intermediary checking of the electrometer's calibration.

5. References

[1] Oliveira, E M *et al* 2006, Management quality system implementation in the LNMRI radionuclidegroup basead on ISO/IEC 17025 requeriment. *XVIII IMEKO Wold Congress Metrology for a suistanable*

development, 17-22.

[2] Azeredo, D A, Machado, L, Barbosa, R A, Cruz P 2019 Estudo de viabilidade para implementa umsistema de verificação intermediária na calibração de eletrômetros, *Congresso Brasileiro de Metrologia das Radiações Ionizantes.*

[3] Machado, L, Azeredo, D A, Barbosa, R A, Cruz, P 2019 Aplicação do calibrador FLUKE/5500A para verificação intermediária na calibração dos eletrômetros, *Congresso Brasileiro deMetrologia das Radiações Ionizantes.*

[4] ABNT NBR ISO IEC 17025:2017.

[5] Machado, L, Barbosa, R A e Cruz, P 2020 Interferência da função de medição relativa-REL no desempenho de eletrômetros durante a checagem intermediária da calibração, *CongressoBrasileiro de Metrologia das Radiações Ionizantes*, online.

[6] Keithley 6517A Electrometer, User's Manual.

[7] INTERNATIONAL ORGANIZATION FOR STANDARDIZATION / AMERICAN SOCIETYFOR TESTING AND MATERIALS. 2002, Standard Guide for Selection and Calibration o Dosimetry Systems for Radiation Processing. ISO/ASTM 51261(E).

Virtual Laboratory of Ionizing Radiation Metrology: Validation by spectrometry

E M Macedo[1,2], M R Nascimento[3], L C Pacífico[4], I F M Garcia[1,2], J S Ferreira[1], W S Santos[5], W Belinato[6], M V T Navarro[1] and J G P Peixoto[2].

[1] Laboratório de Produtos para a Saúde do Instituto Federal de Educação, Ciência e Tecnologia da Bahia (Labprosaud/IFBA), 41.745-715, Salvador, Bahia, Brasil
[2] Instituto de Radioproteção e Dosimetria, 22.783-127, Rio de Janeiro, Rio de Janeiro, Brasil
[3] Departamento de Física, 20550-900, Rio de Janeiro, Rio de Janeiro, Brasil
[4] Departamento de Ciências Radiológicas LABMETRO/DCR/IBRAG/UERJ, 20550-900, Rio de Janeiro, Rio de Janeiro, Brasil
[5] Universidade Federal de Uberlândia, 38408-100, Uberlândia, Minas Gerais, Brasil
[6] Instituto Federal de Educação, Ciência e Tecnologia da Bahia - Campus Vitória da Conquista, 45078-900, Vitória da Conquista, Bahia, Brasil

E-mail: ematosmacedo@gmail.com

Abstract. The digital revolution is a reality. Many areas of industry, services as well as other applications fields, are migrating part of their jobs - even the totality - to the digital environment. Ionizing radiation metrology cannot be out of this world. For this objective, the creation of a virtual environment from the physical systems of radiation meter calibration has been proposed. The methodology tools used consist of describing all the key elements of this laboratory in Monte Carlo, using EGSnrc and *egs_phd* application, to simulate them and to validate the system comparing it with experimental data in the same condition. In this context, the N60 radiation quality, from ISO 4037-1, was chosen for this purpose. An AMPTEK XR-100T CdTe was used on spectrometry, and the spectra were corrected based on the Stripping method. The results showed a good agreement between experimental and simulated spectra, with a difference in their mean energy of 1.4% within the ±5% tolerance range required on ISO 4037-1:1996. After being corrected, both had mean energy of 47.2 keV and a spectral resolution of 33%, which complies with the requirements to produce the ISO N60 reference beam. This indicates that, at first analysis, the virtual system can be exposed to new metrological tests.
Keywords: Digital twin. ISO 4037. Calibration.

1. Introduction

Digital tools are even more present in our life, especially in our work environment. Digital transformation of Industry 4.0 is happening also on Metrology 4.0 [1]. Personal dosimetry, dose prediction in radiotherapy, clinical imaging, among other systems with innovative technology as Artificial Intelligence (AI), Deep Learning, have been widely studied worldwide [2].

Calibration laboratories of radiation meters are a spot of well-known environment, instruments, and practices, once all of them are under quality management and technical requirements compliance [3,4]. Those features are a fertile ground to apply technics of environment virtualization, as Digital Twin (DT) [5].

Digital Twin is based on the development of a faithful copy, in digital way, of any environment. According to the Center for Digital Built Britain, DT consists of "a realistic digital representation of assets, processes or systems in the built or natural environment" [6]. With a DT system, it is possible to evaluate preventive maintenance, equipment life cycles, and previous testing of upgrades [7].

A twin metrology laboratory is the target. To analyze virtually the result of physical phenomena of radiation interaction with matter in those virtual environments, Monte Carlo (MC) simulation is widely used. Some different codes have, most of the time, convergences in their results, which indicates the tool reliability for such applications [8,9].

The aim is to start a virtualization process of a radiation meters calibration laboratory (Labprosaud/IFBA) based on Monte Carlo simulation, using EGSnrc. The first step is to simulate and validate the spectrometry system from Labprosaud/IFBA.

2. Materials and Methods

The virtualization of the laboratory is based on three steps: to establish ISO N60 radiation quality, perform a spectrometry of this beam and simulate the same spectrometry setup in EGSnrc. The first validation is based on the comparison between two spectra (experimental and virtual). The mean energy deviation will be used as a quantitative analysis, taking into account the ±5% tolerance required on ISO 4037-1:1996 [3].

2.1. ISO N60 radiation quality

This quality is considered as the standard quality for calibration and comparison with ISO narrow-spectrum series. It is based on the requirements of ISO 4037 [3,10,11].

The instrumentation used on establishing was the (position 1) ISOVOLT Titan E 160M2, which consist of an X-ray tube with a tungsten anode and inherent filtration of 1 mm Be, and a monopolar generator of 160 kV of maximum tube voltage. It was placed a (position 2) primary collimator, (position 3) shutter, (position 4) additional filtration, and a (position 6) monitor chamber (PTW TM-786) between a couple of collimators. The standard dosimeter was a (position 7) PTW 1-liter spherical ion chamber (TM32002), calibrated in terms of ambient equivalent dose H*(10) at PTB. They are disposed on laboratory setup as illustrated below in figure 1.

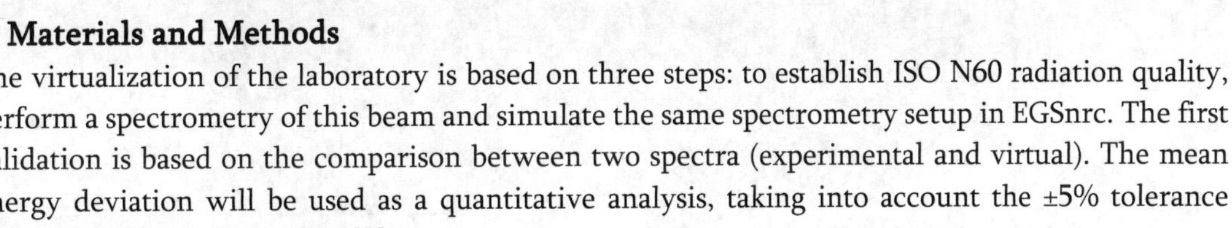

Figure 1. Recommended setup (left) and the real setup on Labprosaud/IFBA (right) [12,13].

The reference value of H*(10) was obtained with the reference chamber 2.5 m far from the source. This setup was explained and discussed in two previous works of the authors [12,14].

2.2. Spectrometry with AMPTEK CdTe

The AMPTEK XR-100T CdTe with PX4 Digital Pulse Processor was used to perform the spectrometry tests [15]. It was used the same setup as in figure 1, and the same distance as the ionization chamber, 2.5 m. The acquisition parameters were 1 mA of tube current, time 240 s, and a 400 μm tungsten collimator. Dead time ratio was 4.5%.

The XR-100T can be seen on the laboratory setup in figure 2.

Figure 2. AMPTEK XR-100T CdTe placed on Labprosaud/IFBA's setup.

Mean energy, $\bar{E}$, is calculated by [3]:

$$\bar{E} = \frac{\int_0^{E_{MAX}} \phi_E E dE}{\int_0^{E_{MAX}} \phi_E dE} \quad (1)$$

Where ϕ_E is the derivative of the fluence of the primary photons of energy E.

To quantify the difference between spectra, a Relation ($R_{\%}$) will be calculated using equation 2 below.

$$R_{\%} = \left| \frac{\bar{E}^r_{exp} - \bar{E}^r_{sim}}{\bar{E}^r_{exp}} \right| \times 100 \quad (2)$$

Where $\bar{E}^g_{exp}$ is the mean energy of the experimental (exp) raw (r) spectrum and $\bar{E}^r_{sim}$ is the mean energy of the simulated (sim) raw (r) spectrum.

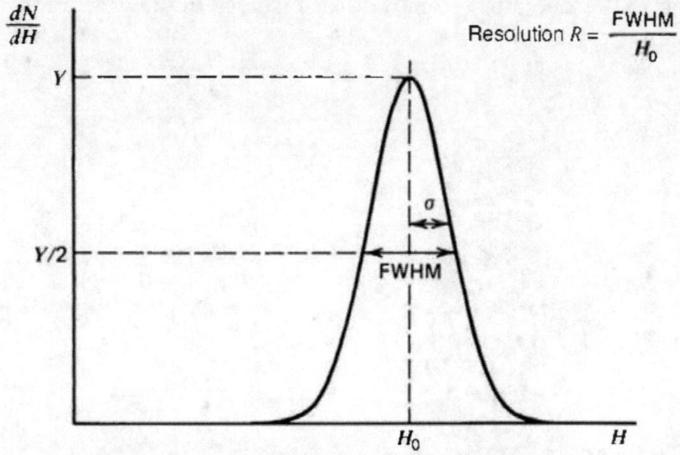

Figure 3. Resolution determination [16].

The Resolution (R) is calculated through full width at half maximum (FWHM) value in the energy (illustrated as H) versus counts by energy (dN/dH), as shown in figure 3. It will be calculated only for corrected spectra, which are used for producing the standard radiation beam.

2.3. EGSnrc simulation

The EGSnrc was chosen to simulate the laboratory environment. Specifically, the *egs_phd* application was used to compute pulse height distribution of the photons and electrons interaction on the simulated CdTe detector [17].

A standard spectrum (generated on SRS78 generator software [18]), with only tungsten anode tube 20 degrees angled and 1 mm Beryllium of inherent filtration was used as parameters, once that the virtual setup will filtrate this radiation with aluminium and copper thin filters. The spectrum is shown in figure 4.

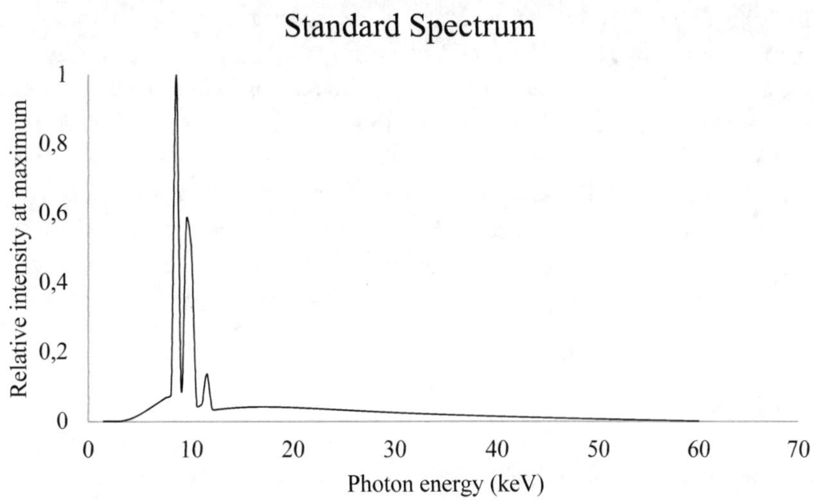

Figure 4. Standard spectrum generated with SRS78 [18].

The key elements of the laboratory scenario were included in the simulation, as can be seeing in figure 5.

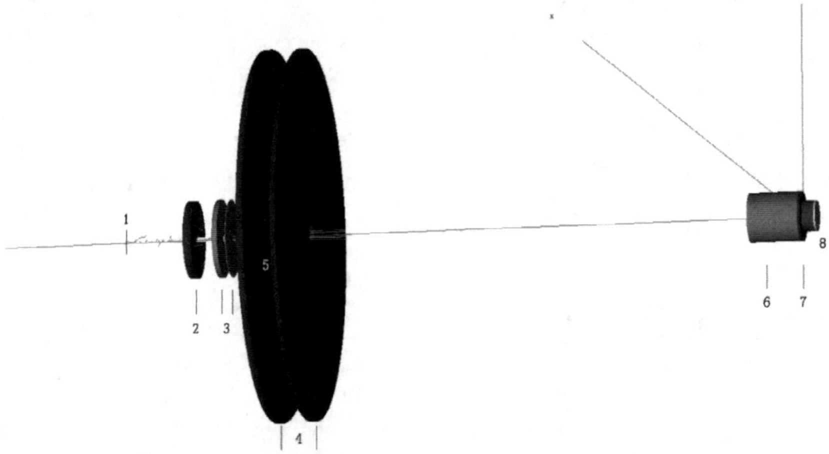

Figure 5. Simulated scenario of N60 spectrometry, on EGSnrc.

Elements in figure 5: 1. X-ray focus; 2. primary collimator; 3. aluminium and cooper filtration; 4. monitor chamber collimators; 5. monitor chamber; 6. brass spacer; 7. 400 μm tungsten disk collimator.

The highlights of the simulation code are:

- Source definition with *egs_collimated_source*;
- Score options with 1024 bins for spectrum output (same as selected Amptek output);
- Spectrometer mount with a Ni cylindrical structure, Be window, Pt and In electrodes, CdTe dead layer and a CdTe depleted layer set as the active detector [19].

2.4. Spectrum correction

Some physical phenomena distort the measured X-ray spectra, causing the softening of its mean energy with values ranging out of the accepted norm value [20]. In that context, those physical issues must be addressed by a correction methodology looking to reduce its disparities to ensure the quality of what is been measured. Consequently, this work makes use of the Stripping method to correct scape peaks, Compton plateau, and intrinsic efficiency issues generated by the CdTe diode instrumentation [20–22]. It is worth mention that this methodology approaches the main three phenomena that distort the measured radiation quality but do not handle all the effects on this subject. Perhaps this could sound undesirable, but it is suitable for the needs of this paper and possess good agreement with norm values and other references as could be also seen in further discussion [23].

3. Results and Discussion

The results of experimental and simulated spectra will be shown, at first separately, and then, adding all possible spectra to compare and discuss them.

3.1. Experimental spectrum
Experimental spectrum was obtained from the N60 radiation quality established on Labprosaud/IFBA.

The "non-spectral" beam features are showed in table 1.

Table 1. ISO N60 beam features.

ISO N60	Tube Voltage [kV]	Additional Filtration [mm]	1st HVL @1m [mm]	2nd HVL @1m [mm]	h
Labprosaud/IFBA	60	4.0 Al + 0.61 Cu	0.240 Cu	0.264 Cu	0.91
ISO requirements	60	4.0 Al + 0.6 Cu	0.24 Cu	0.26 Cu	0.75 to 1

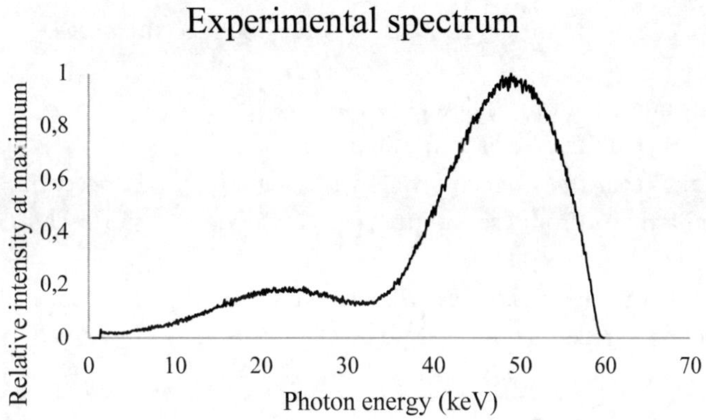

Figure 6. ISO N60 experimental spectrum.

Figure 6 shows the gross experimental spectrum. This means that no correction was applied. The mean energy calculated was 42.7 keV.

3.2. EGSnrc simulated spectrum
The experimental spectrum was corrected by DSc Josilene Santos, based on Tomal e al. 2015. [19] It is shown in figure 7.

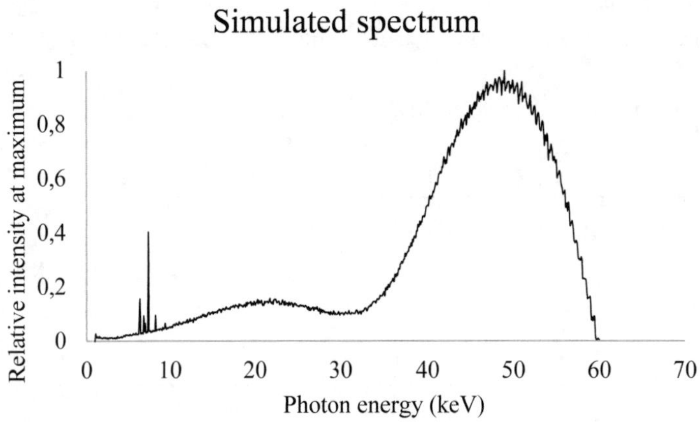

Figure 7. ISO N60 EGSnrc simulated spectrum.

The mean energy and resolution calculated was 43.3 keV.

It's possible to realize that exists a region between 6 to 10 keV where appears some "characteristics peaks" that does not appear on the experimental spectrum. Those peaks can be related to the standard spectrum (see figure 4), which present those characteristics peaks very preeminent.

Figure 8 shows the comparison between experimental and simulated spectra, besides a curve that represents the difference in counts by energy.

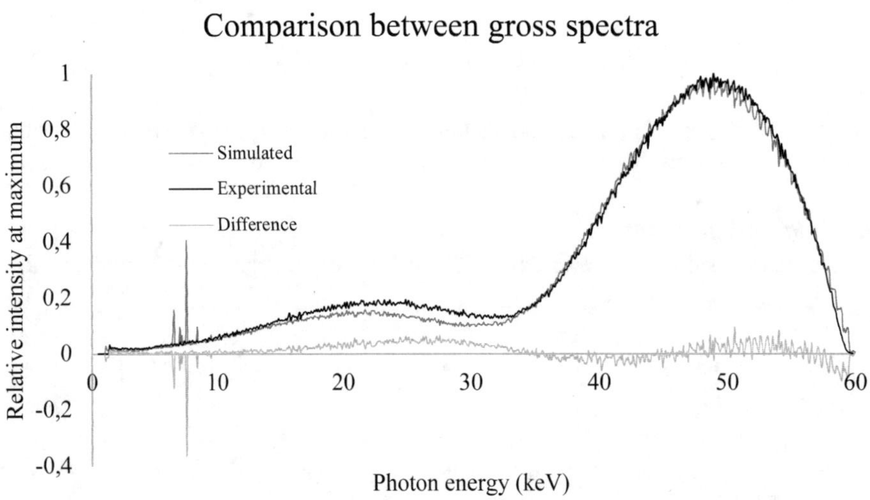

Figure 8. Comparison between experimental and simulated spectra.

The value of $R_{\%}$ was 1.4%, which represents a good agreement, within the ± 5% tolerance range required on ISO 4037-1:1996 [3].

3.3. Corrected spectrum

The experimental and simulated spectra were corrected based on the Stripping method. Both were compared and printed graphically as is shown in figure 6.

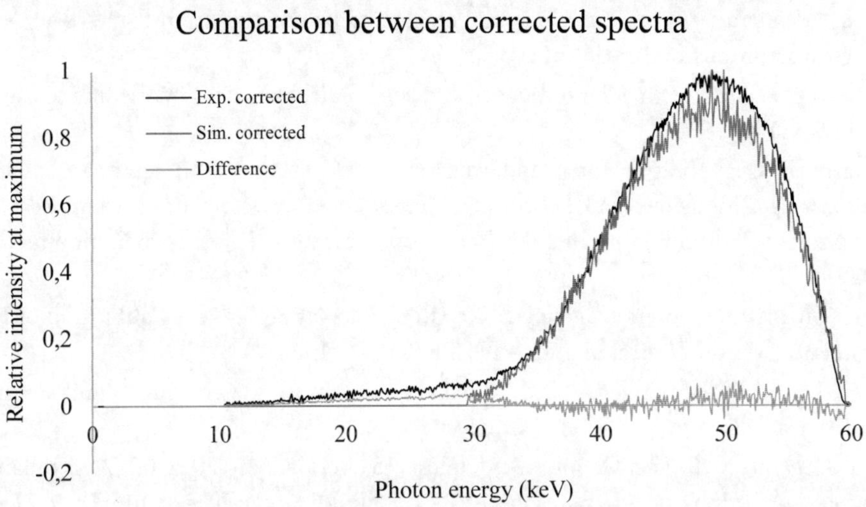

Figure 9. Comparison between experimental and simulated spectra.

The mean energy and spectral resolution calculated for both were 47.2 keV and 33%, respectively. Cut-off energy of 30 keV was used to calculate those two quantities to isolate the main peak.

For producing the ISO N60, ISO 4037-1:1996 requires the mean energy between 45.6 keV and 50.4 keV (5% of tolerance relating to the 48 keV mean energy) and resolution between 30.6 and 41.4% (15% of tolerance relating the 36% of resolution). This means that both the simulated and experimental corrected would be able to produce the standard radiation beam, based on ISO N60 radiation quality. [3]

4. Conclusion

The virtual system created was able to perform a spectrometry test as it is performed on a physical system.

The value of the relation $R_{\%}$ calculated was 1.4%, which represents a good agreement between the mean energy of the experimental and simulated spectra, within the ± 5% tolerance range required on ISO 4037-1:1996.

When both spectra were corrected, to produce the standard radiation beam, based on ISO N60 radiation quality, the mean energy and spectral resolution calculated for both were 47.2 keV and 33%, respectively. These values comply with the ISO 4037 requirements.

The uncertainties analysis of physical and virtual spectrometry will take place in upcoming work.

The results of the first tests performed to virtualize the physical laboratory environment indicate that is possible to move on to the next levels of virtualization, such as validation of radiation meters for the air kerma measurements.

Journal of Physics: Conference Series **2606** (2023) 012023 doi:10.1088/1742-6596/2606/1/012023

References

[1] REMESP 2019 Metrologia 4.0: um caminho longo, mas promissor [webpage] Acessed on 2021-07-09 Available from: https://www.remesp.org.br/noticias/metrologia-40-um-caminho-longo-mas-promissor/1684.html

[2] Avanzo M *et al* 2021 Artificial intelligence and the medical physicist: Welcome to the machine *Appl Sci* **11(4)** 1691

[3] International Organization for Standardization 1996 X and γ Reference Radiations for Calibrating Dosemeters and Dose Rate Meters and for Determining their Response as a Function of Photon Energy - Part 1: Radiation characteristics and production methods **ISO 4037-1**

[4] International Atomic Energy Agency 2000 Safety Reports Series 16 - Calibration of Radiation Protection Monitoring Instruments STI/PUB/**1074** (Vienna)

[5] Bradley D and Hehenberger P 2016 Mechatronic Futures: Challenges and Solutions for Mechatronic Systems and Their Designers Springer

[6] Bolton A *et al* 2018 The Gemini Principles. 15 Acessed on 2021-07-09 Available from: https://www.cdbb.cam.ac.uk/Resources/ResoucePublications/TheGeminiPrinciples.pdf

[7] Madni A *et al* 2019 Leveraging Digital Twin Technology in Model-Based Systems Engineering. *Systems* **7(1)** 7

[8] Teixeira G D J *et al* 2019 Facilidades de códigos de Monte Carlo para obter CSR *Brazilian J Radiat Sci* **7(3B)** 891

[9] Nascimento M *et al* 2021 Monte Carlo method and their codes applied to ionizing radiation. *J Phys Conf Ser.* 1826

[10] International Organization for Standardization 1997 X and γ Reference Radiations for Calibrating Dosemeters and Dose Rate Meters and for Determining their Response as a Function of Photon Energy - Part 2: Dosimetry for radiation protection over the energy ranges from 8 keV to 1,3 MeV and 4 MeV to 9 MeV **ISO 4037-2**

[11] International Organization for Standardization 1999 X and γ reference radiation for calibrating dosemeters and doserate meters and for determining their response as a function of photon energy - Part 3: Calibration of area and personal dosemeters and the measurement of their response as a function of energy and angle of incidence **ISO 4037-3**

[12] Macedo E M 2020 Avaliação da resposta dos equipamentos medidores de radiação utilizados em levantamento radiométrico em mamografia no Brasil *Master's Thesis, Instituto de Radioproteção e Dosimetria* (Rio de Janeiro, RJ)

[13] International Organization for Standardization 2019 Radiological protection -- X and gamma reference radiation for calibrating dosemeters and doserate meters and for determining their response as a function of photon energy -- Part 3: Calibration of area and personal dosemeters and the measurement of their response as a function of energy and angle of incidence **ISO 4037-3**

[14] Macedo E M *et al* 2019 Radiometric Survey in Mammography: Review of regulations, recommendations, field and laboratories practices and first tests. *Braz. J. Rad. Sci.* **7** 3B

[15] AMPTEK Inc. 2018 User Operating Manual XR-100 CdTe X-Ray Detector & Preamplifier RevB2 (Bedford, MA) Acessed on 2021-07-09 Available from: http://www.amptek.com

[16] Knoll G F 2010 Radiation Detection and Measurement *John Wiley & Sons I* 4th ed

[17] GitHub 2021 Release notes · nrc-cnrc/EGSnrc Wiki · GitHub [Internet]. Acessed on 2021-07-09 Available from: https://github.com/nrc-cnrc/EGSnrc/wiki/Release-notes

[18] Cranley K *et al* 1997 Catalogue of Diagnostic X-Ray Spectra and Other Data *IPEM Report No 78 J Radiol Prot*

[19] Tomal A *et al* 2015 Monte Carlo simulation of the response functions of CdTe detectors to be applied in x-ray spectroscopy *Appl Radiat Isot* **100** 32–7

[20] Nascimento MR *et al* 2020 Intrinsic Challenges in X-Ray Spectrometry Instrumentation With CdTe Diode Detector **1–15** Acessed on 2021-07-09 Available from: https://cbmri.org.br/site/intrinsic-challenges-in-x-ray-spectrometry-instrumentation-with-cdte-diode-detector/

[21] Seelentag W W *et al* 1979 Stripping of X-ray bremsstrahlung spectra up to 300 kVp on a desk type computer *Phys Med Biol* **24(4)** 767–80

[22] Di Castro E *et al* 1984 The use of cadmium telluride detectors for the qualitative analysis of diagnostic x-ray spectra *Phys Med Biol* **29(9)** 1117–31.

[23] Santos J C *et al* 2016 Evaluation of conversion coefficients relating air-kerma to H*(10) using primary and transmitted x-ray spectra in the diagnostic radiology energy range *J Radiol Prot* **36(1)** 117–32

Acknowledgments

Acknowledgments to the Instituto de Radioproteção e Dosimetria – IRD/CNEN and Laboratório de Produtos para a Saúde Labprosaud/IFBA. This study was financed in part by the Coordenação de Aperfeiçoamento de Pessoal de Nível Superior - Brasil (CAPES) - Finance Code 001. Thanks also all participants of the IRTech research group.

A simple Kibble balance approach for principle demonstration at Inmetro

F L Cacais[1], V M Loayza[1], A Beatrici[1] , D Quintão[1] and A S Rebelo[1]

[1] Laboratório de Massa, Instituto Nacional de Metrologia, Qualidade e Tecnologia (Inmetro), Duque de Caxias/Xerém, RJ 25250-020,BR

Email:facacais@inmetro.gov.br

Abstract. This work shows the results obtained by Inmetro's mass lab implementation of a simple Kibble balance approach for demonstration of its mass measurement principle based on the Planck's constant. It is based on some modifications of Mimimal Watt Balance work reducing the required electrical measurements by using two loudspeakers mechanically coupled so that one of them acts as a linear actuator in order to perform *BL* measurements in the moving phase. The mean *BL* value obtained in weighing phase was 2.66 N/A and in moving phase 2.67 N/A. The maximum error of mass values measured for standard weights in the range from 5 g to 50 g was lower than 5%.

1. Introduction

In May 2019, the New SI came into effect, in which the seven SI base units were defined as a function of physical constants with fixed numerical values, without uncertainty [1], figure 1. The kilogram, the unit of mass, is now defined as a function of Planck's constant with the value $h = 6.62607015 \times 10^{-34}$ J s [2]. Realizing the unit of mass from this definition requires the application of primary methods such as the Kibble balance and the ^{28}Si ultrapure silicon sphere experiments [3]. However, the acceptance of the results of these experiments as international references for the mass unit within the Mutual Recognition Agreement (MRA) is subject to complying with the compatibility requirements established by the Mass Consultative Committee (CCM). This way of proceeding with the adoption of the primary methods is a consequence of the discrepancy observed in the adjustment of Planck's constant.

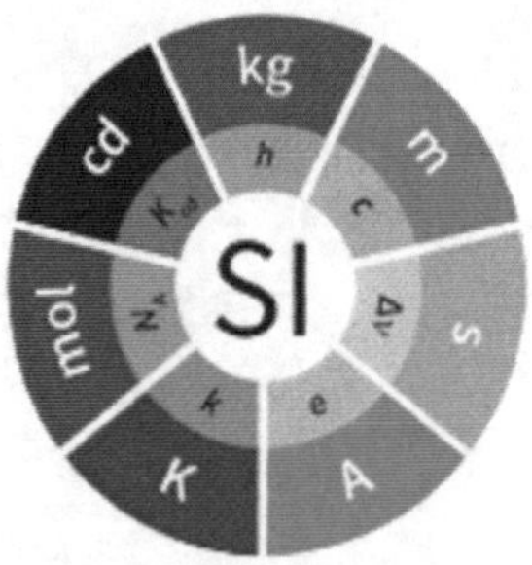

Figure 1. SI Illustration [4].

Following the promulgation of the New SI, the key comparison of the primary methods CCM.M-K8 was carried out with the purpose of evaluating the compatibility between the results as well as the determination of a consensus value for the realized kilogram and its uncertainty. Despite the applied improvements on the experiments a discrepancy still remains between the results of the primary methods. As a consequence, the participant NMIs having primary methods should disseminate the consensus value of the comparison, not their individual results, and the CCM also maintains the call for new primary methods for realizing the unit in the range of 1 kg or less to be developed [5].

In line with this call and as a first step in evaluating the development of a Kibble Balance, the Mass Laboratory (Lamas) developed a simple experimental bench for carrying out weighings directly related to Planck's constant. The basic mechanical setup using two loudspeakers was inspired by the proposal of a possible 'watt' balance based on the 'seismometer' suspension mechanism by Kibble and Robinson [6]. The electrical measurements involve some modification of the Minimal Watt Balance proposed by Professor Ward [7] with the aim of generating the voltage induced in the speaker's coil directly from the mechanical vibration of the loudspeaker in the moving phase. For this, a second loudspeaker is used as a system for driving the mechanical vibration in the first one when it is excited by alternating voltage. This way of proceeding eliminates the need for some electrical measurements required for the Minimal Watt Balance. This work shows the mass measurements results obtained for a set of standard weights from 5 g to 50 g from electric quantities without requiring a mass standard as a reference for calibrating the balance.

2. Theoretical background

2.1. Kibble balance

The Kibble balance is based on the principle of equality between the mechanical and electrical power respectively arisen in the steps of weighing (figure 2 [8]) and moving (figure 3 [8]). In the realization of these two steps the magnetic flux gradient is considered constant. The moving phase is performed in order to directly determine the *BL* parameter required in the weighing phase for mass determination. This is the key characteristic of the all Kibble balances.

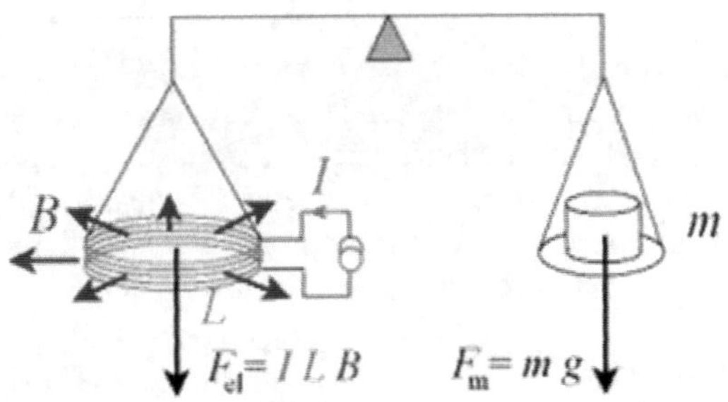

Figure 2. In the weighing phase, the Kibble balance works as a mass comparator with an electromagnetic restoring force. The mass to be measured is balanced by a Lorentz force produced by the circulation of a current *I* in a coil of length *L* immersed in a radial magnetic field *B*. The current is measured from the Josephson and Hall Quantum references [8].

Journal of Physics: Conference Series **2606** (2023) 012024 doi:10.1088/1742-6596/2606/1/012024

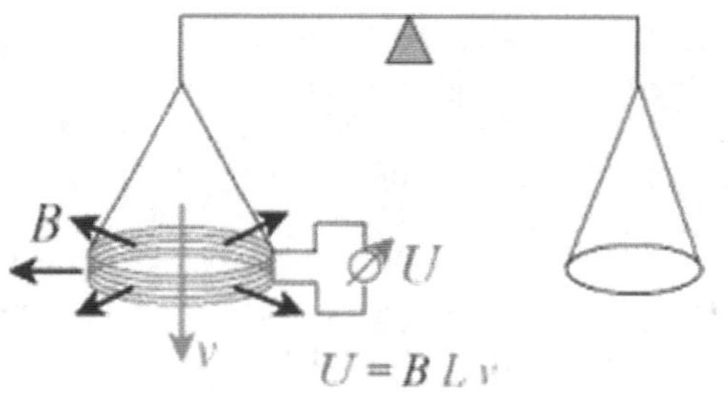

Figure 3. Moving phase of the Kibble balance. The coil is moved vertically with constant velocity v through the same magnetic field B, which leads to an induced voltage U [8].

The mass m in kilograms and the Planck's constant h are related as equation (1) which includes the values of voltages U_1 (weighing phase) and U_2 (moving phase) measured in volts with traceability to Josephson references, the value for resistance R_1 (weighing phase) measured in ohms with traceability to Quantum Hall standards, the fixed electrical constants $K_J\ (=2\frac{e}{h})$ and $R_H\ (=\frac{h}{e^2})$, the local gravity acceleration g and the velocity v of the coil in the moving phase.

$$m = \left[\frac{U_1 U_2 K_J^2 R_K}{4 R_1\, g\, v} \right] h \qquad (1).$$

At this point, it is worth noting that depending on the intended accuracy of the Kibble Balance implementation it may not required to have available on the bench the Josephson or Quantum Hall standards. The assurance of that voltage and current measurements are related to these standards and consequently mass measurements are linked to the Planck's constant is provided by electrical calibrations using standards from a traceability chain which is ultimately based on these quantum electrical standards.

2.2. *Simple Kibble balance approach*

In this approach a common loudspeaker is used as the core element of the Kibble balance. Loudspeakers are made up of a coil which when energized produce an unidirectional movement within a permanent magnetic field. These are the required elements all integrated in a single device. A second identical loudspeaker is used as an actuator in order to produce the oscillatory movement on the coil of the main loudspeaker, figure 4. This is required in order to overcome the disadvantage of the short amplitude of movement of the coil not allowing to drive a constant velocity movement and a direct measurement of the induced voltage.

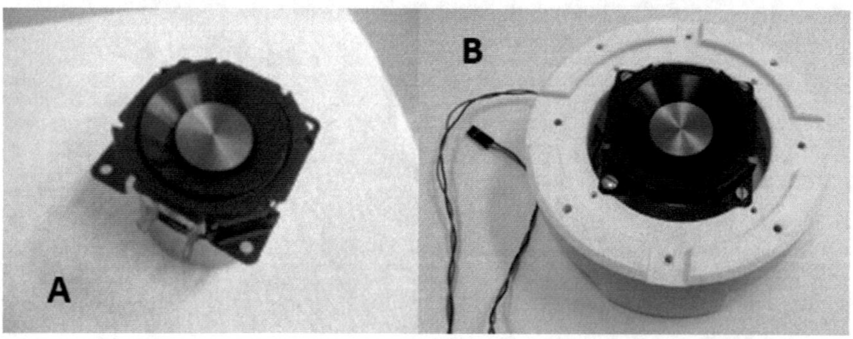

Figure 4. A. Loudspeaker used for this experiment. B. Weighing loudspeaker mounted on its support.

2.2.1. Weighing phase. In the weighing phase the mass under measurement m is placed on the circular plate which is fixed to the speaker's cone which lowers due to the weight force. Then the loudspeaker is powered by a DC voltage supply. The voltage is adjusted until the current I which passes through the speaker coil restores the speaker to its initial vertical position.

In this situation, with the measurement of the electric current I and knowing the value of the acceleration of gravity g, we have, according to Lorentz's law, that:

$$m = {}^{I(BL)}\!\big/_{g} \tag{2}$$

The parameter BL is measured in the moving phase.

2.2.2. Moving phase. In this phase two loudspeakers are connected by means of a rigid rod (plastic tube) shown in figures 5. The upper loudspeaker acting as a linear actuator is powered by an AC voltage generator so due to the mechanical connection, the two speakers move synchronously at the same frequency. However due to the effect of the induced voltage in the upper coil the frequency provided by the generator of AC voltage to the upper loudspeaker is different from the mechanical frequency of the coil's movement causing phase difference between AC voltage and induced voltage in the upper coil [9].

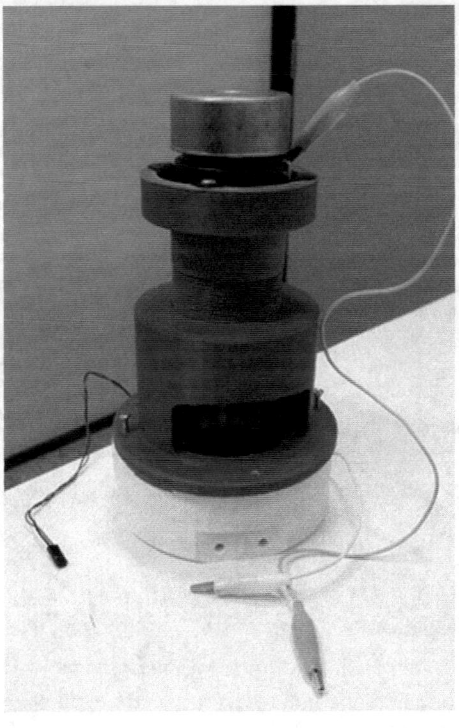

Figure 5. The simple Kibble balance prototype mounted for the moving phase.

However, since the voltage in the lower coil is arisen due to the induction by the synchronized mechanical movement with the upper coil, the velocity v of the lower coil is in phase with the voltage U.

$$U = (BL)v \quad (3).$$

To determine the velocity it is necessary to realize that the resonance the frequency f_0 of the AC generator voltage is the same for the induced voltage for both coils. So the maximum velocity of the oscillatory movement of lower coil is proportional to amplitude of movement A:

$$v_{\max} = 2\pi f_0 A \quad (4).$$

In this resonance condition, the term BL of the lower coil can be determined by measuring the induced voltage amplitude U and by knowing $v_{\max}$:

$$BL = U / 2\pi f_0 A \quad (5).$$

The mass to be measure m can be determined from equations (2) and (5) by:

$$m = IU / 2\pi f_0 A\, g \quad (6).$$

3. Experimental setup

The experimental bench for application of this simple approach of Kibble balance is shown in figure 6.

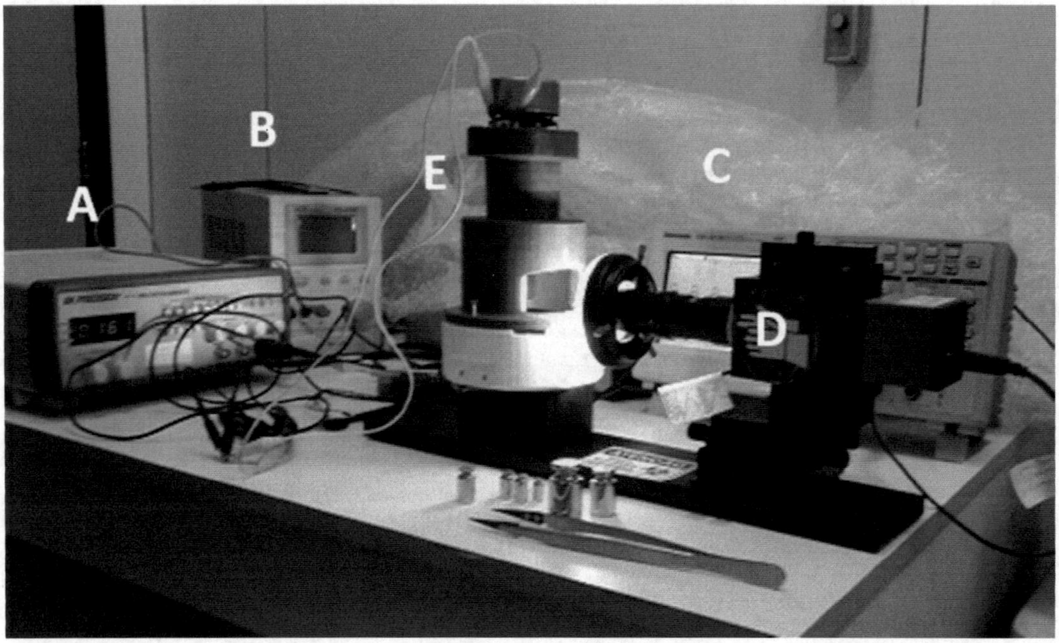

Figure 6. The experimental bench. A frequency generator, B variable DC supply, C digital oscilloscope D digital microscope, E Kibble balance prototype.

In the weighing phase the standard weight (mass to be measured) is put on the loudspeaker circular plate fixed to the cone, figure 7. A variable voltage DC supply provides the current which restores the loudspeaker to the initial vertical position. This current is increased up to the speaker's cone to reach the initial reference vertical position. A digital camera capable of focusing the reference position mark on the cone is used to assure the initial vertical position is reached within ± 5 µm and so the current is measured with the aid of an ordinary digital multimeter.

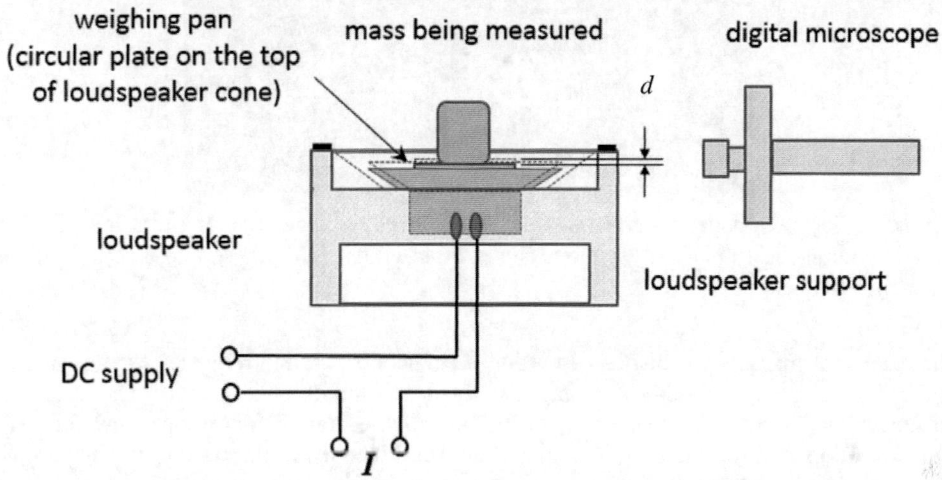

Figure 7. Weighing phase. *d* is the displacement of the cone from reference position produced by the weight force of the mass. *I* is the current required to restore the cone to the reference position using the digital microscope.

A set of standard weights was used to perform weighing phase. This set content is 5 g, 10 g, 10 g*, 20 g e 50 g weights, figure 8

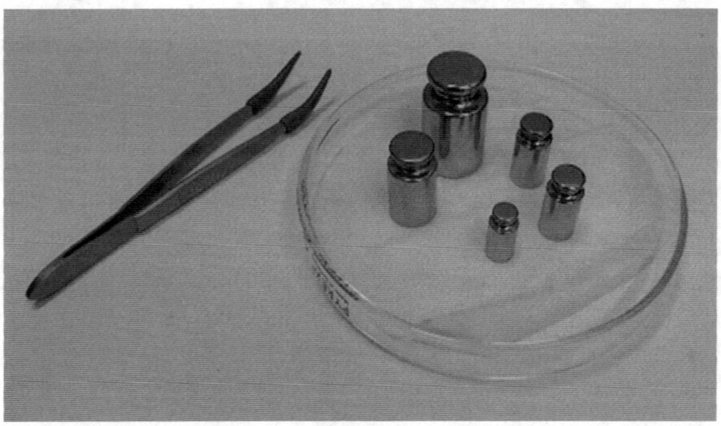

Figure 8. A set of standard weights was used to perform weighing phase.

In the moving phase a two channels digital oscilloscope is used to measure the AC generator voltage and the induced voltage in the lower loudspeaker's coil simultaneously. The XY configuration of oscilloscope allows to make a Lissajous figure, a ellipse which collapses to a straight line right, figure 9, at the resonance frequency.

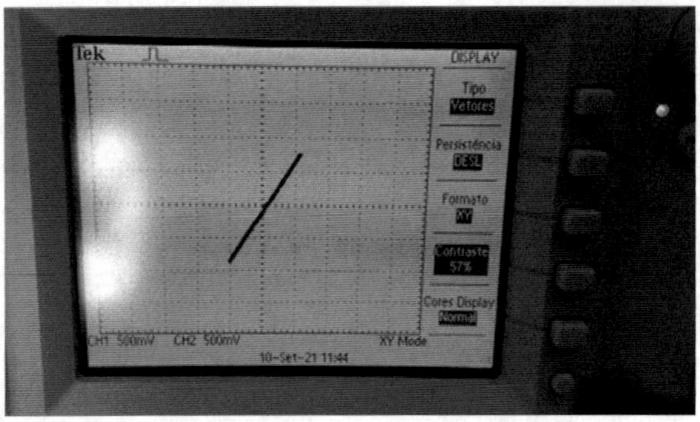

Figure 9. The AC generator voltage and the induced voltage in the lower loudspeaker's coil measured simultaneously make a Lissajous figure of an ellipse to collapse into a straight line right at the resonance frequency.

The resonance frequency is displayed by the AC generator, at this frequency are measured the AC generator voltage and the induced voltage in the lower loudspeaker's coil. The cone's vertical displacement amplitude A is measured with the aid of the digital microscope now focused in the reference position mark on the rod whose oscillation corresponds to the cone oscillation which is the oscillation of the coil, figure 10.

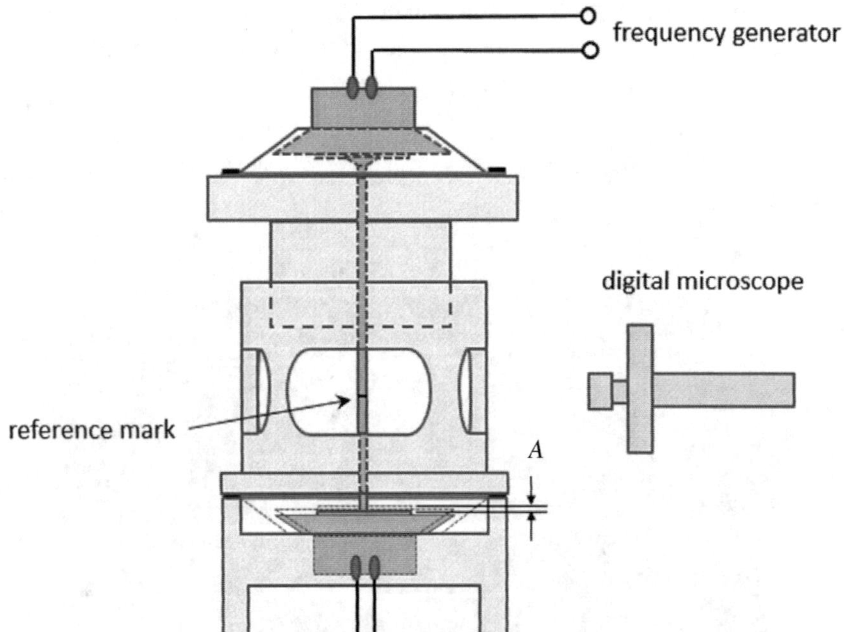

Figure 10. Moving phase. The upper speaker generates the sinusoidal oscillation of its circular plate (cone) by applying a voltage to its coil by the frequency generator at the resonance frequency f_0. This movement is transmitted to the circular plate (cone) of the lower speaker which drives its coil

inducing the voltage U (equation 5). The amplitude of oscillation of the lower speaker is measured on the calibrated screen of the digital microscope and U is measured by a oscilloscope.

4. Results

Table 1 shows the weighing phase current measurements. The estimated *BL* values (in N/A which is equivalent to V.s/m) are also presented.

Table 1. Currents measured for each mass standard nominal weight (assumed $g = 9.8\ m/s^2$).

Mass Identification	Weight N	Restoring current mA	*BL* N/A
5 g	0.049	18	2.78
10 g	0.098	35	2.80
20 g	0.196	74	2.66
10 g + 20 g	0.294	111	2.66
10 g + 10 g*+ 20 g	0.392	155	2.53
50 g	0.49	193	2.54

Table 2 shows the moving phase measurements for induced voltage and displacement amplitudes also for the resonance frequency. Each amplitude measurement was performed twice by different technicians in order to avoid operator's bias. The mass values for the standard weight is obtained by the average of the determined values for *BL* as specified in equation (6).

Table 2. Measurements required for mass determination by the simple Kibble balance.

U V	A µm	f_0 Hz	*BL* N/A	m g
0.147	52.5		2.75	5
0.147	56.4		2.56	10
0.137	49.5		2.72	20
0.137	50.7	162	2.65	30
0.127	47.8		2.61	42
0.127	46.3		2.69	52

The mean *BL* value obtained in the weighing phase 2.66 N/A is very close to the average estimated in moving phase 2.67 N/A. The 2 g error at the highest weights is supposed to be caused by the warm up of the loudspeaker to higher currents which causes reduction in *BL* values.

Journal of Physics: Conference Series **2606** (2023) 012024 doi:10.1088/1742-6596/2606/1/012024

5. Conclusion

This work has shown the implementation of the simple Kibble balance. The measured mass values are close enough to the nominal ones of the standard weights used for demonstration of the Kibble balance principle.

Acknowledgments

An author would like to thank Pronametro/Inmetro. Also the authors would like to thank to Dinam/Dimci/Inmetro, Diele/Dimci/Inmetro and Dimav/Inmetro.

References

[1] INMETRO 2021 Sistema Internacional de Unidades: SI available on https://www.gov.br/inmetro/pt-br/centrais-de-conteudo/publicacoes/brochura-si_portugues_inmetro-ipq_rev01_2021-03-10_conspublica.pdf
[2] Newell D B *et al* 2018 *Metrologia* 55 L13
[3] BIPM 2021 available on https://www.bipm.org/en/publications/mises-en-pratique
[4] BIPM 2021 SI Illustration Guidelines available on https://www.bipm.org/en/measurement-units/si-promotion
[5] Stock M *et al* 2020 *Metrologia* 57 07030
[6] Kibble B P and Robinson I A 2014 *Metrologia* 51 S132
[7] Ward Richard J 2014 *Phys. Educ.* 49 277
[8] Stock M 2011 *Phil. Trans. R. Soc. A* 369 3936
[9] Kraftmakher Yaakov 2009 *Eur. J. Phys.* 30 567

Glass campanula effects on measurement results in high capacity mass comparator

A. Beatrici
Laboratório de Massa (Lamas) – Instituto Nacional de Metrologia, Qualidade e Tecnologia (Inmetro). Duque de Caxias – RJ/BR.

E-mail: abeatrici@inmetro.gov.br

Abstract. The calibrations or checking of mass standards with nominal values greater than 10 kg and high accuracy class require special attention to the environmental conditions stability due to the their volume. The equipment used to calibrate this kind of mass standards usually has a weighing chamber (in this case a glass campanula or bell jar) to avoid, at least to reduce, the effects of environmental changes. In this study will be present the effects of these weighing campanulas on the standard deviations and in mass measurements results using only the internal tare (without weights in the balance plate) and using two 10 kg test weights.
Keywords: Weighing chamber, mass calibration, thermic stabilization.

1. Introduction

Calibration, or checking, of high accuracy class weight collections (OIML E1 accuracy class) [1][2][3] need a high quality environmental conditions control as well as a high stability weighing system. In this study, it will be performed tests in the AX64004 mass comparator using or not the glass campanula. This equipment is designed to calibrate mass nominal values up to 64 kg and have four calibrations positions all of them with a glass campanula (bell jar) (Fig 1 and 2).
The provider data sheet to this kind of mass comparator, in normal use, is:

Maximum Charge:	64.260 kg
Resolution:	0.1 mg
Weighing range:	0 g – 260 g
Granted repeatability:	0.4 mg
Typical repeatability:	0.2 mg (ABA)

Sometimes this kind of equipment need to be used to calibrate special artefacts (not mass standards) an it´s shape cannot allow the glass campanula use. Thus is important to know the effects in these conditions on the measurement results must be considered in the uncertainty budget and mass difference determinations.

Figure 1. Mass comparator and the glass campanula.

Figure 2. Mass comparator without glass campanula.

This approach consists in comparing the measurements results and the standard deviation of weighing using only the internal weights (empty weighing plate) and the same procedure using two 10 kg mass standards. Thus, for the first case, the mass measurements results expected to be null end the standard deviation will provide the effects of this campanula in the stability of the equipment indication and, for the second case, the mass difference between two mass standards that will show the campanulas effects. The measurements will be performed using the double mass substitution scheme (ABBA [4] [5][6]), usually applied in mass comparators measurements. The results show the importance of these stabilizations chambers to obtain better measurement results, in fact, even with the high stability environmental control in the laboratory it was possible to show the importance of the campanula effect.

2. Methodology

To check the effect of each case, with or without weights and with or without campanulas, will be simulated a full mass comparison in both cases. The AX64004 mass comparator has 0.1 milligram of resolution. The usual measurement calibration parameters in the laboratory are:

- Stabilization time: 35 s.
- Integration time: 10 s.
- 8 weighing series with 10 ABBA cycles weighing scheme for each condition.

Thus, using these parameters, we can assume that only the thermic effects and equipment inner heating will be responsible for the changes in the measurement results and standard deviations. It will be studied four cases:

I - Without campanula and without mass standards (only internal tare).
II - With campanula and without mass standards (only internal tare).
III - Without campanula and with two 10 kg mass standards.
IV - With campanula and with two 10 kg mass standards.

The ABBA design corresponding to the weighing sequence (1) and equation (2):

$$A_1\, B_1\, B_2\, A_2\, ;\, A_3\, B_3\, B_4\, A_4\, ;\, A_5 \ldots \quad (1)$$

$$\Delta I_{ABBA} = \frac{\left(I_{B(i+1)} - I_{A(i+1)}\right) + \left(I_{Bi} - I_{Ai}\right)}{2} \quad (2)$$

Where:

I_{Bi} is the balance indication for standard weighing Bi;
I_{Ai} is the balance indication for standard weighing Ai;
i is the weighing order, corresponds to a constant time interval and it´s an integer odd number.

3. Environmental laboratory control

In the Inmetro´s Mass Laboratory, we have a system that keeps the temperature and air relative humidity (*RH*) under stable condition, while the atmospheric pressure is only measured.
Figure 3 shows the laboratory temperature stability. The temperature changes are about ± 0.2 °C in 24 h. The standard deviation in the temperature measurements was 0.1 °C.

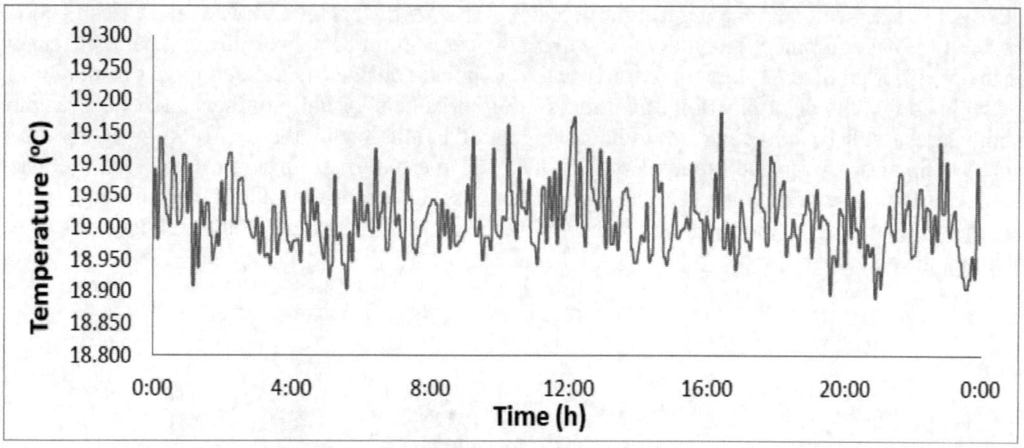

Figure 3. Thermal stability in the laboratory.

The air relative humidity is controlled. The mean value of the *RH* is about 55 % and the range is 50 % up to 60 %. The atmospheric pressure is just measured (not controlled) and the historical range is 1000 hPa up to 1020 hPa.
For the case II and III, we used the mass standards PP019 and PP079, its respective densities are:

PP019 7,888.8 kg/m^3
PP079 7,862.8 kg/m^3

The difference between those two values represents about 0.3 % in volume, thus can be assumed that any difference greater than 1 % is caused by the campanula effects and not by the air density changes. The mean differences in temperature (with or without the glass campanula) is about 0.3 Celsius degrees.

4. Results

Figure 4 shows the mean *SD* obtained in all comparisons for all analyzed cases.

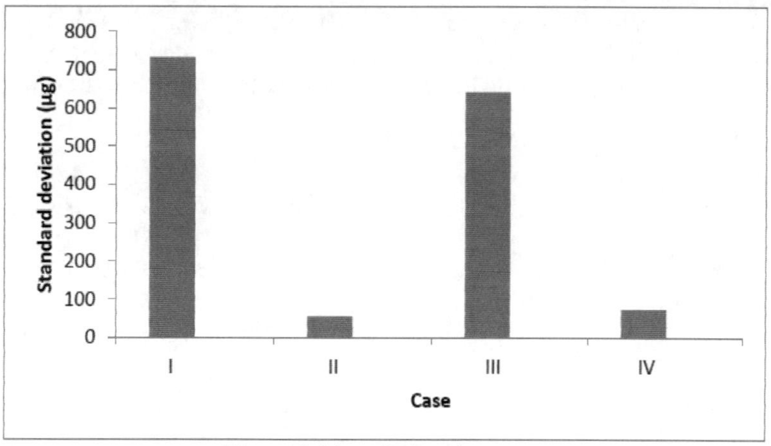

Figure 4. Mean *SD* in micrograms.

Both cases (I and III), corresponding to those measurements without the campanula, the *SD* was greater than the provider reference data (*SD* = 200 micrograms). This is an important result to be consider in mass determination and uncertainty budget. The *SD* differences between the cases I and III are covered by the equipment resolution. However, the cases II and IV (with the campanula) remain under the provider granted *SD* limit and are virtually identical considering the equipment resolution. It can be realized that the determination of the mass differences gets virtually to the same values, Figure 5, in both cases (with 10 kg weights or without weights). Obviously with the two 10 kg weights (cases III and IV) the mass difference (about 3 mg) is due to a true real mass difference between the mass standards but with or without the glass campanula the mass difference was about the same (0.07 mg). In the case without weights the measured mass difference was about 0.06 mg between both cases (with and without glass campanula).

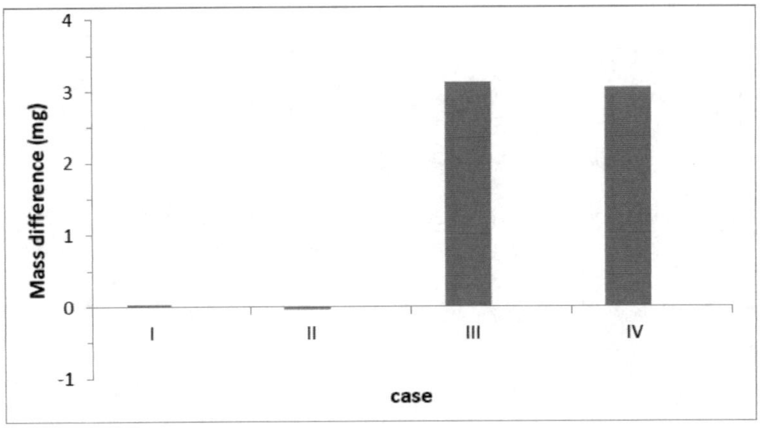

Figure 5. Mass differences in milligrams.

5. Conclusion

The remarkable result that using or not the glass campanula do not causes error in the mass determination can be explained by the excellent environmental control in the laboratory, it could be expected that in worse conditions this result may be not true.

The expected result of the *SD* increasing in measurements results with no glass campanula shows that unusual artefacts when it is not possible to use a campanula the measurement capabilities of the laboratory can be analyzed with care.

Both results can be understanding due to the low differences between the temperature inside and outside the glass campanula, about 0.3 °C, this is due to the equipment design that avoid the heating in the weighing plate zone and the size of the campanula area that allow the heat dissipation more efficiently.

6. References

[1] OIML Recommendation R111 (Parts-1/2) 2004. *Weights of classes E_1, E_2, F_1, F_2, M_1, M_{1-2}, M_2, M_{2-3} and M_3*. International Organization of Legal Metrology.

[2] OIML D28 2004. *Conventional value of the result of weighing in air*. International Organization of Legal Metrology.

[3] EURAMET/cg-18 (version 4.0) 2015. *Guidelines on the Calibration of Non-Automatic Weighing Instruments*. Euramet.

[4] A Beatrici et al. *Analysis of drift correction in different simulated weighing schemes*. J. Phys.: Conf. Ser. v. **648**, 2015. doi:10.1088/1742-6596/648/1/012025.

[5] Sutton C M and Clarkson M T 1993/1994. *A General Approach to Comparisons in the Presence of Drift*. Metrologia **30**, 487-493.

[6] Gläser M 2000. *Cycles of comparison measurements, uncertainties and Efficiencies*. Meas. Sci. Technol. **11**, 20-24.

Inmetro's participation in SIM kilogram dissemination project (SKDP) Mass stability analysis of the SIM21 mass standard (1 kg) over three years in the SIM Kilogram Dissemination Project (SKDP)

F L Cacais[1], V M Loayza[1], A Beatrici[1] , D Quintão[1] and A S Rebelo[1]

[1] Laboratório de Massa, Instituto Nacional de Metrologia, Qualidade e Tecnologia (Inmetro), Duque de Caxias/Xerém, RJ 25250-020,BR

Email:facacais@inmetro.gov.br

Abstract. Inmetro's mass laboratory (Lamas) participated in SIM Kilogram Dissemination Project (SKDP). The assessment of mass stability of the SIM21 1 kg stainless steel mass standard was carried out between 2018 and 2021. The results show a typical variability for a weighing system but not clear trend in measurements. Therefore one can say no mass drift for SIM21 mass standard was observed along this period.

1. Introduction

In the SIM meeting in Montevideo Uruguay (2018), the SIM kilogram dissemination project (SKDP) was introduced to the participant, National Metrology Institutes (NMIs), and the Brazil's participation was agreed. The (SKDP) is a research project involving a set of highly correlated stainless steel mass artifacts standards provided to the participating NMIs across the SIM region aiming to assess their stability in use. Each mass standard is made of the same material and has similar surface properties. They have each been manufactured, characterized, and handled in a similar manner prior to distribution by NIST/NRC. Also the mass drift of each standard was determined after washing over a period of approximately 8 month, and a calibration report was generated for each mass containing its final mass calibration value. When distributed, the mass artifacts were hold in a sealed high purity enclosure to ensure that transport environments for each mass be equal. The enclosure is itself housed during transport within a watertight carrying case that also contains a data logger which records continuously the environmental conditions temperature, pressure, and relative humidity. During transport and subsequent storage in the participant's laboratory the data logger is kept with the mass at all possible times to maintain a record of the ambient conditions the mass standard was subject to during storage.

The mass measurements for the S21 standard were performed at Inmetro between 2018 and 2021 and this work shows the results of mass stability checking.

2. Mass artifact description

The cylindrical mass standards were manufactured from one of two stainless steel rods with each artifact numbered based on the position within the rod from which it was cut. The density of each artifact was determined at NIST (USA) or NRC (Canada) using two different methods (aerostatic, hydrostatic). At both laboratories the mass artifacts were cleaned (ethanol degreaser and distilled water rinse) and five mass calibrations performed over an approximately 8 month period before distribution. The magnetic susceptibility and permanent magnetization of a sample subset of the mass ensemble was determined using a susceptometer. Mass artifacts was found to comply with requirements for magnetic properties of OIML E_1 accuracy class [1]. Data of S21 characterization are shown in tables 1 – 3 [2]. Figure 1 shows the receipt of the SIM21 mass standard.

Table 1. Volume, Density and Coefficient of Volumetric Expansion.

Mass Identification	Volume [cm^3] at 20 °C	Volume Uncertainty [cm^3] (k = 1)	Density [g/cm^3] at 20 °C	Density Uncertainty [g/cm^3] (k = 1)	Volumetric Coef. of Exp. [°C^{-1}]
SIM21	124.4236	0.0037	803.706	0.00024	0.000045

Table 2. Mass Corrections and Uncertainties.

Date	Nominal [g]	True Mass Correction [mg]	True Mass Uncertainty mg (k = 1)	Conventional Mass Correction [mg]	Conventional Mass Uncertainty [mg] (k = 1)
Before 20 May 2019	1000	0.4675	0.0123	1.1592	0.0114
After 20 May 2019	1000	0.4675	0.0158	1.1592	0.0152

Table 3. Measurement Environment.

	Temperature [°C]	Humidity [%RH]	Pressure [kPa]	Computed Air Density [mg/cm^3]
Average	20.270	45.5	100.425	1.1879

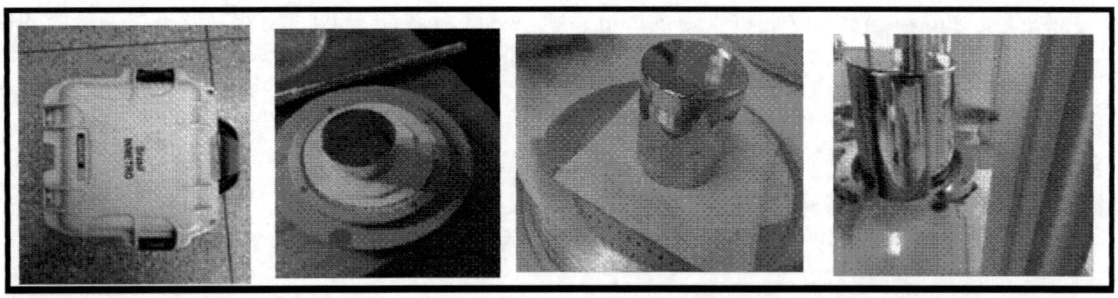

Figure 1. Receipt, storage and placing upon the weight handler of AT1006 mass comparator of S21 standard.

Journal of Physics: Conference Series **2606** (2023) 012026 doi:10.1088/1742-6596/2606/1/012026

3. Mass measurements

Weighings are performed in a 1 kg automatic mass comparator Mettler Toledo AT1006 which works integrated with the climate station Klimet A30. The 1 kg mass standards used to check the stability of SIM/NIST S21 standard in mass and conventional mass are shown in Inmetro's traceability chain is shown in figure 2 [3].

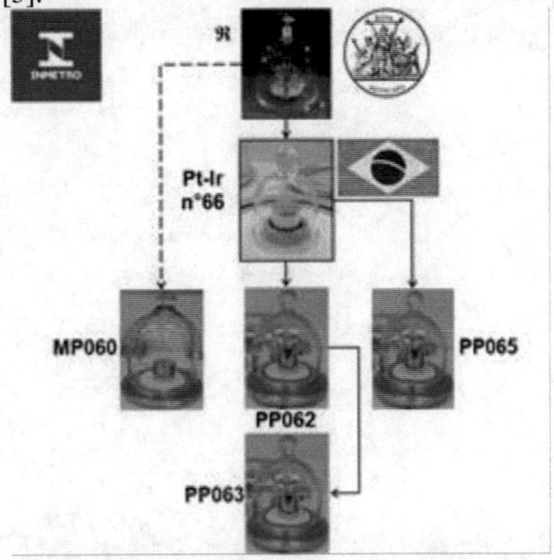

Figure 2. Inmetro's traceability chain

Mass measurements were executed by either weighing designs or direct comparison methods. Air density were determined by CIPM equation [4]

4. Environmental and use conditions

The typical environmental conditions in a period of a week (May 26th to June 4th 2021) in Module 1 of Lamas are shown in figures 3, 4 and 5.

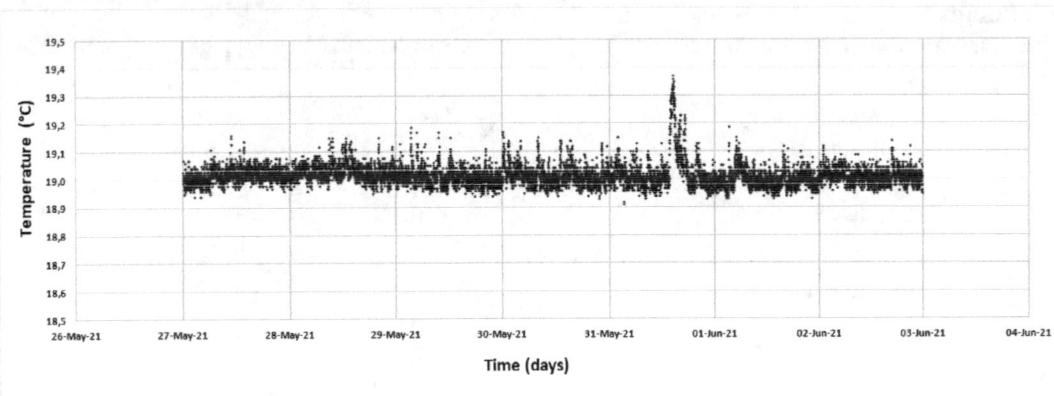

Figure 3. Air temperature.

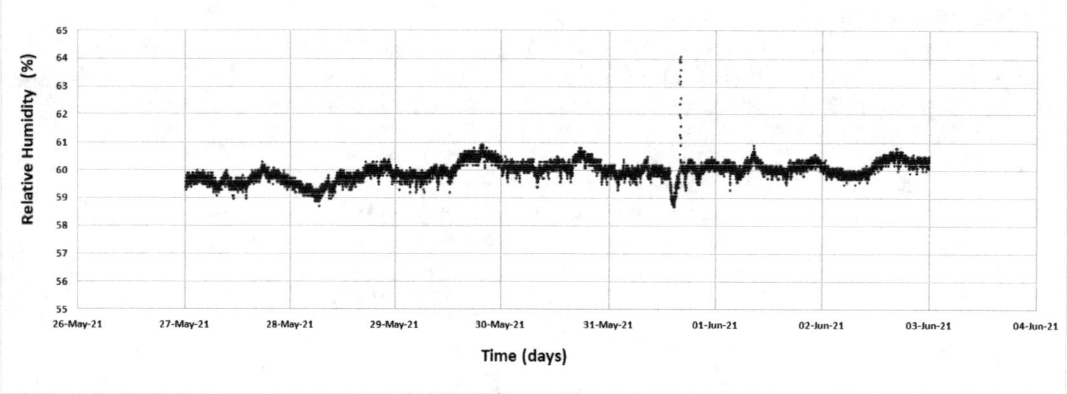

Figure 4. Air relative humidity.

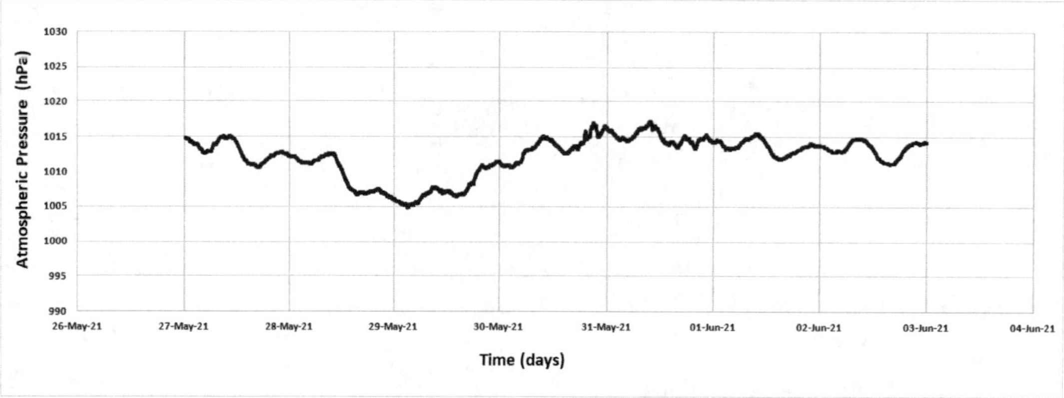

Figure 5. Atmospheric pressure.

5. Results
The mass measurements for the S21 standard performed between 2018 and 2021 are shown in table 4.

Table 4. Measurement results from 2018 up to 2021.

Date	(Mass – 1 kg)/mg	$u(k=1)$/mg	Ref Std
10/29/2018	0.477	0.018	MP060
12/03/2018	0.480	0.033	PP062
12/09/2018	0.492	0.033	PP062
06/14/2019	0.477	0.020	MP060
11/29/2019	0.480	0.020	PP062
12/03/2019	0.478	0.020	PP062
01/14/2020	0.475	0.020	PP065
05/29/2021	0.474	0.016	MP060

The measurements performed in conventional mass are shown in table 5.

Table 5. Measurement results in conventional mass.

Date	(Conv. Mass - 1 kg)/mg	$u(k=1)$/mg	Ref Std
10/29/2018	1.168	0.018	MP060
12/02/2018	1.167	0.034	PP063
12/03/2018	1.171	0.033	PP062
12/09/2018	1.184	0.033	PP062
12/09/2018	1.176	0.034	PP063
06/14/2019	1.169	0.020	MP060
11/29/2019	1.172	0.020	PP062
12/03/2019	1.170	0.020	PP062
01/14/2020	1.167	0.020	PP065
07/24/2020	1.186	0.023	PP063
07/28/2020	1.194	0.023	PP063
08/05/2020	1.189	0.024	PP063
08/12/2020	1.190	0.023	PP063
05/29/2021	1.166	0.016	MP060

The deviations $d_i = M_{Inmetro} - M_{NIST}$ between Inmetro's mass or conventional mass measurements and that provided by NIST allow to evaluate the progress of measurements over time. The deviation in mass and conventional mass are shown, respectively, on figures 6 and 7. Since Inmetro and NIST measurements can be considered uncorrelated, uncertainty bars u (d_i) means the root of the sum of the variances associated to the respective u (k=1).

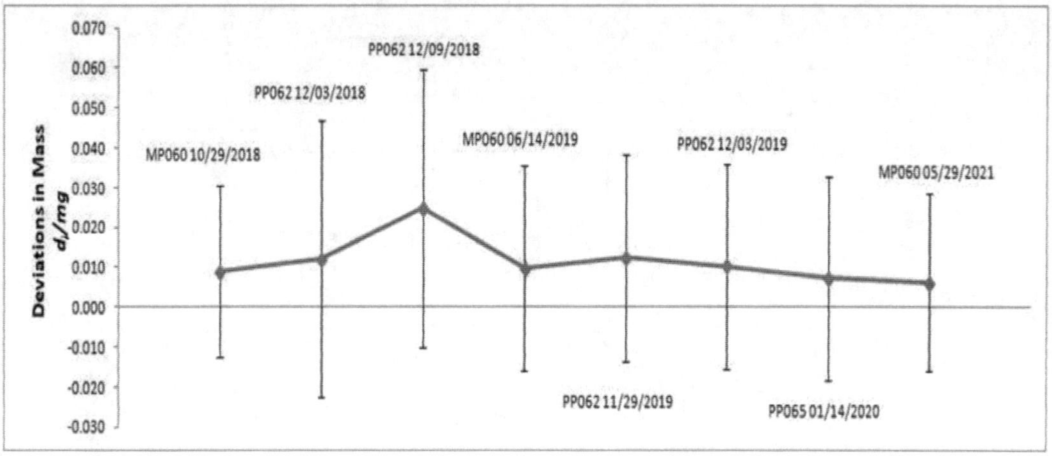

Figure 6. Deviations in mass between Inmetro and NIST measurements over time.

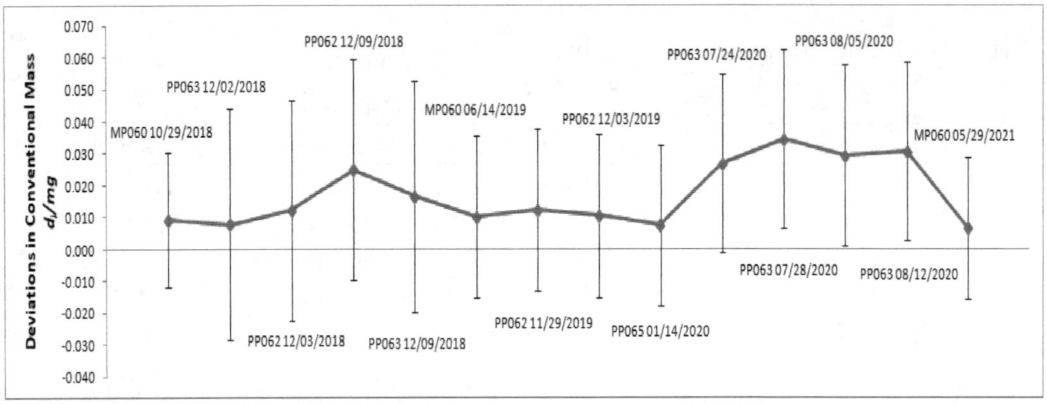

Figure 7. Deviations in conventional mass between Inmetro and NIST measurements over time.

Since mass measurements are correlated with correlation coefficient 1 due to systematic effects as buoyancy or traceability, mass variations are affected just for random effects. Thus, at consecutive times i and j ($j > i$) the mass variation $M_{\text{inmetro}\,j}$ - $M_{\text{inmetro}\,i}$ can be evaluated with reduced uncertainties. Inmetro's typical repeatability for 1 kg comparisons at 10 ABBA weighing cycles is about 0.002 mg yielding a 95% confidence interval from Student-t of 0.007 mg for mass variations. Mass and conventional mass variations between consecutive measurements are shown on figures 8 and 9.

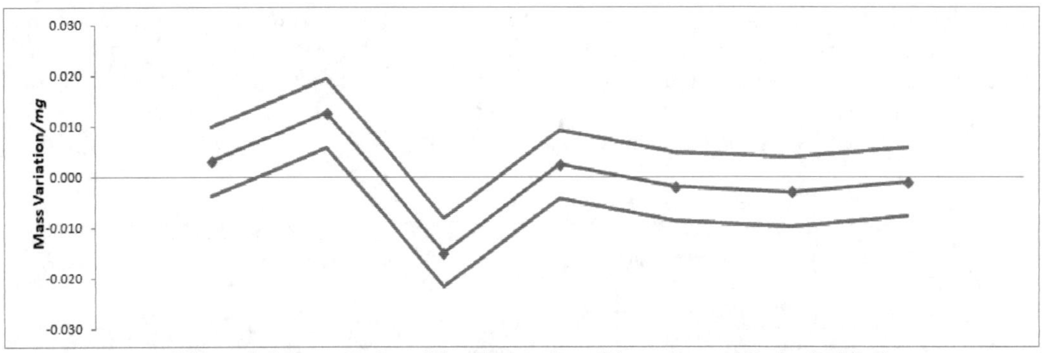

Figure 8. Mass variations (blue line) and confidence interval limits (red line).

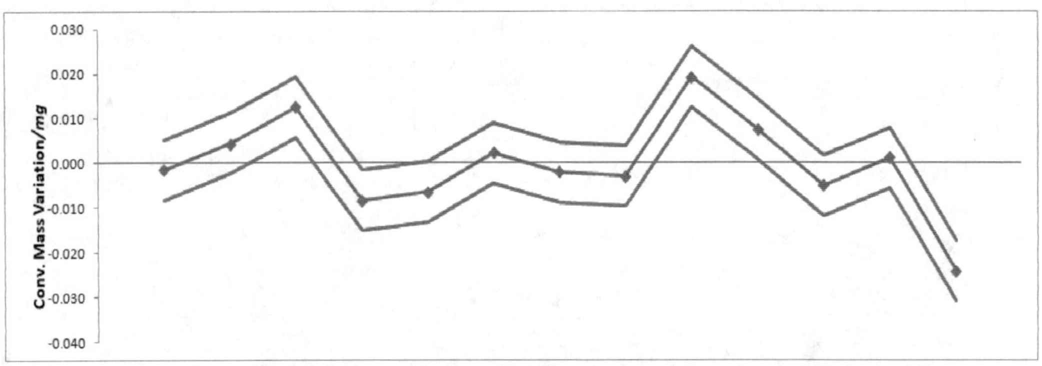

Figure 9. Conventional mass variations (blue line) and confidence interval limits (red line).

6. Conclusion

Inmetro has participated in SIM Kilogram Dissemination Project (SKDP). The mass measurements for the SIM21 standard were carried out at Inmetro between 2018 and 2021.

Although most mass and conventional mass measurements are compatible with the measurement provided by NIST, some are not. However a trend in the measurements is not clear. The observed mass variations demonstrate typical random fluctuations compatible with the variability of a measurement system. Therefore, it can be considered that within 30 μg no mass variation of the SIM21 standard was observed.

Acknowledgments

An author would like to thanks Pronametro/Inmetro.

References

[1] OIML, R111 2004 *Weights of Classes* E_1, E_2, F_1, F_2, M_1,$M_{1\text{-}2}$, M_2, $M_{2\text{-}3}$, M_3 Part 1: metrological and technical requirements edition (E) (Paris: International Organization of Legal Metrology) (https://www.oiml.org/en/publications/recommendations/publication_view?p_type=1&p_status=1).
[2] Nist, Report of Calibration 68407/SIM21-2018, 09/27/2018.
[3] Inmetro, Test Report Dimci 0656/2021, 07/20/2021.
[4] Picard A,Davis R S,Gläser M and Fujii K 2008 *Metrologia* 45 149.

Journal of Physics: Conference Series **2606** (2023) 012027 doi:10.1088/1742-6596/2606/1/012027

A reference system to calibrate low-current meters in the range of picoamperes to microamperes

M V V Pinto[1], R V F Ventura[1], R P Landim[1] and R P David[1]

[1] Instituto Nacional de Metrologia, Qualidade e Tecnologia (Inmetro), Av. Nossa Senhora das Graças, 50, Xerém, Duque de Caxias - RJ - Brasil

E-mail: mvviegas@inmetro.gov.br

Abstract. The calibration of low current meters in the range of picoamperes up to microamperes is important for the electronics industry, for the photometry area, and especially for the ionizing radiation area. In Brazil, traceability is limited to ten microamperes. In this paper, a reference system using Ohm's law method is presented. It was possible to calibrate an electrometer between 2 picoamperes and 20 microamperes, with uncertainties between 0.70 % and 0.000060 %, respectively, as well as to validate the developed system and Ohm's law method by comparing it with the method used to calibrate values above 10 microamperes, in an intralaboratory comparison.

1. Introduction

The calibration of currents in the range below a few nanoamperes is important for semiconductor testing, photometry, and in the area of dosimetry [1,2]. In addition to the areas already mentioned, areas of environmental particulate monitoring and characterizing electrical insulators also require traceability at low currents, in the range of nA and below [3].

At the Brazilian National Metrology Institute (Inmetro), a system providing traceability in currents below the present limit of Inmetro's calibration and measurement capability (CMC) is under development. In order to develop this system, many approaches are possible. The Consultative Committee for Electricity and Magnetism (CCEM) of the International Bureau of Weights and Measures (BIPM) [4] shows three possible methods to achieve the realization of the ampere, a) Ohm's law method [3,5], b) the capacitor-charging method [6–8], and c) the single-electron transport method [9–11]. Another method, the active transresistance (ULCA), is accepted and, despite of not being mentioned by the BIPM in [4], its traceability is based on the quantum Hall effect and the Josephson effect. Nevertheless, it is different from Ohm's law method and it is considered by some authors [3] as another traceability route for low currents. The ultra-low current amplifier (ULCA) [12–14] was developed by the PTB and can be considered the state-of-art for calibration of low currents due to its stability and very low uncertainties (around 10 μA/A). Ohm's law method and the ULCA method can provide traceability in ranges below the μA range, and the other two methods are limited only in case of currents below the nA range. The principal method to provide traceability in below the nA range low direct current is the capacitor-charging method [3] that is the method used by most national metrology institutes participating in key comparison at low currents [15]. This method is used mostly in the range of attoampere to picoampere, and, when the current is increased to the nanoampere range, the voltage ramp needs to have its maximum voltage increased because the largest commercially available gas-filled capacitor available has a capacitance of around 10 nF. This makes it impossible to directly compare the present method (direct current calibration using a current source) with the capacitor-charging method, because the range of values does not overlap. In order to make it possible the comparison of the present method with one of the methods used to provide traceability in a low current range, we need to use a method capable of providing a current of about 10 μA. Only two methods are possible, the ULCA method, and Ohm's law method. As there is no ULCA available at Inmetro, it was selected Ohm's law method. With the use of Ohm's law method and the capacitor-charging method, in the future, we will be able to provide traceability also for current ranges below 10 μA. Additionally, with Ohm's Law method, we can calibrate low current meters from the pA range to the μA range, as we showed in this paper. Ohm's law method can reach current values of the order of picoamperes, however, its uncertainty in this range of picoamperes is greater than the uncertainty obtained with the capacitor-charging method. Despite the best uncertainty, the capacitor-charging method is not capable of reaching currents above the nA range, which would leave an undesirable gap in the traceability of direct current. As a first step to enlarge the range of direct current, the objective of this study is to compare the two methods (the present method, direct calibration using a current source; and the alternative method, Ohm's law) in some points above 10 μA, applying current in direct and reverse mode, and making a complete calibration of one electrometer from 2 pA up to 20 μA.

2. Materials and methods

In order to validate the application of Ohm's law method, as well as the developed automation software [16], an intralaboratory comparison was performed. A multifunction calibrator Fluke model 5720A was used as a standard current and voltage generator to perform this comparison. Were used a set of standard resistors Guildline model 9334A, a PC to control the equipment, and a thermo-hygrometer to monitor the environmental condition. An electrometer Keithley model 6514A was used as unity under test (UUT). All measurements were made in a Faraday cage and the communication with instruments was made with an optically isolated general-purpose interface bus (GPIB).

A software was developed to make this calibration semi-automatically. A complete description of the software used to control and manage the calibration, and the equations for uncertainty evaluation are presented in [16].

Firstly, the multifunction calibrator was connected directly to the electrometer, and the test points minus and plus 10 μA, 15 μA, and 19 μA were calibrated. Each measurement was made with a measuring interval of 60 seconds, with 30 seconds between each measurement for stabilization, and it was repeated eight times. Each read value was the mean value of 10 measurements made by the electrometer. Secondly, the calibrator was connected in series with the standard resistor and the electrometer, and the same test points were calibrated, in the following sequence: Zero, Point, Zero, repeating this sequence eight times, and the result was the average of the eight readings. The results for each point were the difference between the average of the measurement at the calibration point, minus the average of the zeros immediately before and after. This is made to prevent effects caused by changes in the values of zero during the interval of measurements. Figure 1 shows the block diagram of the two compared calibration methods.

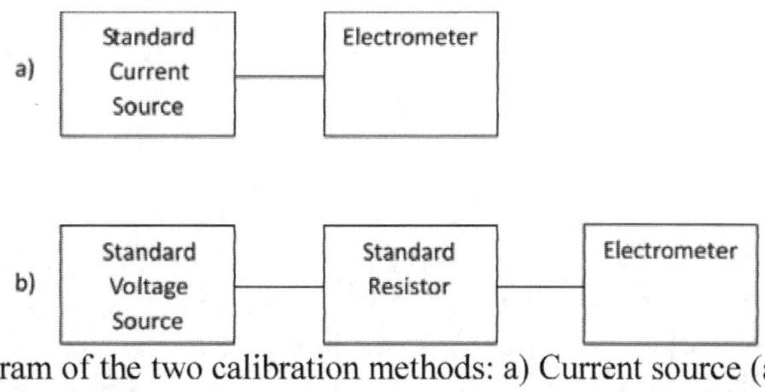

Figure 1 – Block diagram of the two calibration methods: a) Current source (actual method) and b) Ohm's law (alternative method).

In the calibration of the ranges from pA to μA, we used different resistors. The same range was calibrated with different resistors and different applied voltage levels, to verify the reproducibility. Calibration was also performed several times on different days, and in the final uncertainty, the reproducibility contribution was considered. It was possible to calibrate the ranges of 20 pA, 200 pA, 2 nA, 20 nA, 200 nA, 2 μA, and 20 μA with the different resistors used. In order to validate the proposed system, the 20 μA range was calibrated using the actual method (Current source) and Ohm's law method. After the measurements, we estimated the uncertainty of each method, to make the comparison feasible. The criteria used to consider the two methods equivalent was the normalized error criteria ($|e_n| \leq 1$).

3. Results

Seven different calibration rounds were made. Five using Ohm's law method, and two using the current source method. We compare six different points, +10 μA, +15 μA, +19 μA, -10 μA, - 15 μA and -19 μA. All values had normalized errors far below the limit, as can be seen in table 1.

Table 1. Normalized error results.

Measured Value (μA)	-19	-15	-10	10	15	19
Normalized error $|e_n|$	0.21	0.26	0.18	0.21	0.12	0.07

The results obtained for each method for the measurement above 10 μA in the direct and reverse modes are summarized in figure 2:

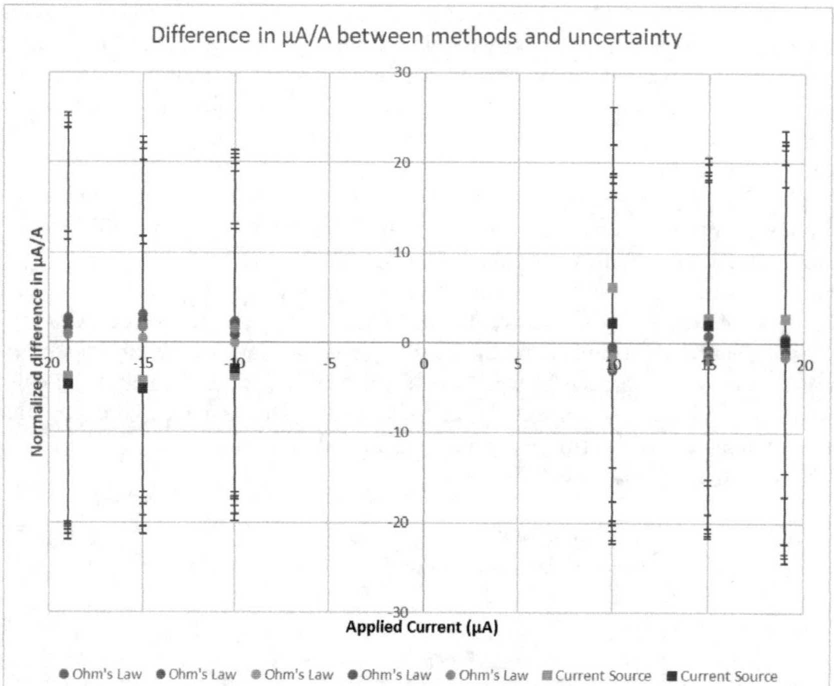

Figure 2 – Uncertainties and the normalized difference between both methods at +/- 10 μA; +/- 15 μA; +/- 19 μA. The five calibrations were performed on different days to verify the reproducibility of Ohm's law method. Two calibrations were performed using the current source method to assess the compatibility between the methods.

As it can be seen in figure 2, the uncertainty obtained with ohm's law method is, in this preliminary evaluation, quite similar to the uncertainty obtained by the current source method. Table 2 shows the relative uncertainties obtained in the range from 2 pA up to 20 μA.

Table 2. Range of relative uncertainties obtained for Ohm's Law method.

Ranges	Uncertainty
>2 μA up to 20 μA	6.0 μA/A up to 23 μA/A
>200 nA up to 2 μA	48 μA/A up to 7 μA/A
>20 nA up to 200 nA	0.096 % up to 0.010%
>2 nA up to 20 nA	0.038 % up to 0.0044%
>200 pA up to 2 nA	0.041 % up to 0.0046%
>20 pA up to 200 pA	0.31 % up to 0.032%
≥2 pA up to 20 pA	0.66 % up to 0.069%

In the normal use of electrometers, the calibration certificate presents a calibration coefficient [17,18], defined by

$$ECC = I_{supplied}/I_{readout} \quad (1)$$

The calibration coefficient must be very close to 1 (one), where the value 1 means that the value read by the instrument is equal to the applied value, that is, the error is zero. Because of this, we chose ECC to show the calibration results, as well. Figure 2 shows the calibration results using the developed system (Ohm's Law-based) for ranges from 20 pA up to 20 μA.

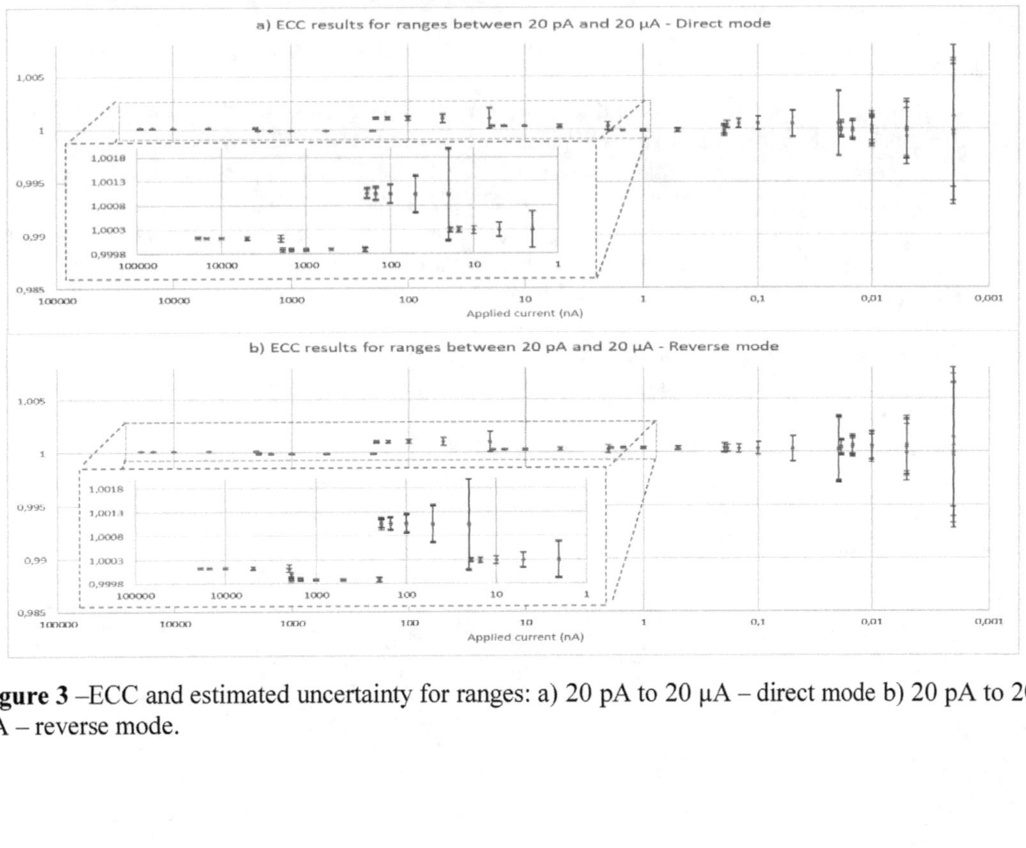

Figure 3 –ECC and estimated uncertainty for ranges: a) 20 pA to 20 μA – direct mode b) 20 pA to 20 μA – reverse mode.

It is possible to observe that the value changes when the range changes, which shows that it is necessary to evaluate each range individually and to correct the read values considering the selected range. The values and uncertainties obtained are satisfactory, with the uncertainties and errors being smaller than the specification of the electrometer manufacturer in all calibrated ranges. The calibration coefficient values for each range remain constant within the ranges, meaning that there is good linearity in each range. It is possible, as recommended by the International Atomic Energy Agency (IAEA) in [17], to define a correction coefficient for the ranges, with one range being the reference range (coefficient equal to 1) and the other ones being corrected related to the reference range.

4. Conclusion

A reference system using Ohm's law method was presented. This system is semi-automatically controlled by software, whose details and uncertainty evaluation details are presented in [16]. In order to validate the application of Ohm's law method, as well as the developed automation software [16], an intralaboratory comparison was successfully performed. It was possible to calibrate an electrometer between 2 picoamperes and 20 microamperes, with uncertainties between 0.70 % and 0.000060 %, respectively. The objective of this work was achieved since the comparison between the two methods was successfully made. The initial steps for validation of the use of Ohm's law method for calibrating low current electrometers at Inmetro were made.

The normalized error results are satisfactory. Each different method was compared, and the criteria were attended. Calibration of direct current between 20 μA and 2 pA was made, and uncertainties are satisfactory. The calibration coefficient values for each range remain constant within the ranges, meaning that there is good linearity in each range. Also, the calibration coefficient value changes when the range changes, which shows that it is necessary to evaluate each range individually and to correct the read values considering the selected range. We can utilize Ohm's law method to enlarge the range of traceability of Inmetro and can offer this service to our clients, mainly industry, research labs, and calibration labs. The next steps are to perform calibrations of different electrometers, gaining experience to later participate in a BIPM key-comparison to validate the calibration and measurement capability.

5. References

[1] Willenberg G-D, Tauscher H N and Warnecke P 2003 A traceable precision current source for currents between 100 aA and 10 pA *IEEE Trans. Instrum. Meas.* **52** 436–9

[2] Fletcher N E, Giblin S P, Williams J M and Lines K J 2007 New Capability for Generating and Measuring Small DC Currents at NPL *IEEE Trans. Instrum. Meas.* **56** 326–30

[3] Giblin S and Tompkins J 2020 From Counting Electrons to Calibrating Ammeters: Improved Methodologies for Traceable Measurements of Small Electric Currents

[4] BIPM 2019 Mise en pratique for the definition of the ampere and other electric units in the SI *SI Brochure-Appendix 2* 3–8

[5] Jarrett D G, Payagala S U, Fitzgerald R, Bergeron D E, Cessna J T, Perera C J W and Zimmerman N M 2020 Ohm's Law Low-current Calibration System for Ionization Chambers *CPEM Digest (Conference on Precision Electromagnetic Measurements)* vol 2020-Augus

[6] vandenBrom H E, delaCourt P and Rietveld G 2005 Accurate Subpicoampere Current Source Based on a Differentiating Capacitor With Software-Controlled Nonlinearity Compensation *IEEE Trans. Instrum. Meas.* **54** 554–8

[7] Ehtesham B, Bist P S and John T 2017 Development of an Automated Precision Direct Current Source for Generation of pA Currents Based on Capacitance Charging Method at CSIR-NPL *MAPAN* **32** 17–22

[8] Kim W-S, Kim M-S, Kim K-T, Chong Y, Kim H-J, Kim M and Tauscher H N 2008 A high precision calibration setup for low-current meters in the range of 10 pA to 1 fA *2008 Conference on Precision Electromagnetic Measurements Digest* (IEEE)

[9] Callegaro L 2020 A quantum ampere *tm - Tech. Mess.* **87** 258–65

[10] Scherer H and Schumacher H W 2019 Single-Electron Pumps and Quantum Current Metrology in~the~Revised SI *Ann. Phys.* **531** 1800371
[11] Giblin S P, Kataoka M, Fletcher J D, See P, Janssen T J B M, Griffiths J P, Jones G A C, Farrer I and Ritchie D A 2012 Towards a quantum representation of the ampere using single electron pumps *Nat. Commun.* **3**
[12] Scherer H, Drung D, Krause C, Gotz M and Becker U 2019 Electrometer Calibration with Sub-Part-Per-Million Uncertainty *IEEE Trans. Instrum. Meas.* **68**
[13] Drung D, Gotz M, Pesel E and Scherer H 2015 Improving the Traceable Measurement and Generation of Small Direct Currents *IEEE Trans. Instrum. Meas.* **64**
[14] Fitzgerald R, Bergeron D E, Giblin S P, Jarrett D G, Judge S M, Michotte C, Scherer H and Zimmerman N M 2020 The next generation of current measurement for ionization chambers *Appl. Radiat. Isot.* **163**
[15] Willenberg G-D 2013 EUROMET.EM-S24: Supplementary comparison of small current sources *Metrologia* **50** 1002
[16] Pinto M V V, Ventura R V F, Landim R P and David R P A software to calibrate low-current meters in the range of picoamperes to microamperes *Submitt. to Metrol. 2021*
[17] INTERNATIONAL ATOMIC ENERGY AGENCY 2009 *Calibration of Reference Dosimeters for External Beam Radiotherapy* (Vienna: INTERNATIONAL ATOMIC ENERGY AGENCY)
[18] Becker P and Meghzifene A 2002 *Pilot study to verify the calibration of electrometers* (International Atomic Energy Agency (IAEA))

A software to calibrate low-current meters in the range of picoamperes to microamperes

M V V Pinto[1], **R V F Ventura**[1], **R P Landim**[1] **and R P David**[1]

[1] Instituto Nacional de Metrologia, Qualidade e Tecnologia (Inmetro), Av. Nossa Senhora das Graças, 50, Xerém, Duque de Caxias - RJ - Brasil

mvviegas@inmetro.gov.br

Abstract. Among the main pillars of industry 4.0 are big data, artificial intelligence, and the industrial internet of things (IIoT) which, together with industrial automation, allow for significant operational advances. In the field of calibration, where processes and procedures must be standardized, automation diminishes the variability caused by human intervention. In addition, the process is no longer limited to the time the technician is present, speeding up deliveries and improving laboratory efficiency and productivity. In this work, a software for the calibration of low current meters is presented. Also, a mathematical model and uncertainty analysis related to the adopted calibration method is presented. It was possible to calibrate an electrometer in a fully automated way, with minimal intervention by the technician. This software was successfully validated through an intralaboratory comparison.

1. Introduction

The calibration of current detectors, electrometers, electrochemical transducers, and dosimeters for ionizing radiation demands accurate measurements of low current [1].

At the Brazilian National Metrology Institute (Inmetro), a system to provide traceability in currents below the present limit of its calibration and measurement capability (CMC) is under development. Inmetro's present limit is 10 μA, and a calibration system needs to be developed in order to achieve the pA range. This system, as presented in [2], can be used to calibrate low current meters from 2 pA up to 20 μA in direct current using Ohm's law method. In this paper, the details of the software and the uncertainty evaluation were discussed, while in [2] we present the validation of the system and the results of the calibration of an electrometer in this range of values. In the ionizing radiation area, calibration of electrometers is very important and, in 2019, the Consultative Committee for Ionizing Radiation (CCRI) and the Consultative Committee for Electricity and Magnetism (CCEM) of the International Bureau of Weights and Measures (BIPM) started a joint Task Group on this topic (CCRI-CCEM, 2020) [3], which Inmetro joined this year.

Many challenges are involved in low current measurements. According to Babita et al. [1], any charged body with some potential, like the human body, can induce a noise current in the order of nA, caused by electrostatic coupling. Because of that, it is important to take some precautions to execute low current measurements, like putting the system in a Faraday cage, maintaining all connections free of contaminants, using low noise cables, and putting the cables on a surface free of vibrations, to minimize triboelectric effects. To enable this, software which can manage all the calibration process, control the equipment, and take the readings without intervention from the staff are needed. This permits the calibration to be made without interference or with minimal interference from the calibration technic and prevents human errors in the measurement process. The developed software was designed to allow flexibility in using standards, and in communicating with different equipment from different manufacturers, as is usually necessary for modern calibration laboratories [4].

In order to make the measurements, Ohm's Law method was used. According to [5,6] this method can be used to achieve the realization of the ampere (amongst other two methods).

2. Materials and methods

Ohm's law method consists of one voltage source, and one standard resistor in series with the electrometer or low current meter, the unity under test (UUT). We use a PC running the software, which makes all the control of equipment, records the measurement data, processes the data to estimate the uncertainty, and generates the measurement report. All measurements were made in a Faraday cage and the communication with instruments was made with an optically isolated general-purpose interface bus (GPIB).

The software also controls the measurement process, the time of measurement, the time to stabilization, and the change of measurement points and ranges. The user only needs to make the connections and, as requested by the software, change the standard resistor. All information about the standards, like the calibration data, temperature coefficients, voltage coefficients, and all data needed to estimate the uncertainty are stored in a database. The measurement data are stored in the same database to allow quick interaction between the developed software and the other tools used in the laboratory. The software was developed using LabVIEW® 2014, and the use of the software minimizes human intervention and other influences which can cause measurement errors. This feature improves the repeatability and reliability of measurements.

3. Uncertainty evaluation

Uncertainty evaluation and control of equipment are made by the software. The mathematical model is shown below:

$$I = ((V_{cal} + \delta vb_{el}) / (R_{std} + \delta V_r + \delta T_r + \delta S_r)) + \delta r_{st} + \delta n_{cb} + \delta r_{el} \qquad (1)$$

Where:

I is the value of the direct current being measured,

V_{cal} is the voltage applied by the multifunction calibrator,

δvb_{el} is the correction due to the voltage burden of the electrometer, obtained from the manufacturer's manual,

R_{std} is the resistance of the standard resistor,

δV_r is the correction due to the voltage coefficient of the standard resistor,

δT_r is the correction due to the temperature coefficient of the standard resistor,

δS_r is the correction due to the standard resistor stability,

δr_{st} is the correction due to measurement reproducibility,

δn_{cb} is the correction due to the cables, for triboelectric and other effects, from the literature,

δr_{el} is the correction due to the electrometer resolution.

All calculations were performed using the methodology presented in [7].

4. The software

The interface to the user is similar to the other tools used in the laboratory, with administrative information, like the client, number of the administrative process and the name of the technician responsible to perform the calibration. The information about the unity under test (UUT) is available on the same screen, like the manufacturer, the model, serial number, and ID. The ranges available, GPIB address and calibration points in each range are on the same screen. Information about the standards used to generate the voltage values and the resistors are available too. While opening the software, the screen is loaded with information from the database. The user can then select the customer, enter the calibration process number, and the name of the technician responsible to perform the calibration. They can also select the manufacturer and model of the equipment to be calibrated, which will automatically have its ranges, configuration, and reading commands loaded into the software. The user can inform the identification code (ID) of the equipment, if any, and the GPIB address. The loaded commands can be freely modified by the user, to configure the equipment being calibrated in the most suitable way, or according to the customer's request. The user can store the calibration points and ranges to use the same set of calibration points for different equipment of the same client, or different clients. On the measurement screen, the user can configure the stabilization time, the reading time, the number of readings for each point, and choose whether to calibrate all points at once, starting from a certain point, or whether to calibrate just a single point. The time count, current reading, and the range and point currently being calibrated are available for the technician to view. The software informs, through system messages, what is being done at the moment. All individual readings are available for the technician to view, and zero-corrected values are also shown, together with the absolute error. The relative error is shown in a graph, where the technician can check the stability of the measurements. Figure 1 shows two software screens, and figure 2 shows an auxiliary screen with resistance standards information.

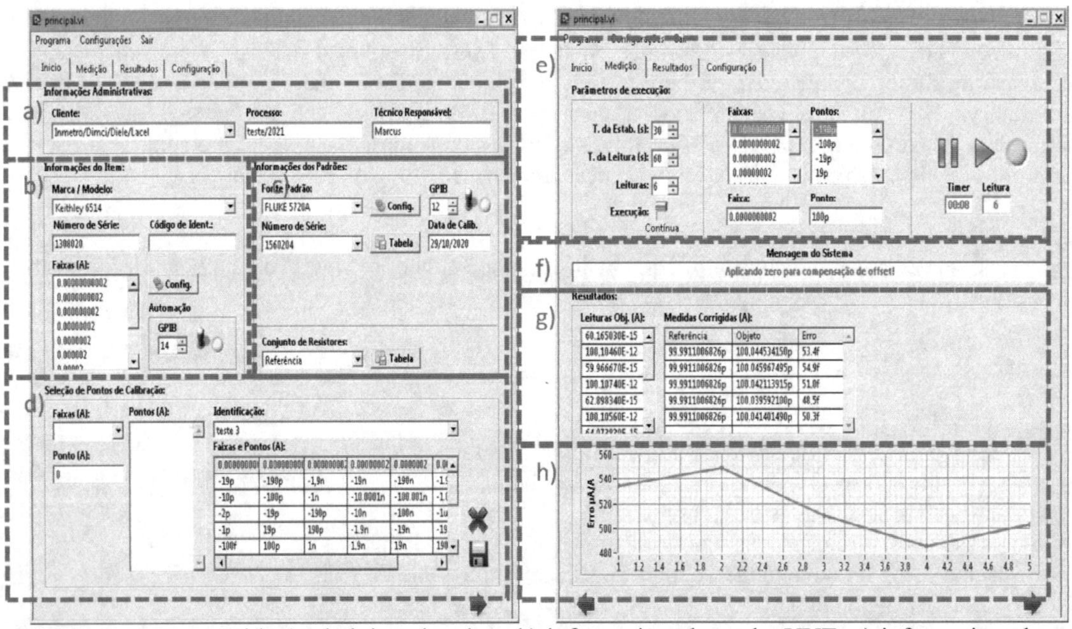

Figure 1 – Screens with a) administrative data; b) information about the UUT; c) information about standards; d) information about calibration points and ranges; e) calibration parameters, like stabilization time, reading time, number of readings, and ranges and calibration points; f) System messages; g) readings and calculated values; and h) graphic showing the calculated relative error for each measurement.

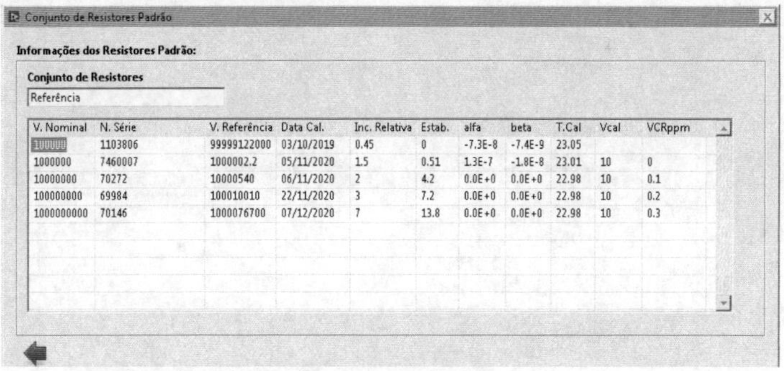

Figure 2 – Auxiliary screen showing the resistance parameters.

Figure 3 shows an auxiliary screen with the communication commands used by the software to control and configure the UUT and it can be changed by the user. The commands used in each calibration were registered in a log file. The commands have been separated into start commands, range selection commands, measurement execution commands, mean read, and standard deviation read commands, and end commands. This structure is flexible enough to allow different equipment from different manufacturers to be calibrated, requiring only the appropriate commands to be entered by the user.

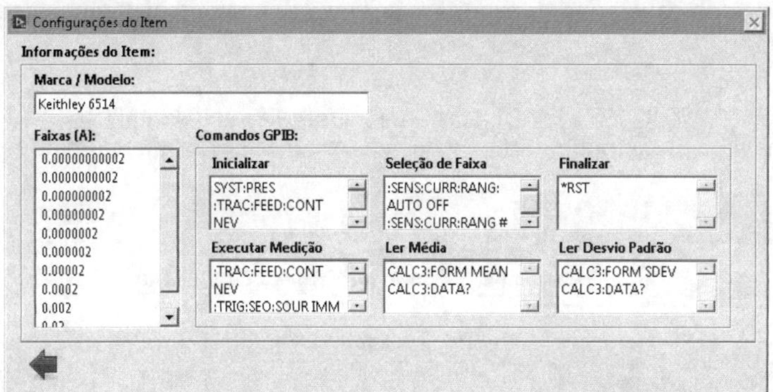

Figure 3 – Auxiliary screen with the communication commands to control the UUT.

Figure 4 shows the screen where the user can select the calibration data and each measurement archive, as well to export the data to generate the calibration report. The ".ltr" file brings all raw calibration data (for statistical analysis and audit purposes) plus the UUT information, like its serial number (which is also in the "ltr" file name, right after the client name). Files can be filtered by customer, by item, and by serial number.

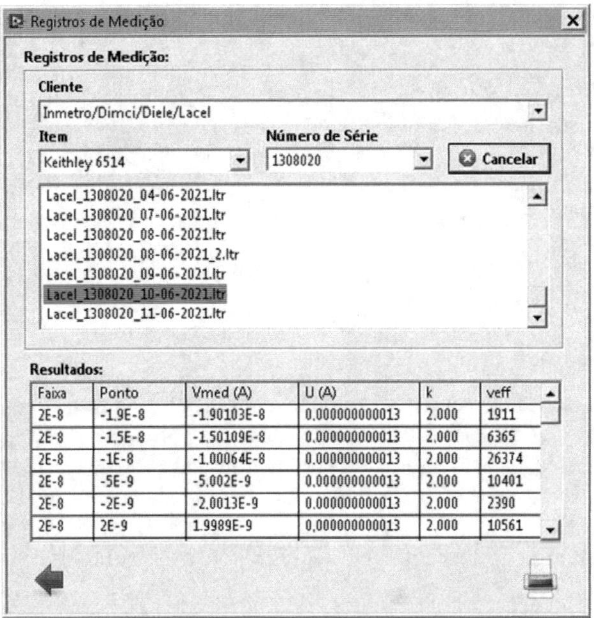

Figure 4 – Screen with calibration data. Archives show all recorded calibration data for each selected client and the UUT.

Figure 6 shows an example for the calibration report, bringing the client's name, the UUT complete identification, the date of calibration, the technician responsible for the calibration execution, and so on.

REGISTRO DE MEDIÇÕES - LOW IDC
Laboratório de Calibração em Metrologia Elétrica - Lacel

Cliente: Inmetro/Dimci/Diele/Lacel
Objeto Calibrado: Amperímetro Digital - Fabricante: Keithley - Modelo: 6514 - Nº Série: 1308020
Cod. Identificação: Não Identificado
Número do Processo : teste/2021
Data da Calibração : 10/06/2021
Técnico Responsável: Marcus
Registro: Lacel_1308020_10-06-2021.ltr

Padrões Utilizados:
Fonte Dc Padrão: FLUKE 5720A - Nº de série: 6695201
Conjunto de resistores padrão: Referência

Faixa [nA]	Ponto [nA]	Valor Medido [nA]	Erro [μA/A]	Incerteza [nA]	k	veff
0.02	-0.019	-0.01901	542.1	0.000013	2.00	1000
0.02	-0.015	-0.015011	726.7	0.000013	2.00	1000
0.02	-0.01	-0.010006	640.0	0.000013	2.00	1000
0.02	-0.005	-0.005002	400.0	0.000013	2.00	1000
0.02	-0.002	-0.002001	650.0	0.000013	2.00	1000
0.02	0.002	0.001999	-550.0	0.000013	2.00	1000
0.02	0.005	0.004999	-200.0	0.000013	2.00	1000
0.02	0.01	0.01	-10.00	0.000014	2.00	1000
0.02	0.015	0.014997	-180.0	0.000013	2.00	1000
0.02	0.019	0.018996	-189.5	0.000013	2.00	1000
0.2	-0.19	-0.190066	347.4	0.000061	2.00	1000
0.2	-0.15	-0.150049	324.7	0.000061	2.00	1000
0.2	-0.1	-0.100038	377.0	0.000061	2.00	1000
0.2	-0.05	-0.050017	334.0	0.000061	2.00	1000

Figure 5 – Example of calibration report generated by the calibration software.

5. The software validation

This software was successfully validated through an intralaboratory comparison. A complete calibration over the entire current range covered by the method was performed, using the software, with satisfactory results. Details of this calibration and intralaboratory comparison are described in [2].

6. Conclusion

The objective of this work was fully achieved since the software to control and manage the calibration of low current meters in the range of picoamperes to microamperes was made and it is working properly.

The developed software was presented, and the results are fully satisfactory. The software can minimize the necessity of staff intervention, and it can be used without the need for constant monitoring and can even continue measuring outside the technicians' office hours, increasing efficiency and productivity.

Calibration of direct current between 20 μA and 2 pA was made, and uncertainties are satisfactory. We can utilize the developed software at Inmetro and can offer this service to our clients, especially industry, research labs, and calibration labs.

7. References

[1] Babita, Sharma D K, Satish, Ansari M A and Saxena A K 2014 A versatile automation program using LabVIEW for low dc current measurement *J. Sci. Ind. Res. (India).* **73**

[2] Pinto M V V, Ventura R V F, Landim R P and David R P A reference system to calibrate low-current meters in the range of picoamperes to microamperes *Submitt. to Metrol. 2021*

[3] Fitzgerald R, Bergeron D E, Giblin S P, Jarrett D G, Judge S M, Michotte C, Scherer H and Zimmerman N M 2020 The next generation of current measurement for ionization chambers *Appl. Radiat. Isot.* **163**

[4] Tasic T, Bojkovski J, Pusnik I and Drnovsek J 2000 Reliability design of a calibration laboratory automation software *Conference on Precision Electromagnetic Measurements. Conference Digest. CPEM 2000 (Cat. No.00CH37031)* pp 600–1

[5] Callegaro L, D'Elia V, Capra P P and Sosso A 2007 Techniques for Traceable Measurements of Small Currents *IEEE Trans. Instrum. Meas.* **56** 295–9

[6] BIPM 2019 Mise en pratique for the definition of the ampere and other electric units in the SI *SI Brochure-Appendix 2* 3–8

[7] BIPM 2008 *JCGM 100: Evaluation of Measurement Data - Guide to the Expression of Uncertainty in Measurement*

Bilateral comparison of dc voltage at 10 V between UTE and INMETRO

D Slomovitz [1], R P Landim [2], H de Souza [3], L Trigo [3]

[1] UDELAR, Uruguay
[2] INMETRO, Brazil
[3] UTE, Uruguay

E-mail: dslomo@gmail.com

Abstract: A bilateral comparison between the Instituto Nacional de Metrologia, Qualidade e Tecnología (INMETRO) and the Administración Nacional de Usinas y Trasmisiones Eléctricas (UTE) was performed during 2019 and 2020. The National Laboratory Standards used were two Josephson Voltage Standards at INMETRO (a CJVS and a PJVS), and a bank of multiple Zeners at UTE, both at 10 V level. The traveling standards were two commercial Zener based sources. The INMETRO was the pilot laboratory and the comparison error was (-1.79 ± 2.55) parts in 10^7, k=2.

1. Introduction

The discovery of the Josephson effect, in 1962, lead to the use of Josephson Voltage Standards (JVS) as the volt representation in the International System of units (SI) and to the first practical 1 V Josephson standards production in 1985 [1]. The SI redefinition, in 2019, allowed the use of the JVSs as the volt realization [2].

In the early 1990s, shortly after the production of the first 1 V JVS, the Bureau International des Poids et Mesures (BIPM) started the first JVSs key-comparison, at 1.018 V (BIPM.EM-K10.a) [3]. Such key-comparisons are performed on-site, directly comparing two JVSs and allow to reach a few parts in 10^9 of expanded uncertainty. In this case, the pilot laboratory needs to ship its entire JVS system to the participant(s) laboratory(ies). However, when shipping an entire JVS is not an option, another way that NMIs can check the coherence of JVSs is using Zener diode-based dc voltage references (Zeners) as travelling standards in international comparisons. In this case, pressure and temperature corrections need to be made, in order to allow reaching a few parts in 10^8 of expanded uncertainty [4]. The National Metrology Institutes (NMI) which do not have a JVS usually rely on Zeners as travelling standards to assure traceability to the SI volt via JVSs calibrations and comparisons with other NMIs. For both situations, the BIPM also provides specific key-comparisons (BIPM.EM-K11.a&b, for 1.018 V and 10 V, respectively) [4].

Three key-comparisons, linked to the above mentioned ones were performed in the Inter-American Metrology System (SIM) Regional Metrology Organization, so far: the SIM.EM.BIPM-K11.b (in 2006, using Zeners for an indirect on-site comparison, resulting in 4.32 parts in 10^9 of expanded uncertainty, k=2) [5], the SIM.EM.BIPM-K10.b (in 2007, in a direct on-site JVS comparison, reaching 2.07 parts in 10^{10} of expanded uncertainty, k=2) [6] and the SIM.EM.BIPM-K10.b.1 (in 2009, in a direct on-site JVS comparison, reaching 1.48 parts in 10^{10} of expanded uncertainty, k=2) [7]. There are no records of off-site indirect (via Zeners) JVS key-comparison within the SIM, like the most recent one, performed between the BIPM and the KEBS (Kenya), in 2017 (BIPM.EM-K11.a&b, which resulted in 1.55 parts in 10^7 of expanded uncertainty, k=2) [8]. This kind of comparison is important because most of South America NMI base their National Voltage Standard on this type of source.

This report presents the results of a bilateral comparison on dc voltage performed by the Uruguayan Designed National Metrology Institute (UTE) and the Brazilian National Metrology Institute (INMETRO). The nominal value was 10 V and the traveling standards were two Zener based electronic references. The measurements were conducted from August 2019 to January 2020. The responsible for the measurements were Heriguatí de Souza at UTE and Regis Pinheiro Landim at INMETRO. This comparison was a pilot study-case, since it was a training exercise for a future dc voltage key-comparison within the CIPM MRA [9], intended to happen with more NMIs.

The objective of this comparison was to compare the measurement capabilities of the mentioned two NMIs, in the field of calibration of dc voltages. These NMIs establish the degree of equivalence between their national measurement standards by performing this bilateral comparison.

2. Traveling Standard

The traveling standards were two Zener-type electronic dc reference standards, manufactured by Fluke, model 732B, property of UTE, labeled as #03 and #04. This model has a 1.018 V and a 10 V output voltages, but in this comparison only the 10 V output voltage was considered. The pressure dependences of each traveling standards were determined by UTE. These values were 7.4 nV/hPa and 5.7 nV/hPa, respectively. The influence of the ambient temperature was also determined by UTE for both standards, around -0.1 μV/K, with 0.04 μV/K of uncertainty. No correction was done due to this influence factor, but it was included in the uncertainty budget.

During transport, the standards were battery powered. Additional external batteries were added to extend the time up to five days. This was enough to avoid loss of calibration status (and the many years of calibration data). After arrival at each participant's laboratory, the front panel LOW BAT and IN CAL indicators were checked to assure that there was not power loss.

Before starting the calibrations, the standards were stabilized at room temperature for at least three days in the measuring room. In case the standards were disconnected from the ac power line during the measurements, they were reconnected when measurements were not taking place.

3. Measurement conditions

Participants measured the 10 V output voltage of each traveling standards, together with the resistance of the respective thermistor, the air pressure, the ambient temperature and the humidity.

At INMETRO the measurements were carried out with the standards operated with their internal batteries, i.e., disconnected from the AC line power. To allow the standards to stabilize, battery operated measurements started at least after one hour of disconnection from the ac line power. The duration of the disconnection was restricted to six hours or less.

At UTE, a continuous automatic measuring system was used, taking measurements every fifteen minutes. So, traveling standards stayed connected to ac power. Anyway, studies of the behavior of these two units under ac or battery powered showed very good agreement between both connections (one order lower than UTE uncertainty).

At INMETRO, the measurement shield was made as follows: the Zener GUARD terminal was connected to its CHASSIS, which was connected to the CHASSIS of the JVSs (either the Conventional Josephson Voltage Standard – CJVS or the Programmable Josephson Voltage Standard – PJVS). The 10-V measurement cable copper shield was connected to the same JVS CHASSIS point at the scanner side (not at the Zener side to avoid ground loops). This measurement shield was connected to the ground through the JVS ground (which is connected to the laboratory individual ground network outside the building, through one single wire). More details regarding INMETRO CJVS and PJVS systems can be seen in [10] and [11], respectively.

At UTE, the GUARD terminal was connected to the GROUND of the measuring system, and the CHASIS terminal to GROUND of the power network.

4. Drift of traveling standards

The drift of each of the two traveling standards was calculated from the initial and final values at UTE. Shorter periods cannot allow good estimation of drift because of the intrinsic noise of this kind of sources. With the best linear adjust for the #03 unit, the drift was found to be -160 nV/month, and for the #04 one, it was found to be -30 nV/month.

5. Results of the bilateral comparison on dc voltage

The sequence of measurements was UTE-INMETRO-UTE. All measurements took four months. Each laboratory reported the value for both traveling standards.

Table 1. Average results informed by each laboratory.

Date	Laboratory	Value (V)	Uncertainty (μV)
Standard #3			
30/08/2019	UTE	10.0000447	2.54
19/10/2019	INMETRO	10.00004673	0.25
29/12/2019	UTE	10.00004412	2.54
Standard #4			
30/08/2019	UTE	10.00006901	2.54
19/10/2019	INMETRO	10.00007077	0.25
29/12/2019	UTE	10.00006893	2.54

Figure 1, in Appendix I, show all measurements of traveling standard #04, and table 1 shows the results of the average values measured by each laboratory with their declared uncertainties. The dates in the table are the average date for each case.

For the calculation of the differences between UTE and INMETRO, a linear drift was assumed for both standards. Table 2 shows the differences between the average values of the two measurements at UTE and INMETRO, taking into account the date differences between each measurement.

Table 2. Comparison results.

Laboratory	Average value (V)	Uncertainty (μV)
Standard #3		
UTE	10.00004449	2.54
INMETRO	10.00004673	0.25
DIFFERENCE	-2.24E-06	2.55
Standard #4		
UTE	10.00006898	2.54
INMETRO	10.00007077	0.25
DIFFERENCE	-1.79E-06	2.55

6. Uncertainty

The uncertainty budgets of both laboratories are detailed in table 3 for UTE and table 4 for INMETRO (see Appendix I). INMETRO atmospheric pressure was taken as the reference and UTE results were corrected according that condition. The differences between ambient temperature were inside 1 K, so no correction was done due to this factor, However, it was included in the uncertainty budget.

UTE expanded uncertainty, at approximately 95 % of probability, was 2.54 μV, and INMETRO 0.25 μV. The large difference is mainly due to their reference standards. The normalized error was calculated according to (1).

$$E_n = \left| Vute\text{-}Vinmetro \right| / \sqrt{(Uute^2 + Uinmetro^2)} \quad (1)$$

For the traveling standard #03, E_n=0.88, and for #04, E_n=0.70. As both values are lower than 1, the voltage differences between the two laboratories are consistent with the stated uncertainties.

7. Calibration equipment

The dc voltage national standard, at UTE, is one of two banks of 140 Zeners [12-13]. The set is divided in four groups of 35 each, and the average voltage of each one is available through external terminals. This allows comparisons between different parts of the bank to observe differences in the drift and noise of each one of them with respect to the average. Continuous measurements are done each 15 minutes. Its behavior has been observed since 2004, for more than 15 years. It was calibrated in four occasions, being the last one at BIPM in 2015. Therefore, there was not any external comparison since the last five years. The maximum deviation related to its initial value is inside +0 μV, −1.5 μV.

At INMETRO, two Josephson were used. In order to check the consistency between the CJVS (Conventional JVS) and the PJVS (Programmable JVS) systems of INMETRO, the measurements were done using both systems. Figure 2 (see Appendix I) shows the excellent consistency between them. Five measurements were done each day. The Zeners were disconnected from the ac power early in the morning and measurements were initiated at least one hour later. Measurements were taken either in the afternoon, or in the morning, to avoid any tendency due to the time of disconnection from the electrical network.

INMETRO has self-traceability and UTE traceability is from BIPM. The last calibration of its National Standard was five years ago, and since then, no external comparison was performed.

8. Conclusions

A useful interlaboratory comparison of dc voltage between UTE and INMETRO has been conducted. The range was 10 V. Methods and standards were very different for each participant as declared uncertainties. As traveling standards, two FLUKE Zeners, model 732B, were used.

The uncertainty bars in INMETRO measurements demonstrate consistency in all measurements, including between its PJVS and CJVS systems. The uncertainty bars in the UTE measurements also demonstrate consistency in its measurements. Clearly, there is an offset between the set of values measured at UTE and the set of values measured in INMETRO, but this difference is covered by the combined uncertainties. The results obtained support the measurement and calibration capabilities (CMCs) published in the CIPM's key comparisons databases (MRA).

9. References

[1] Benz S P and Hamilton C A, 2004, Application of the Josephson effect to voltage metrology, *Proceedings of the IEEE*, **92**, no. 10, pp. 1617-1629.

[2] Landim R P, Carvalho H R and de Oliveira V C, 2021, The redefined SI and the electromagnetic quantities in detail – part I: current and voltage, *Journal of Physics: Conference Series*, **1826**.

[3] Reymann D and Witt T J, April 1993, International comparisons of Josephson array voltage standards, *IEEE Transactions on Instrumentation and Measurement*, **42**, no. 2, pp. 596-599.

[4] Technical protocol for BIPM.EM-K11a&b comparisons, *BIPM*, 2018.

[5] Hernández D, Navarrete E, Avilés D and Yi-hua Tang, 2007, Final report on bilateral comparison of dc voltage references between CENAM and NIST (SIM.EM.BIPM-K11.b), *Metrologia*, **44**, no. 1A.

[6] Yi-hua Tang and Wood B, 2008, Comparison of the NIST and NRC Josephson voltage standards (SIM.EM.BIPM-K10.b), *Metrologia*, **45**, no. 1A.

[7] Yi-hua Tang, Landim R, Afonso E and Ferreira V, 2010, Comparison of the INMETRO and NIST Josephson Voltage Standards (part of the ongoing regional key comparison SIM.EM.BIPM-K10.b.1), *Metrologia*, **47**, no. 1A.

[8] Solve S, Chayramy R, Aguko G O and Stock M, 2018, Bilateral comparison of 1.018 V and 10 V standards between the KEBS (Kenya) and the BIPM, January to March 2018, (part of the ongoing BIPM key comparison BIPM.EM-K11.a and b), *Metrologia*, **55**, no. 1A.

[9] Measurement comparisons in the CIPM MRA Guidelines for organizing, participating and reporting CIPM MRA-G-11, *BIPM*, ver. 1.0, 11/01/2021.

[10] Landim R , Tang Y, Afonso E and Ferreira V, 2011, A 10 V Josephson Voltage Standard Comparison Between NIST and INMETRO as a Link to BIPM, *IEEE Transactions on Instrumentation and Measurement*, **60**, no. 7, pp. 2353-2358.

[11] Landim R P, Alzamora M, Ferreira V and Afonso E, 2012, Inmetro 10 V Programmable Josephson Voltage Standard implementation, 2012, *Conference on Precision electromagnetic Measurements*, pp. 176-177.

[12] Slomovitz D, Trigo L and de Souza H, 2010, Long term behavior of a multi-zener 10 V voltage standard, *2010 Conference on Precision electromagnetic Measurements* , pp. 211-212.

[13] Souza H, Trigo L, Slomovitz D, 2016, Eleven years of monitoring an ultra-stable 10 V zener-based voltage standard, *Journal of Physics: Conference Series*, ***733***, pp. 1-6.

APPENDIX I

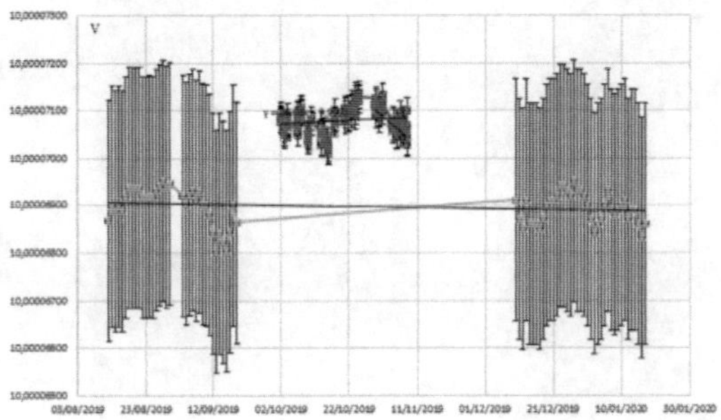

Figure 1. Results of UTE (green) and INMETRO (purple) for the traveling standard #04.

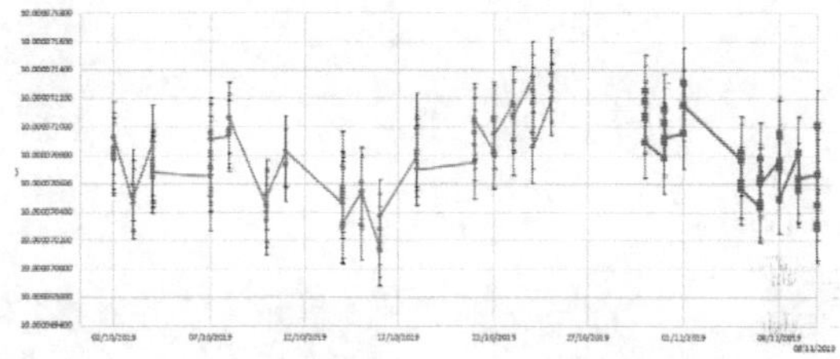

Figure 2. Results of the INMETRO PJVS (blue) and CJVS (red) measurements for the traveling standard #04.

Table 3. Uncertainty budget of UTE

Quantity	Value		Uncertainty		Type	Distribution	Standard uncertainty		Sensitivity coefficient		Uncertainty contribution
Xi									ci		µV
Reference standard	10.0006084	V	0.65	µV	B	Normal	0.65	µV	1		0.65
Drif of reference standards	0.0	µV	0.9	µV	A	Normal	0.9	µV	1		0.9
UUT-REF	-539.5	µV	0.31	µV	A	Normal	0.31	µV	1		0.31
Offset voltages	0.02	µV	0.5	µV	A	Normal	0.5	µV	1		0.5
DMM accuracy	0	µV	0.3	µV	B	Rectangular	0.17	µV	1		0.17
DMM resolution	0	µV	0.01	µV	B	Rectangular	0.0058	µV	1		0.006
Atmosferic presure	1009	hPa	10	hPa	B	Rectangular	5.8	hPa	0.0074	µV/hPa	0.04
Temperature	22.6	°C	1	K	B	Rectangular	0.58	K	0.04	µV/K	0.02
UUT power voltage source	0	µV	0.1	µV	B	Rectangular	0.058	µV	1		0.06
Combinded uncertainty											1.27
Rounded expanded measurement uncertainty (k = 2)											2.54

Table 4. Uncertainty budget of INMETRO

Uncertainty Components	Type	10 V	Deg. Of Freedom
Null voltage variability (noise, nV)	A	119.4	37
Zero Offset (u_z, nV)	A	11.6	50
Microwave frequency (u_f, nV)	B	1.2	∞
Uncorrected leakage (u_L, nV)	B	0.0	∞
Combined Standard Uncertainty (u_c, nV)	AB	120.0	
v eff		37.7	
k		2.07	
Overall Uncertainty (U, nV)		248	

Journal of Physics: Conference Series **2606** (2023) 012030 doi:10.1088/1742-6596/2606/1/012030

Development of an upgraded E-field strength probe to be used as a transfer standard for metrological proposes

Juan C. Mateus Sánchez[1], Donglin Meng[2]

[1]National Institute of Metrology, Quality and Technology – Inmetro
Duque de Caxias, CEP 25250-020, RJ – Brazil.

[2]National Institute of Metrology of China – NIM China
No. 18, Bei San Huan Dong Lu, Chaoyang District, Beijing, 100029, China.

[1]jcsanchez@inmetro.gov.br, [2]mengdl@nim.ac.cn

Abstract. This paper describes the development of an E-field strength sensor probe based on schottky diode technology to be used as a transfer standard for metrological proposes. This sensor probe was designed in a joint project between the National Metrology Institutes of Brazil (Inmetro) and China (NIM China) to bring traceability to many facilities generating E-field strength of both Institutes. The developed sensor allows transferring the E-field traceability from a μ-TEM Cell used as a primary standard to secondary standards like TEM and G-TEM Cells in the frequency range of 500 kHz to 2 GHz and at E-field intensities from 3 V/m to 70 V/m, with low uncertainty and a good dynamic range. The working principle of the new measuring system is similar to an old system that is not manufactured anymore, but it was re-designed in order to improve its performance and characteristics obtaining, as a result, a reduction in size of more than 10 times. Stability test results are also presented, showing that the new sensor is suitable for future inter-comparisons.

1. Introduction

Electric field (E-field) strength is one of the quantities used in the radiofrequency area. E-field strength is not a basic SI-Unit and cannot be realized as a material object such as the kilogram; nevertheless, to generate known electromagnetic field strength as a primary standard the national metrology laboratories use different infrastructures and techniques [1].

To compare the electric field strength generated in such facilities around the world, the Consultative Committee for Electricity and Magnetism, Radio Frequency (CCEM) starting in 1999 the comparison "CCEM-RF-K.20" which officially started in September 2000 with 13 participants. The comparison was based on the circulation of one standard field measuring system (E-field probe), to assess the E-field strength generated by many facilities in the frequency range of 10 MHz to 1 GHz.

The comparison was interrupted many times but it continued and ended finally in September 2004. Many of the interruptions were due to the fact that the circulating E-field probe suffered instabilities and presented drift during the 5 years of duration of the key comparison. Anyway, the results were considered consistent and support the equivalence of national standards laboratories.

Today, the measuring system in Figure 1 is used by NIM as a transfer standard to bring traceability to many of its working (secondary) standards. That system called TFS-11 was manufactured in Germany and is basically the same used in the comparison "CCEM-RF-K.20". Nevertheless, as this measuring system is not manufactured anymore, NIM and Inmetro started a project to design and built in China a similar, but improved, measuring system.

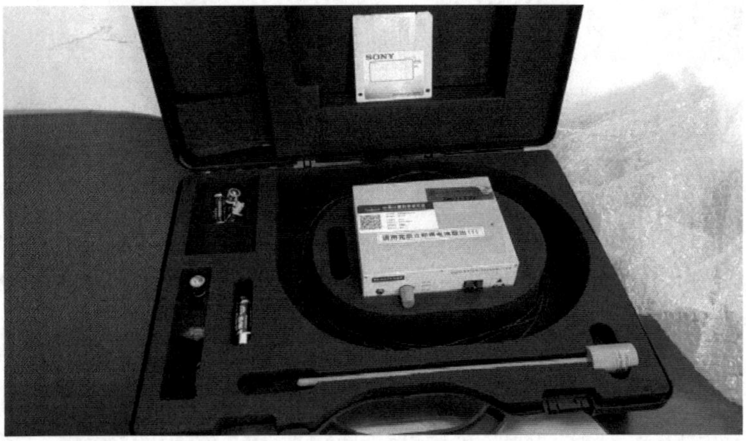

Figure 1. E-field strength measuring system currently used by NIM as a transfer standard.

The study and design of the sensor were in cooperation between Inmetro and NIM. This paper reports the achievements reached during that project, the main characteristics of the sensor that was designed and a series of tests that were conducted in the NIM facilities to evaluate its performance, as well as its improvements and advantages.

Results show that the new sensor probe has better performance than the older one, including better linear behavior at low E-field strength levels, excellent repeatability, good selectivity and good stability. By using this sensor, NIM can extend the E-field strength levels traced to the primary standard and improve its process of bring traceability among its secondary standards.

Although the new sensor is based in an old technology, it is still cheaper than an optical probe; it doesn't require any other instrument to get the reading; is easy to transport; very sensitive and stable; because of this, we consider it suitable to use in future key comparisons.

2. Review on design of small E-Field probes

The TSF-11 was studied in detail by reviewing the technical specifications and the scientific papers regarding the sensing principle. The developing of E-field probes for metrology purposes was initiated in the 80's. A summary of sensing technologies for E-field strength is reported in section B.1.2 of [2]. In the following lines a revision on the reported literature of technologies similar to the TFS-11 will be discussed.

Papers reporting a sensor like TFS-11 were presented by Berger [3] and by Münter [4], but previous work was developed before that papers. In [5] M. Kanda released the theory of a single sensor to perform simultaneous electric and magnetic near-field measurements. This author presents a theory indicating that it is possible to measure the magnetic-loop and electric dipole currents using a loop antenna terminated with identical loads at two diametrically opposite points. The experimental results with a loop antenna with 0.16 m of radius were presented.

The first proposal to use a probe as a transfer standard for metrological purposes was released by Kanda in 1987 [6]. In that paper, a probe body with only 1 cm in diameter and 4 cm long was presented. It is a resistively tapered dipole with a low-barrier, beam-lead Schottky detector diode shunting each dipole at the center gap. The main characteristics are: (i) a useful frequency from 100 kHz to 18 GHz; (ii) a flat frequency response (± 2 dB from 1 MHz to 15 GHz) and (iii) a wide dynamic range between 1 V/m to 1600 V/m. The importance of [6] is because it explains the role of any component of the sensing principle of sensors like the TFS-11, for example the role of the dipole element, the proper choice of the Schottky detector and the high impedance transmission line after the

Schottky diode. Other works about the use of a probe as a transfer standard were subsequently published in 1997 [7] and 2000 [8].

In [9] modifications on the theory proposed by Kanda in [5] were presented, showing practical considerations on the sensor construction like the difficulty to get precisely matched loads to terminate the loop. In this paper the authors presented the experimental results on a 4 cm square wire loop compared with the modified theory. In [10] and [11] the author summarizes the theoretical analysis of various standard antennas for measuring radio-frequency electric and magnetic fields.

2.1. *Design Specifications for a Schottky Diode-Based Sensor*

Figure 2 summarizes the design specifications for a new E-field strength measuring system. As the sensor probe needs to be fitted in the existing NIM's μ-TEM Cell, many dimensional restrictions need to be considered for the new design. Software specific requirements were made in order to improve the operability of the sensor, as well as to allow use it for metrological proposes. Figure 3 shows the final prototype called STT RF06-M and Figure 4 shows the sensor being calibrating in NIM facilities.

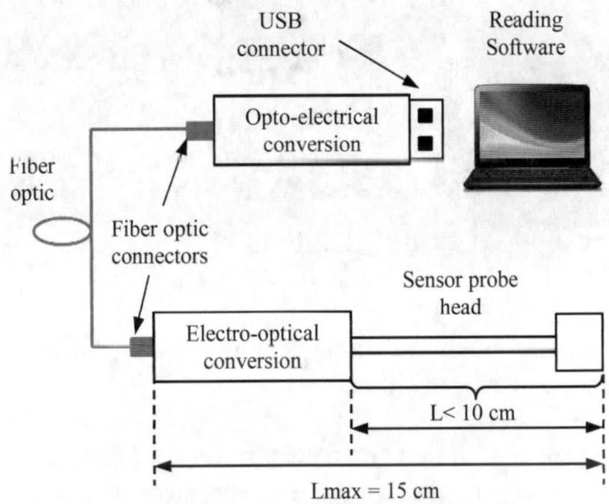

Figure 2. Minimum design specifications for the new E-field strength measuring system.

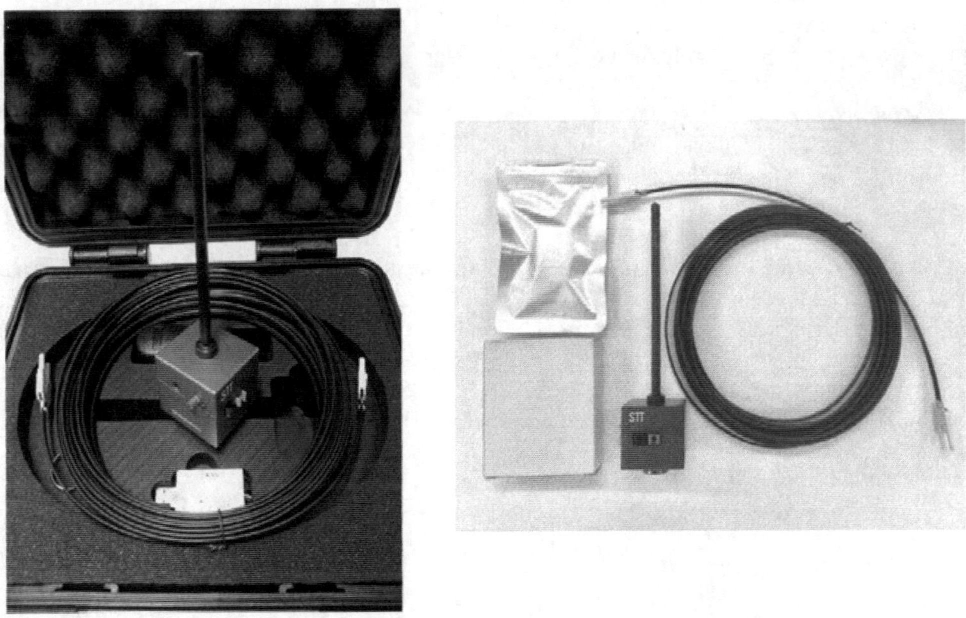

Figure 3. Final prototype called STT RF06-M

Figure 4. Final prototype being calibrated at NIM facilities.

2.2. Software capabilities.
During the development of the sensor probe, we identified many desirable features for the software that were successfully implemented. The following lines describe such features.

2.2.1. Reading Stability. STT RF-06M E-field probe takes instantaneous measurements at each 0.5 s, therefore, any change in the E-field strength is displayed by the computer becoming difficult to register the measurement if it changes too fast. So an interesting feature is to display the measurement average during a period of time defined by the user. It should be useful to determine a measurement result during a calibration or during the process to transfer the traceability.

2.2.2. Readings Records. STT RF-06M E-field probe can record the measurement results in a *.txt file for future analysis. If no average time is set the produced file will show the instantaneous measurement (at each 0.5 s), otherwise the software will record the average values over the specified time in the *.txt file. The user can give a name to the file before record the measurements.

2.2.3. Correction Factors. The measurement reading can also be corrected by a correction factor obtained from calibration against the primary standard. Using this feature no additional calculations need to be done during the traceability transfer. The user can record the individual frequency points and the corresponding correction factor. Frequencies between two frequency points are corrected by the lower correction factor.

In the final prototype an intensive test program was carried out, which will be described in the next section.

3. Performance tests of STT RF06-M

3.1. Dynamic Range – DR [dB]
The procedure described in section 8.3.3.2 of standard IEEE 1309 [2] was followed to obtain the dynamic range (*DR*). For this document the *DR* was considered as the relationship between the largest

and smallest values that the sensor probe can measure accurately. Therefore, *DR* was computed by the following equation:

$$DR = 20 \cdot \log\left(\varepsilon^{\max}(f) - \varepsilon^{\min}(f)\right) \tag{1}$$

Where,
$\varepsilon^{\max}$ is the maximum error on E-field strength at frequency *f* and,
$\varepsilon^{\min}$ is the minimum error on E-field strength at frequency *f*.

Figure 5 shows the accuracy obtained from 10 MHz to 1 GHz for five different E-field strengths. Figure 6 shows values of *DR* calculated by (1). Observe that by the definition given in (1), as lower as the values of *DR* are the accuracy is better, meaning that accuracy doesn't variates too much with E-field strength changes. In Figure 6 is possible to see that *DR* is between 9 dB and 15 dB. In average, we can observe in this plot that *DR* = 12.5 dB with standard deviation of 1.5 dB measured between 3 V/m to 70 V/m.

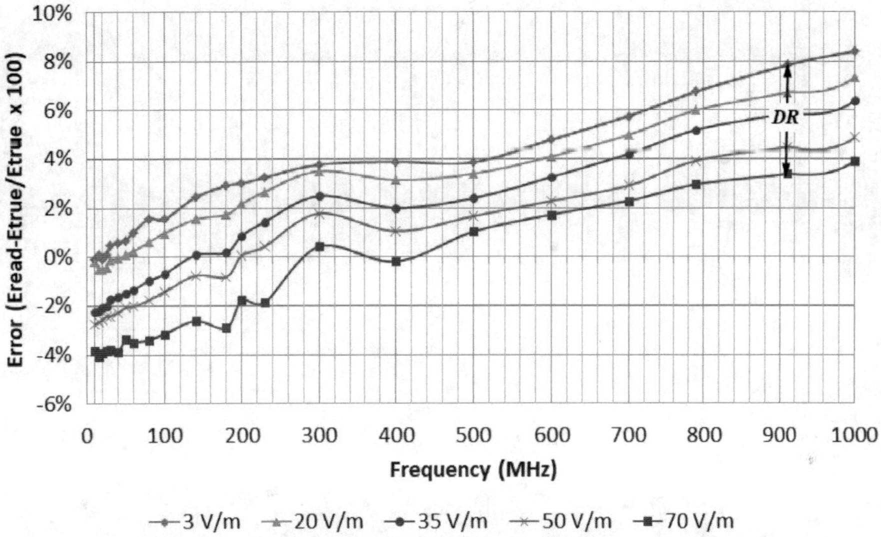

Figure 5. Dynamic Range

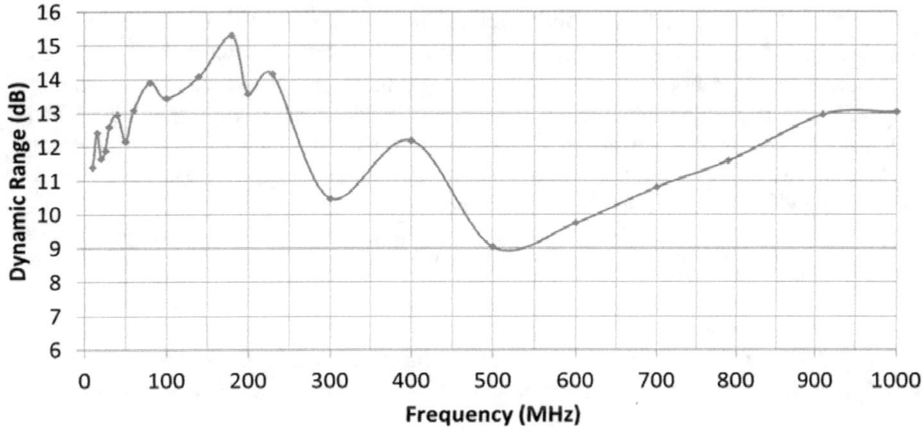

Figure 6. Dynamic Range (dB)

3.2. Selectivity – Sel [dB]

Selectivity is the capacity of the E-field probe to reject orthogonal components of the E-field. This characteristic was measured by radiating a constant E-field (i.e. 60 V/m at 10 MHz) and rotating the STT RF06 around its axis, such as selectivity can be computed with the following equation:

$$Sel = 20 \cdot \log\left(\frac{E^{nom}}{E^{read}(\delta)}\right) \tag{2}$$

Where,

E^{nom} is a nominal E-field value at a specified frequency (i.e. 60 V/m at 10 MHz);

E^{read} is the reading from the E-probe [V/m];

δ is the rotation angle (from 0° to 180° or 360°).

Figure 7 shows the selectivity values computed by (2) for the E-field probe when radiated at 60 V/m at 10 MHz and it is rotated 360° around its axis.

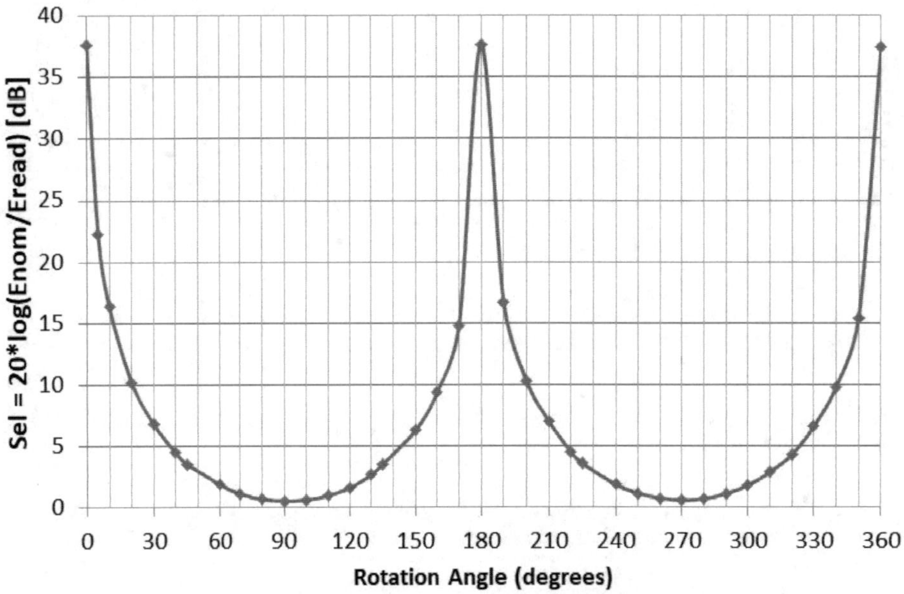

Figure 7. Selectivity values at 10 MHz and 60 V/m

In the figure above we can observe very small changes on selectivity when the probe is misaligned with the E-field vector (or not aligned at 90°). Using this approach we can evaluate the effect of misalignment during the calibration process, for instance, for a large angle misalignment like ±10° around 90° the effect in accuracy is approximately ±1.5%. Considering that a misalignment of 2° can be easily perceived by the eye of the operator, this effect in accuracy can be estimated in ±0.3%.

Evaluating selectivity at many frequencies we concluded that this parameter was between 30 dB (1 GHz) to 38 dB (10 MHz) which is reasonable if compared with the minimum value of 20 dB specified by section B.1.1 of standard IEEE 1309 [2].

3.3. Accuracy and Frequency Response

The accuracy of STT RF-06M was evaluated by calculating the E-field strength error (*e*) with the following equations:

$$e(dB) = 20 \cdot \log\left(\frac{E^{read}}{E^{true}}\right) \tag{3}$$

$$e(\%) = \frac{E^{read} - E^{true}}{E^{true}} \times 100 \tag{4}$$

Where,
E^{read} is the reading of STT RF-06M
E^{true} is the E-field true value generated by the μ-TEM Cell.
Using (3) and (4), was possible to plot the accuracy in function of the frequency range of the STT RF-06M as can be seen in Figure 8 and 9.

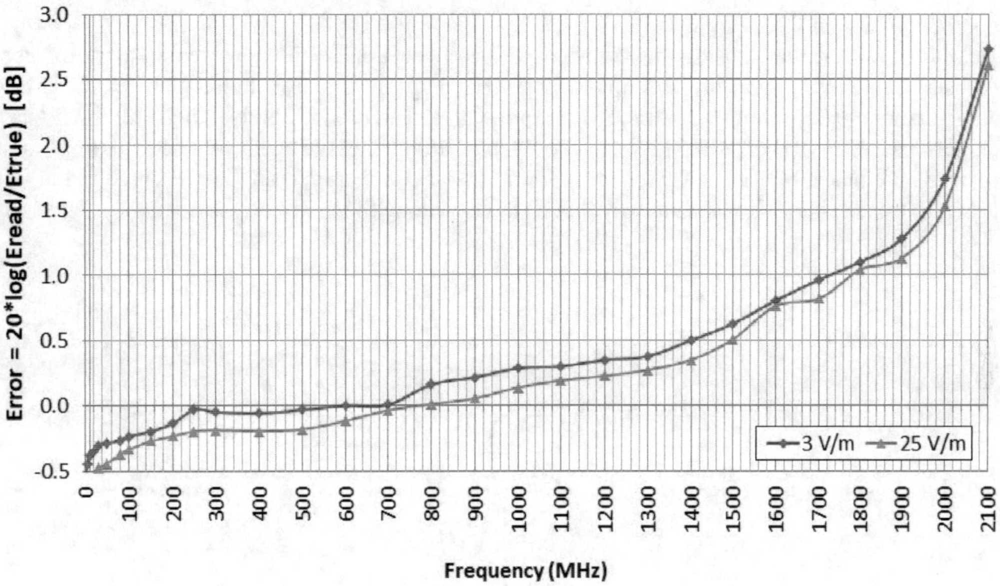

Figure 8. Error vs frequency range of STT RF06M

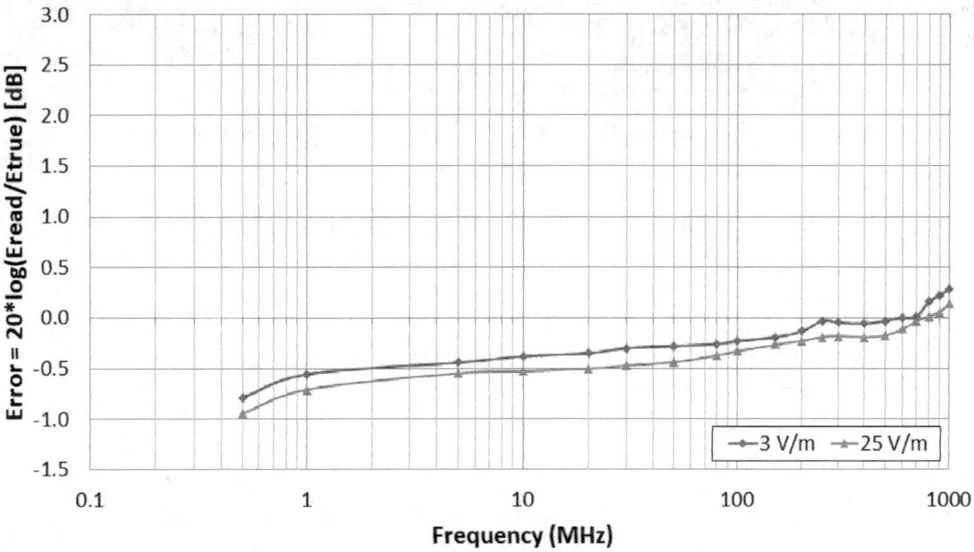

Figure 9. Error vs frequency range up to 1 GHz in logarithmic scale

In the Figure 8, we can observe a very flat frequency response curve from 500 kHz to 2 GHz with error variation less than 2 dB. In the frequency range of 500 kHz to 1.4 GHz the error variation is ±0.5 dB. Observing this plot and using (4), we can say that the error varies between ±10% for the frequency range of 500 kHz to 1.6 GHz.

In Figure 9 we plotted the frequency response in logarithmic scale up to 1 GHz. The results suggest that the STT RF-06M still can be used for frequencies lower than 500 kHz (bellow the manufacturer specification). This is an improvement on the frequency range that can be traced from the NIM's primary standard.

3.4. Repeatability and Stability

Repeatability was evaluated using the feature described in section 2.2.3. Many readings were recorded for constant E-field strength of 20 V/m at 500 MHz. Figure 10 shows more than 31,000 different readings taken in 1 day from 10:30 am to 16:30 pm with an interval of 2 h for lunch. As can be seen in this figure, the STT RF-06M has a very good repeatability measuring in average 19.96 V/m with standard deviation of 0.02 V/m. The power variation was also recorded and is presented in red in this plot. In addition, Table 1 shows data to evaluate stability along the time.

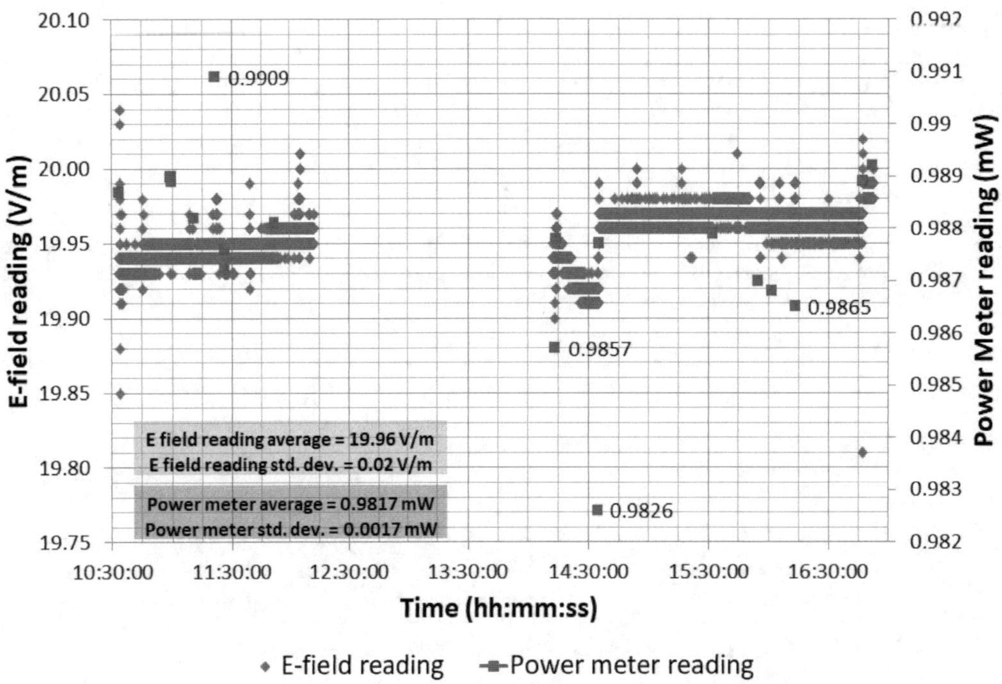

Figure 10. STT RF-06 Repeatability at 20 V/m and 500 MHz

Table 1. STT RF-06M E-field Readings for 20 V/m over the time

f (MHz)	0.5	10	100	500	1000	1500	2000	Average	Std. Dev.
08/11/2019	17.21	18.82	19.38	19.71	20.50	21.32	24.11	20.38	1.90
15/11/2019	17.53	18.66	19.05	19.70	20.09	21.11	22.31	19.92	1.22
06/12/2019	17.84	18.65	18.97	19.40	19.75	20.79	21.68	19.72	1.10

3.5. Linearity – L (dB)

Linearity is the relationship between the applied E-field strength and the reading from the probe over the E-field strength range. Figure 11 shows the linearity characteristic which was obtained from the frequency response curves at many E-field strength levels.

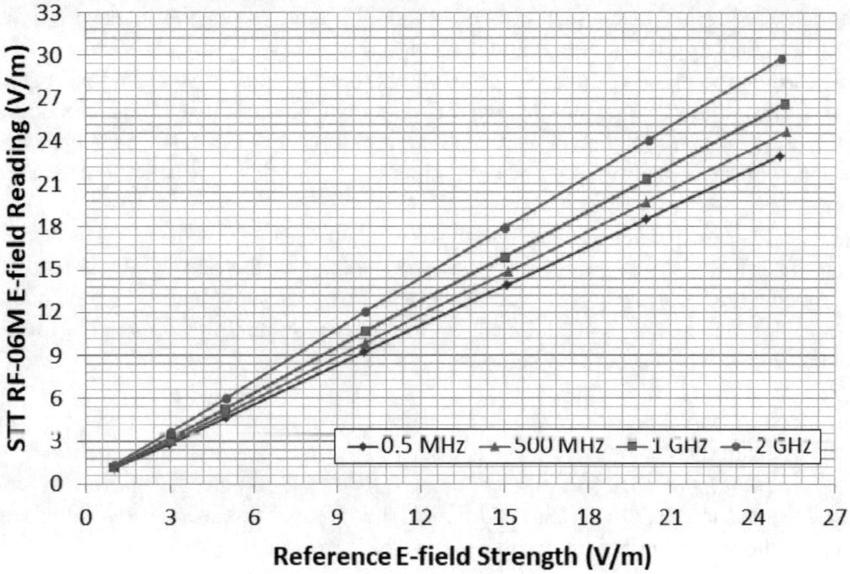

Figure 11. STT RF-06M Linearity characteristic

Figure 12 also shows a comparison between the three probes used in this study: TFS 11, STT RF-06M and the optical based probe called CS1210. As can be seen the TFS-11 deviates from the linear model for low E-field strength values due to limitation on dynamic range. In contrast STT RF-06M and CS1210 presents a linear behaviour for all the dynamic range.

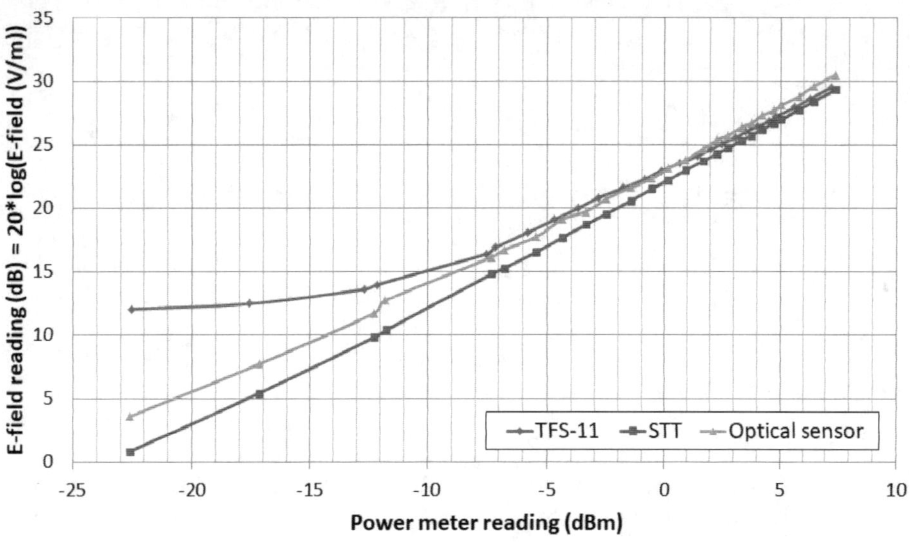

Figure 12. Linearity comparison between 3 E-field probes measured in the G-TEM Cell at 100 MHz

4. Uncertainty Evaluation

4.1. Distortion introduced by the probe when calibrated at a primary standard

The Chinese national primary Standard is a Micro TEM Cell, model MTC1000 (μ-TEM Cell). It is designed to form a complete set with the TFS-11 transfer field strength measurement system. The probe's shape of TFS-11 is a disc which thickness is only 2 mm. It occupies a space less than 1/10 of the length between the MTC1000 septum and its roof.

Normally, loading the probe into the μ-TEM Cell will produce a change in the line impedance, thus changing the field strength. One test to observe this effect was to see if a small difference in VSWR of μ-TEM Cell happens when the TFS-11 was inside the cell or not. After carried out this test, NIM's researchers concluded that such effect can be completely ignored, as the VSWR doesn't exceed the prescribed maximum limits.

However, what would happen if the recently developed sensor STT RF-06M is used instead of TFS-11? Can this effect also be ignored? As the sensor's head of the STT RF-06M is a small square with 11 mm of side, it occupies a bigger space than the TFS-11, but less than 1/3 of the length between the septum and the roof of the μ-TEM Cell (i.e. 34.489 mm), therefore to estimate this uncertainty contribution the following tests were carried out:

4.1.1. Transmission Coefficient Measurements. A similar test done with TFS-11 was conducted with STT RF-06M. By using a Vector Network Analyser (VNA) we measured the S-parameters (S11 and S21) with and without the STT-RF-06M in the μ-TEM Cell. Figure 13 shows the transmission coefficient (S21) measured in 3 different conditions: (i) without any sensor, (ii) with the sensor turned off and (iii) with the sensor turned on. In all 3 cases the largest difference in S21 coefficient was less than 0.02 dB, meaning that presence of the STT sensor does not change significantly the μ-TEM Cell behavior.

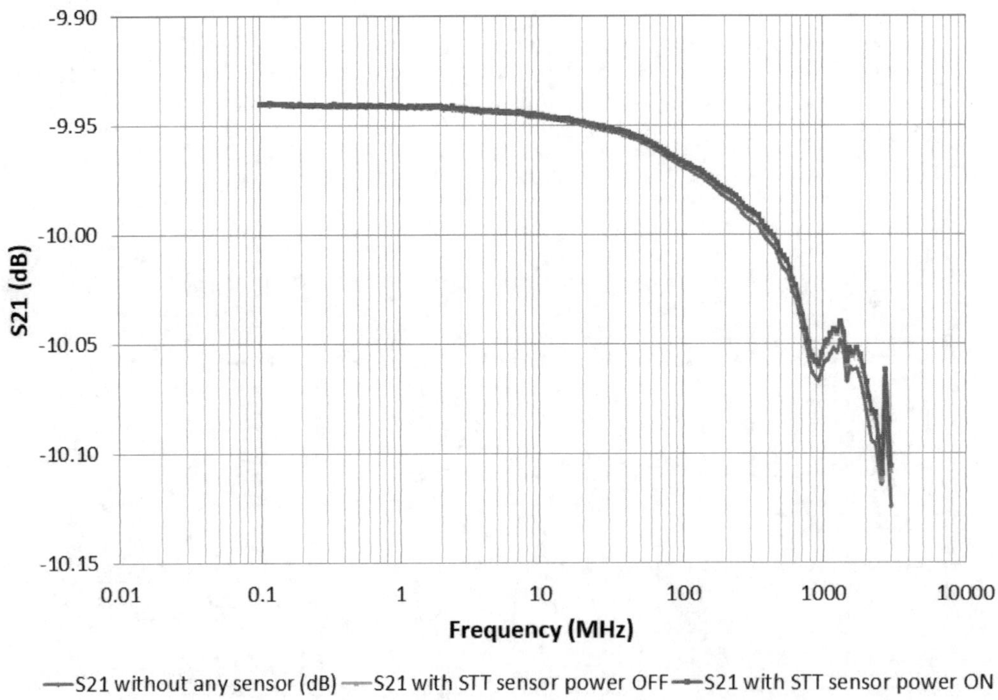

Figure 13. S21 of μ-TEM Cell at 3 different conditions

4.1.2. Calibration comparison in a bigger TEM Cell. The frequency response in the µ-TEM Cell was compared with the response measured in a larger TEM Cell (with a 450 mm length between septum and roof). To do that the accuracy of STT RF-06 was measured in both facilities between 10 MHz to 100 MHz at 20 V/m. First, correction factors were obtained in the µ-TEM Cell and used to correct the readings of STT RF-06. Second, the STT RF-06 was calibrated again in the larger TEM Cell.

Table 2. Calibration results of STT RF06 in a larger TEM cell

Freq. (MHz)	**Standard Value**	**STT RF06M reading**	**Error**		
	V/m	**V/m**	**V/m**	**%**	**dB**
0.5	19.95	19.88	-0.07	-0.35	-0.03
0.8	19.95	20.20	0.25	1.25	0.11
1	19.97	19.60	-0.37	-1.85	-0.16
3	20.01	19.92	-0.09	-0.45	-0.04
5	20.05	19.70	-0.35	-1.75	-0.15
8	19.92	19.57	-0.35	-1.76	-0.15
10	19.99	19.58	-0.41	-2.05	-0.18
20	20.18	19.94	-0.24	-1.19	-0.10
30	20.12	20.22	0.10	0.50	0.04
50	20.10	20.59	0.49	2.44	0.21
80	19.97	19.88	-0.09	-0.45	-0.04
100	20.08	19.78	-0.30	-1.49	-0.13

Table 2 above shows the calibration results obtained in such facility. As we can see calibration errors of STT-RF-06M are less than 0.25 dB, which are inside the uncertainty limits declared to generate the E-field strength of this facility (0.6 dB).

As a conclusion, we can say that there are no significant distortions in the field due to the presence of the sensor STT-RF-06M.

4.2. Measurement Uncertainty of sensor being used as a transfer standard

The main contributors to the measurement uncertainty of the sensor while is calibrated in the µ-TEM Cell are:

- Repeatability (Computed in section 3.4 as being 0.02 V/m).
- Sensor misalignment regarding the E-field (Computed in section 3.2 as being 0.06 V/m).
- Uncertainty of the generated E-field (Estimated in 0.44 V/m).

The combined standard uncertainty gives 0.45 V/m.

5. CONCLUSIONS

A new developed E-field strength sensor is presented. This sensor was developed to attend requirements of the National Metrology Institute of China to be used as a transfer standard instead a measuring E-field probe that is not manufactured anymore. The sensor is based on Schottky diode technology but it was upgraded with the latest technology available allowing keeping characteristics like low cost, excellent dynamic range, frequency response and repeatability.

The use of this sensor as a transfer standard allows extending the traceability services using a µ-TEM Cell as primary reference standard for E-field Strength. In fact, before this project the old TFS-11 worked at a frequency range from 10 MHz to 1 GHz; now the new transfer standard can calibrate facilities from 500 kHz to 2 GHz. The E-field strength, that before this project were limited to just 20 V/m, now was extended allowing transferring the traceability from 3 V/m to 70 V/m still keeping the uncertainty around 1 dB.

After some stability tests, we consider that STT RF-06M can be used as a transfer standard for key comparison of E-field Strength.

Acknowledgments

This work was supported in part by the Talented Youth Scientist Program of the Dept. of International Cooperation of the Ministry of Science and Technology (MOST), China and under the project AKY1940 of the National Institute of Metrology – NIM China.

References

[1] Final Report on GT-RF Key comparison CCEM.RF-K20: *Comparison of Electrical Field Strength Measurements*. Swiss Federal Office of Metrology and Accreditation, Bern-Wabern – Switzerland, August 2005.
[2] IEEE Standard 1309:2013 *Calibration of Electromagnetic Field Sensors and Probes (Excluding Antennas) from 9 kHz to 40 GHz.*
[3] Berger H. S., Kumara V and Matloubi K. "Considerations in the design of a broadband E-field sensing system". *IEEE 1988 International Symposium on Electromagnetic Compatibility.*
[4] Münter K., Pape R. and J. Glimm "Portable E-Field Strength Meter and its Traceable Calibration Up to 1 GHz Using a μ-TEM Cell" *IEEE Transactions on Instrumentation and Measurement*, vol. 46, no. 2, April 1997.
[5] Kanda M. "An Electromagnetic Near-Field Sensor for Simultaneous Electric and Magnetic-Field Measurements". *IEEE Transactions on Electromagnetic Compatibility*, vol. Emc-26, no. 3, August 1984.
[6] Kanda M and Driver L. D. "An Isotropic Electric-Field Probe with Tapered Resistive Dipoles for Broad-Band Use, 100 kHz to 18 GHz". *IEEE MTT-S International Microwave Symposium Digest*, Baltimore, MD, USA, 1986.
[7] Glimm J., Münter K., Pape R., Schrader T. and Spitzer M., "The new national standard of EM field strength; realization and dissemination," *International Zurich Symposium on Electromagnetic Compatibility*, pp. 611–13, Zurich, Switzerland, Feb. 1997.
[8] Loader B., Liang W., Alexander M., and Gregory A., "An optically modulated field sensor to make TEM cells traceable to the NPL calculable standard dipole," *IEEE Conference on Precision Electromagnetic Measurements*, pp. 283–284, Sydney, Australia, May 2000.
[9] Miles E. G. Upton and Andrew C. Marvin. "Improvements to an Electromagnetic Near-Field Sensor for Simultaneous Electric and Magnetic Field Measurement". *IEEE Transactions on Electromagnetic Compatibility*, vol. 35, no. 1, February 1993.
[10] Kanda M. "Standard Probes for Electromagnetic Field Measurements". *IEEE Transactions on Antennas and Propagation*, vol. 41, no. 10, October 1993.
[11] Kanda M. "Standard Antennas for Electromagnetic Interference Measurements and Methods to Calibrate Them". *IEEE Transactions on Electromagnetic Compatibility*, vol. 36, no. 4, November 1994.
[12] Schwerdt P.I., Berger J. and Petermann I. "E-Field Measurements in the Near-Field of Transmitting Antennas using an Integrated Electro-Optical Sensor". *Proceedings of Symposium on Electromagnetic Compatibility*, 1996.
[13] Loader B., Liang W., Alexander M. and Gregory A. "An optically modulated field sensor to make TEM cells traceable to the NPL calculable standard dipole antenna". *Conference on Precision Electromagnetic Measurements CPEM* 2000.
[14] "Standard uncertainty budget for the E-field strength in the range of 10 MHz to 1 GHz". Electromagnetic Compatibility (EMC) Laboratory, Division of Electronics and Information Technology, National Institute of Metrology, China. Dec, 2008.
[15] Glimm J., Munter K., Pape R., Spitzer M and Schrader Th. "Traceability and uncertainty of field-strength measurements" *Conference on Precision Electromagnetic Measurements CPEM* 1998.

Journal of Physics: Conference Series **2606** (2023) 012031 doi:10.1088/1742-6596/2606/1/012031

Digitalization of an Electrical Metrology Laboratory: Hybrid-remote calibrations

Gean Marcos Geronymo

Instituto Nacional de Metrologia, Qualidade e Tecnologia - Inmetro

E-mail: gmgeronymo@inmetro.gov.br

Abstract. This paper gives a summary of the recent improvements in the digitalization process of the Electrical Standardization Metrology Laboratory (Lampe) of Inmetro and presents a solution for hybrid-remote calibrations with remote control of calibration systems from anywhere using the Internet.

1. Introduction

The laboratories of the Electrical Metrology Division of Inmetro are responsible for providing traceability of electrical quantities in Brazil. The Electrical Standardization Metrology Laboratory (Lampe) is responsible for providing traceability to standard resistors (both ac and dc), standard inductors, standard capacitors, capacitance bridges and ac-dc transfer standards (for both voltage and current).

Since the late 90's the laboratories of Inmetro's Electrical Metrology Division have been focusing on automating its calibration systems in order to keep up to the increasing demanding of calibration services [1].

In this paper we present our last improvements on the digitalization of the laboratory processes, including the development of a web-based system with a structured SQL database and a remote access solution to remote control our calibration systems. With those improvements, we were able to maintain the execution of some calibrations services during the social distancing period imposed by the COVID-19 pandemic in 2020 and 2021.

2. Recent Improvements

In 2016 we started upgrading one of our calibration services (DC resistors) to a web-based application, using open source software tools [2]. The concept was inspired by a technical paper from IBM proposing the integration of the efficient mathematical processing of Scilab to web pages using PHP scripting language [3]. The first step was to rewrite an old Visual Basic application used for uncertainty calculation and measurement report generation that was becoming obsolete and hard to maintain. A new application was developed using Scilab powerful math functions. Then, with the integration of Scilab and PHP we were able to build a friendly user interface. In addition, we started to use a database management system (PostgreSQL) to store all the measurement and administrative data in a structured way. Using the same concepts presented in the IBM technical paper, Gnuplot, an open source data plotting tool, was integrated

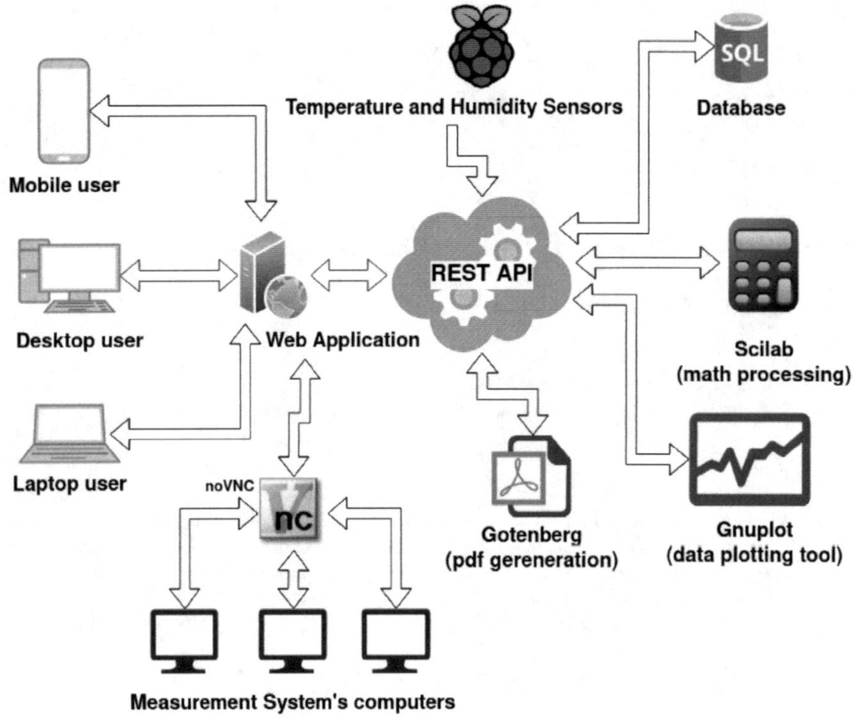

Figure 1. System architecture.

to our web application. Using Gnuplot we can dynamically generate plots of measurement data and historical calibration data, speeding up the measurement analysis.

In 2019 we proposed a "Smart Lab" architecture following the Industry 4.0 design principles [4]. In the Smart Lab architecture we propose the use of REST APIs for the interconnection between different systems, allowing the integration of custom made hardware and software with commercially available solutions. By following this architecture we can comply with the four design principles of Industry 4.0: interconnection, information transparency, decentralized decisions and technical assistance [5].

During 2020 we started the migration of other laboratory processes, including some custom made software, electronic spreadsheets and paper forms to a common web application based on the Smart Lab architecture. The system is called "SYS-LAMPE" and incorporates the main processes of the daily routine of the laboratory.

3. System Architecture

Figure 1 shows the system architecture as a block diagram. The system runs on a Linux server and is based on Docker containers. Containers are a standardized unit of software that allows developers to isolate their application from its environment, solving most of the compatibility issues [6]. By using containers we can integrate multiple different software pieces in our system in an efficient and reproducible way.

The first calibration service module implemented in the new system is for DC Resistor calibration and is presented in Figure 2. This module can read measurement data from the text files generated by the commercial resistance bridges from Measurement International that we have in the laboratory and extract all the relevant information. We can do all the measurement

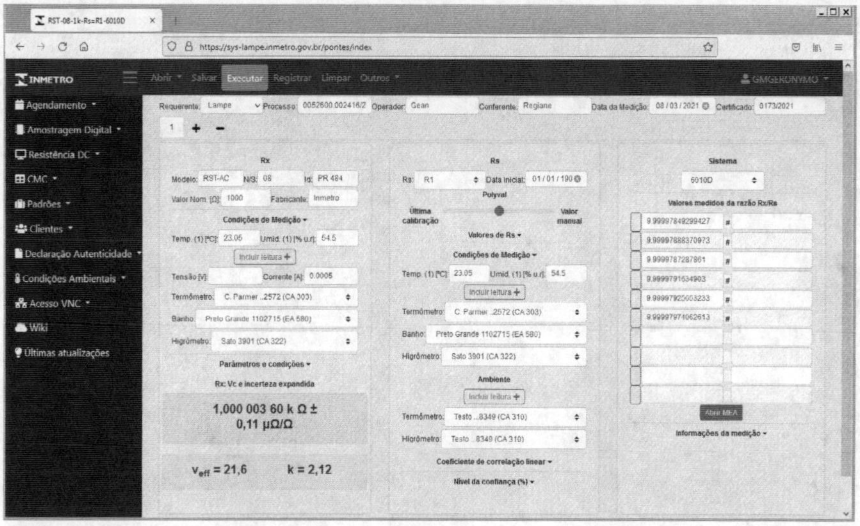

Figure 2. DC Resistor calibration.

analysis, check historical calibration data, do uncertainty calculation and generate measurement reports, and, finally, generate the calibration certificates automatically.

The automatic generation of calibration certificates was implemented using phpWord [7]. In 2021 Inmetro started to issue electronic calibration certificates, in PDF format, accompanied by a declaration of authenticity with a SHA256 hash code of the issued PDF, digitally signed in Brazilian federal government's electronic information system (SEI). Our system is fully compliant with this new policy, generating both the calibration certificate in PDF format and the hash code to be included in the declaration of authenticity, speeding up the workflow of the laboratory technicians.

Our PDF generator engine uses an application called Gotenberg, that implements a docker-powered stateless API for converting HTML and Office documents do PDF [8]. The PDF generator tool is used to convert the calibration certificates do PDF format automatically and to export the measurement reports.

An useful feature included in our system is remote access to the measurement system's computers using VNC, directly on the web application (Figure 3). This feature was implemented with noVNC, an open source VNC client built using JavaScript, that runs well in any modern web browser (including mobile web browsers for iOS and Android) [9]. To use this feature, we need to install a VNC server in each measurement computer. We use TightVNC, an open source VNC server.

Currently we have temperature, humidity and pressure sensors connected to Raspberry Pi tiny credit card sized computers to monitor the environmental conditions of the five physical laboratory rooms of Lampe. We use both commercially available digital termohygrometers and custom build sensors using DHT22 and BME280 sensors. The Raspberry Pi computers send the measurement data directly to the REST API that stores them in a structured SQL database. Then, in the web application, we can easily access the information as time series charts (Figure 4).

Our next step is to extend this automatic monitoring to our temperature controlled air and oil baths used for DC Resistor calibrations (currently this is still done manually).

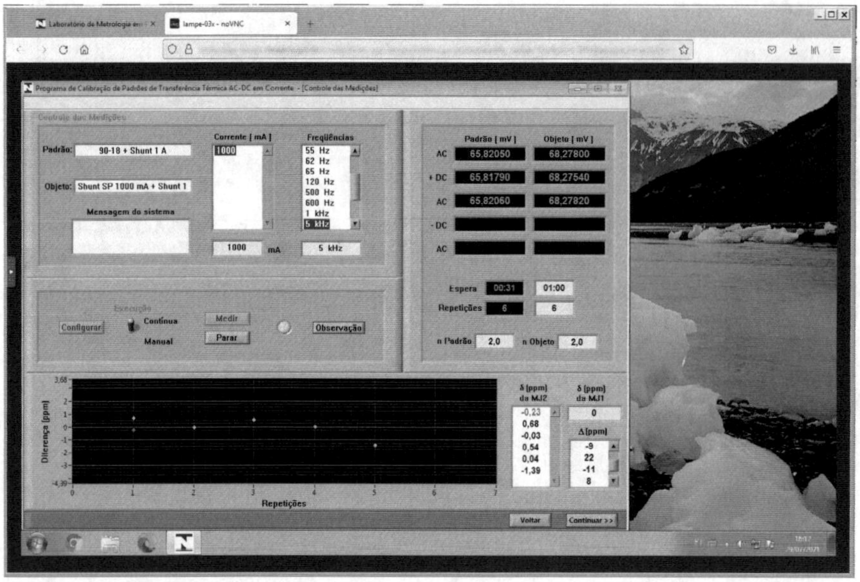

Figure 3. VNC access to measurement computers directly in the web application.

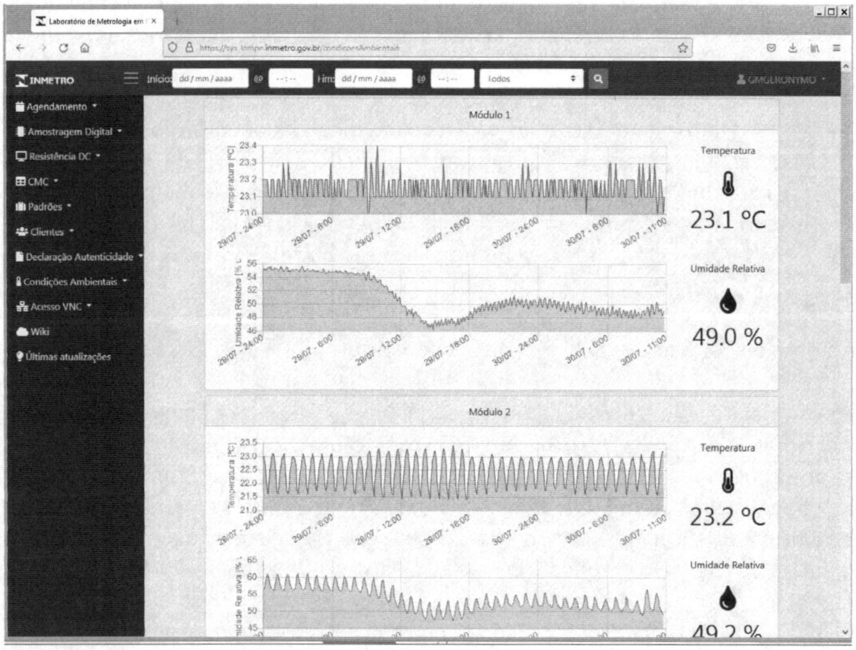

Figure 4. Environmental conditions logging of the physical laboratories.

4. Hybrid-Remote Calibrations

We can find in the literature several references to internet-based calibrations, internet enabled calibrations and remote calibrations. One of the first works in this area was presented by NIST, in 2000 [10]. We have reports of similar works in NPL and JV [11] and IEN [12]. Most of the works in the remote calibrations subject are focused on "on-site" calibrations, where the customer's instrument is remotely calibrated in its original environment. In those approaches, the logistics of the calibration are reversed: the laboratory that will perform the calibration sends the standards to the customer's laboratory, and the calibration process can be partially or fully controlled via internet.

To overcome the limitations imposed by the social distancing imposed by the COVID-19 pandemic, we tried a different approach of remote calibration, that we will call "hybrid-remote" calibrations.

With the integration of our automated measurement systems with SYS-LAMPE, a technician working from home just need remote access via Virtual Private Network (VPN) to be able to perform a calibration service. This technician has access to all relevant data of the laboratory, including the access to the calibration schedule, quality system documentation, laboratory equipment management and to the environmental conditions (temperature, humidity and ambient pressure) data log. In this scenario, with only one technician working physically on the laboratory, to make the physical connections of the device under test and the calibration system, the technicians working from home can continue with the execution of some calibration services.

5. Conclusion

A large portion of Lampe's customers is made up of Inmetro's own laboratories, which need traceability in electrical quantities such as DC resistance. With hybrid-remote calibrations we were able to perform calibrations during the social distancing period in a secure way, reducing negative impacts to the calibration schedule of other laboratories.

References

[1] Afonso E *et al.* 2001 An automated system for the calibration of high accuracy multifunction instruments *IMEKO TC4 Symp. - Trends in Electrical Measurement and Instrumentation* (Lisbon, Portugal)

[2] Geronymo G M and Silva M C 2016 *Journal of Physics: Conference Series* **733** 012067 URL `https://doi.org/10.1088/1742-6596/733/1/012067`

[3] Lewis D J 2010 Efficient mathematical processing using scilab through php URL `http://www.ibm.com/developerworks/library/os-php-scilab/os-php-scilab-pdf.pdf`

[4] Geronymo G M 2021 *Journal of Physics: Conference Series* **1826** 012027 URL `https://doi.org/10.1088/1742-6596/1826/1/012027`

[5] Hermann M, Pentek T and Otto B 2016 Design principles for industrie 4.0 scenarios *2016 49th Hawaii International Conference on System Sciences (HICSS)* pp 3928–3937

[6] Merkel D 2014 *Linux J.* **2014** ISSN 1075-3583

[7] Phpword URL `https://github.com/PHPOffice/PHPWord`

[8] Gotenberg – a docker-powered stateless api for converting html, markdown and office documents to pdf URL `https://thecodingmachine.io/gotenberg-a-stateless-api-for-converting-files-to-pdf`

[9] novnc - the open source vnc client URL `https://novnc.com`

[10] Baca L, Duda L, Walker R, Oldham N and Parker M 2000 Internet-based calibration of a multifunction calibrator (Proc. 2000 National Conference of Standards Laboratories Workshop and Symposium (NCSL), Toronto, 1, CA)

[11] Sand A, Stevens M and Parkin G 2007 *IEEE Transactions on Instrumentation and Measurement* **56** 1986–1991

[12] Cassiago C, Callegaro L and La Paglia G 2004 Internet calibration for electrical metrology: first application at ien *Proceedings of the 21st IEEE Instrumentation and Measurement Technology Conference (IEEE Cat. No.04CH37510)* vol 1 pp 102–105 Vol.1

Journal of Physics: Conference Series **2606** (2023) 012032 doi:10.1088/1742-6596/2606/1/012032

Evaluation of the effects of leakage current in the calibration of standard resistors between two dc resistance bridges and a digital sampling impedance bridge

Márcio Cândido da Silva and Gean Marcos Geronymo

Instituto Nacional de Metrologia, Qualidade e Tecnologia – Inmetro, Rio de Janeiro, Brasil

mcsilva@inmetro.gov.br

Abstract. This article discusses calibrations results of standard resistors performed with two commercially available equipment and a digital sampling impedance bridge developed at Inmetro. One of the commercial bridges uses the current comparison method and the other the potentiometric method with a binary voltage divider. Calibration was performed with standard resistors of 10 kΩ and 100 kΩ with significant leakage current. This study showed divergent results between each methodology. A specific methodology was used to estimate the leakage resistance of these resistors. This study contributes to establish criteria for correction and/or inclusion of uncertainty components in resistors calibrated in these methodologies.

1. Introduction

The Electrical Standardization Metrology Laboratory (Lampe) calibrates standard resistors in direct current (dc) and alternating current (ac) from primary and secondary laboratories, industries, electric power companies, research centers, and universities. Lampe has two commercial bridges to calibrate dc resistors and a digital bridge (DigBrid) for ac resistors.

Lampe has two models of commercial bridges: the 6010D bridge which uses the current comparison method and the 6000B bridge which uses the potentiometric method using the binary divider principle, both of Measurements International (MIL).

The 6010D bridge switches the direction of currents during resistor calibration to reduce the effects of eddy currents and voltages. Its calibration range is from 1 mΩ to 100 kΩ.

The 6000B bridge was built with a measurement methodology that compensates[1] the effect of leakage currents[2,3] from resistors and connection cables[4] during calibration. Its calibration range is from 10 kΩ to 1 TΩ [5].

The DigBrid can perform comparisons in phase (impedance of the same nature) or in quadrature (ac resistances with capacitors) of four terminal pair impedances. It operates in a wide frequency range (50 Hz to 3 kHz) and impedance (1 nF to 1 μF or 10 Ω to 100 kΩ), with measurement uncertainties around a few parts in 10^6 [6].

The calibration of standard resistors can be affected by several quantities such as: temperature, insulation resistance of the resistor terminals and connection cables, humidity, voltage, pressure, short-term drift, leakage current and low frequency resistance noise due to resistor material and resistance range.

Lampe calibrates standard resistors with equipment and environment at controlled temperature.

In this way, the quantities that have the greatest influence on the calibration of resistors in Lampe are the bridge specifications, the applied voltage, and the leakage current due to the insulation of cables and resistors.

In this work, the influence of the leakage resistance on the resistor calibration when using each one of these bridge models was evaluated.

2. Methodology

Figure 1 shows a simplified standard resistor diagram. Between the High and Low connection terminals and the protection box there is an insulation resistance which, depending on its order of magnitude and the nominal value of the standard, may influence the result of the calibration.

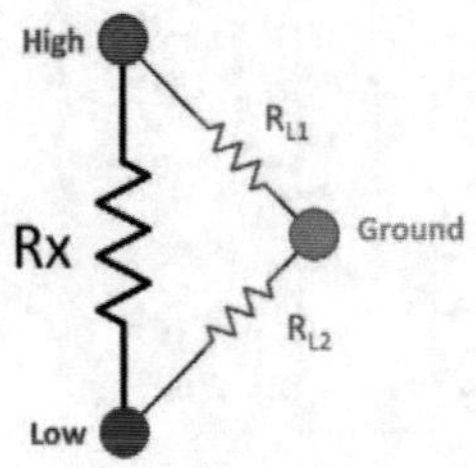

Figure 1. Standard resistor diagram.

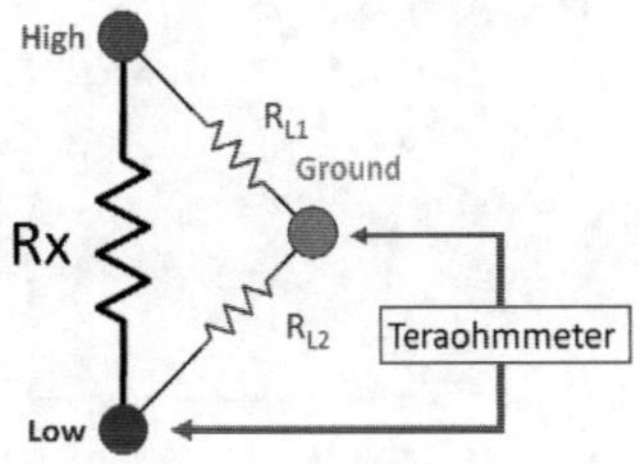

Figure 2. Teraohmmeter conection diagram.

In addition to insulation resistance, there are other types of contribution due to the way it is constructed. In this study, the composition of the insulation resistance of resistors, cables and other forms of resistive loss [7,8] are called leakage resistance (R_L) and the resistor under calibration is called R_X.

Before performing the calibrations, the insulation resistance of the connecting cables was measured with the Sefelec Teraohmmeter M1501P and they presented insulation values greater than 10 TΩ for an applied voltage of 50 V.

The effect of the leakage resistance was analyzed when measuring resistors of 10 kΩ and 100 kΩ with (grounded) and without (floating) the attachment of ground terminal to the connection cable of the bridges. Then, the results were compared to the teraohmmeter measurements.

3. Measurements analysis

3.1. Teraohmmeter measurements

At figure 2, when measuring the isolation resistance using the Teraohmmeter, the measured value will correspond to the parallel of (R_{L1} +R_X) and R_{L2}.

Table 1 shows the Teraohmmeter measurements with different voltages.

Table 1. Teraohmmeter measurements.

Voltage (V)	Measurements (GΩ)	
	A04(10 kΩ)	A05(100 kΩ)
1	0.8	1.6
5	0.5	1.1
10	0.4	0.9
20	0.3	0.8
50	0.3	0.7

3.2. Resistor measurements on current comparator bridge (CCB)

In Figure 1, it is observed that the resistance to be measured without connecting the ground terminal is the parallel of R_X with the sum of R_{L1} and R_{L2}.

Figure 3 illustrates the connection of the resistor grounded to CCB with the representation of the alternating current during the calibration of a standard resistor. When the CCB polarity is positive with respect to the High terminal of the standard resistor, R_{L2} is eliminated, and current flows through R_X and R_{L1}. When the polarity is reversed, current flows through R_X and R_{L2}, eliminating R_{L1}. Thus, the calibration result will be the R_X value in parallel with the average of R_{L1} and R_{L2}.

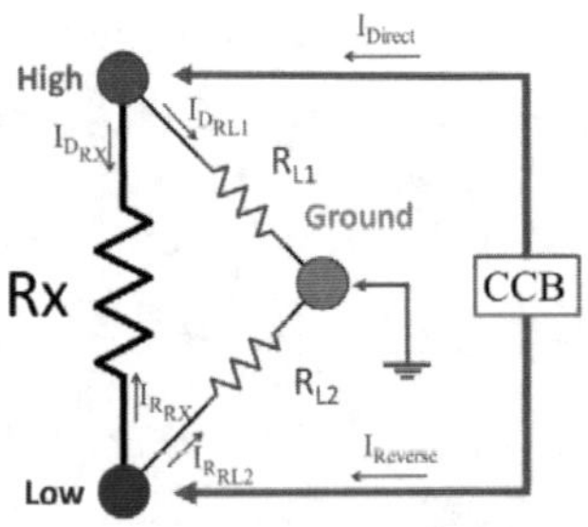

Figure 3. CCB connection diagram.

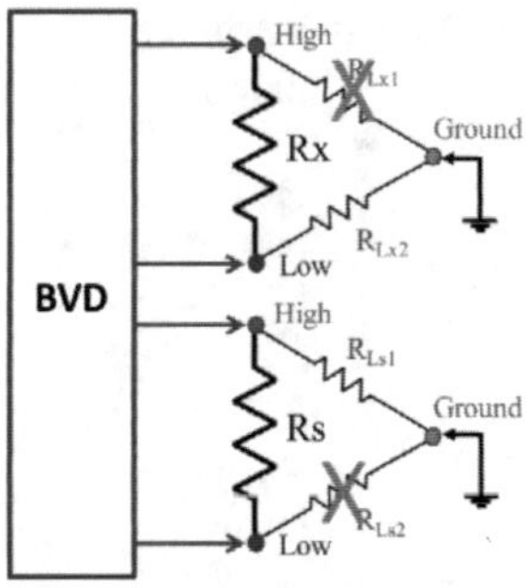

Figure 4. BVD connection diagram.

Table 2 shows the difference (Δ) between floating and grounded values for different values of applied current (I_a).

Table 2. Measurements on CCB. (Resistance values with expanded uncertainty less than 0.60 μΩ/Ω).

Resistor	I_a (mA)	V_{Rx} (V)	Floating value (kΩ)	Grounded value (kΩ)	Δ (μΩ/Ω)
A04	0.5	5	10.0000812	9.9999838	9.7
	1	10	10.0000699	9.9999539	11.6
A05	0.05	5	99.997569	99.993629	39.4
	0.1	10	99.997304	99.992702	46.0

3.3. Resistor measurements on potentiometric binary voltage divider bridge (BVD)

In BVD, if the ground terminal of the resistor to be calibrated is not connected to the ground cable and the same voltage conditions are maintained, the result of its calibration will be close to that obtained by the CCB.

Figure 4 shows the connection of the resistor grounded to BVD, where R_{Lx1} and R_{Ls2} are eliminated, leaving only R_{Lx2} and R_{Ls1} in parallel with standard resistor Rs. In this type of configuration, R_{Lx2} and R_{Ls1} would tend to influence the calibration result by increasing the value of the Rx/Rs ratio, obtaining a result greater than expected for Rx.

However, BVD has a measurement methodology that compensates the influence of the leakage resistance of cables and the resistors.

Table 3. Measurements on BVD. (Resistance values with expanded uncertainty less than 0.60 μΩ/Ω).

Resistor	V_a (V)	V_{Rx} (V)	Floating value (kΩ)	Grounded value (kΩ)	Δ (μΩ/Ω)
A04	10	5	10.0000869	10.0001378	-5.1
	20	10	10.0000729	10.0001372	-6.4
A05	5	4.5	99.997908	99.999809	-19.0
	10	9.1	99.997519	99.999841	-23.2

Table 3 shows the difference (Δ) between de floating and grounded values for different values of applied voltage (V_a).

3.4 Resistor measurements on DigBrid

The A04 resistor was measured by DigBrid with an applied rms voltage of 5 V in different frequencies (f_a). The impedance Z of ac resistor is defined by $Z = R\ (1 + j\ \omega\tau)$ where R is the value of the ac resistor and τ is the time constant.

Grounded values agree with those found on BVD. Floating values were not analyzed in this time because of the dependence of coaxial connections.

Table 4 shows A04 grounded measurements results on DigBrid and the difference (Δ) between BVD results for applied voltage of 10 V.

Table 4. DigiBrid´s Measurements. (Resistance values with expanded uncertainty less than 4.0 μΩ/Ω).

f_a (Hz)	Value (kΩ)	τ (ns)	Δ (μΩ/Ω)
64 Hz	10.000113	1	-2.5
398 Hz	10.000127	5	-1.1
796 Hz	10.000102	6	-3.6
1592 Hz	10.000099	10	-3.9

3.5 Estimating the leakage resistance value

From the results in tables 2 and 3, it is possible to estimate the value of the leakage resistance of each resistor and compare it with the results obtained from teraohmmeter.

Considering that the difference (Δ) found in CCB and BVD are less than one part in 10^4, the calculation of the leakage resistance can be simplified and estimated.

In CCB, by approximation, $(R_{L1} + R_{L2})/2 = Rx/\Delta_{CCB}$.

In BVD, by approximation, $(R_{L1} + R_{L2}) = Rx/\Delta_{BVD}$.

Table 5. Leakage resistance result (R_L).

Resistor	V_{Rx} (V)	Teraohmeter (GΩ)	CCB (GΩ)	BVD (GΩ)
A04	5	1.0	1.0	1.0
	10	0.8	0.9	0.8
A05	5	2.2	2.5	2.6
	10	1.8	2.2	2.2

In table 5 the correlation is made between the applied voltage and results for the leakage resistance calculated in each bridge and the teraohmmeter. For the final calculation, it was considered that $R_{L1}=R_{L2}=R_L$.

4. Conclusion

Through the measurements carried out on the two bridges, it was observed that the BVD bridge leakage current compensation system works well in the ranges of 10 kΩ and 100 kΩ.

The CCB does not have a similar leakage current compensation system, and the characteristics of the resistor to be measured in this range of values must be observed. It is possible to estimate the leakage resistance value and calculate a possible correction and/or uncertainty component.

The difference between DigBrid and BVD bridge are less then DigBrid´s uncertainty

References

[1] Dunn A F 1991 Measurement of resistance ratios in the range to 100 megohms *IEEE Trans. on Instr. and Meas.* **40** n 2 p 278-280.

[2] Cabral V, Ribeiro L and Sousaet J A 2018 J. Phys.: Conf. Ser. **1044** 012073.

[3] Honig R 2010 Practical aspects of high resistance measurements *Cal Lab: The International Journal of Metrology.*

[4] Giblin S P and Drung D 2018 Limitations to the Short-Term Stability of 100 MΩ and 1 GΩ Standard Resistors *Proc. 2018 Conference on Precision Electromagnetic Measurements (CPEM 2018) (Paris).* DOI: 10.1109/CPEM.2018.8501095.

[5] Silva M C 2019 *Projectus* **4**(1) 88–101. DOI: 10.15202/25254146.2019v4n1p88.

[6] Kürte Ihlenfeld W G and Vasconcellos R B 2017 A digital five-terminal impedance bridge *IEEE Transactions on Instrumentation and Measurement* **66** n 6 p 1546-52.

[7] Silva M C 2020 *Projectus* **5**(1) 13–21. DOI: 10.15202/25254146.2020v5n1p13.

[8] Silva M C and Geronymo G M 2021 Evaluation of the effects of leakage current in the calibration of standard resistors between two dc resistance bridges and a digital sampling impedance bridge *Proc. XIV Congresso Internacional de Metrologia Elétrica (SEMETRO) (Rio de Janeiro)* 204823 Available at https://metrologia2021.org.br/?p=2304.

Acknowledgments

The authors would like to thank the team Lampe, Lameq and Lacel interaction to carry out this work.

11th Brazilian Congress on Metrology (Metrologia 2021) IOP Publishing
Journal of Physics: Conference Series 2606 (2023) 012033 doi:10.1088/1742-6596/2606/1/012033

Identification and evaluation of uncertainty contributions to optical current sensor measurement uncertainty

M M Costa[1] **and J C W A Costa**[2]

[1]Faculty of Electrical and Biomedical Engineering, Federal University of Pará, Belém, Brazil
[2]Faculty of Electrical and Biomedical Engineering and Faculty of Computer and Telecommunication Engineering, Federal University of Pará, Belém, Brazil

marcelo.melo@eletronorte.com.br [1], jweyl@ufpa.br [2]

Abstract. Optical current sensors can replace traditional current sensor due to their several advantages, as electromagnetic immunity, low dimensions, low cost and others. One of the current main challenges is to develop optical current sensors with good accuracy and stability. This paper presents identification and evaluation of uncertainty contributions to measurement uncertainty evaluation of optical current sensors based on Faraday effect.

1. Introduction
Optical current sensors (OCSs) offer attractive features such as wide bandwidth, lightweight and small size, adjustable turn-ratio, integrated functions (e.g., voltage and current sensing in one device), as well as safety and environmental benefits when used in power systems (elimination of oil or gas insulation, no open secondaries, and no ferro-resonance) [1].
Due to these key characteristics, OCSs can replace conventional current sensors in power systems, as current transformers (CTs). However, measurement accuracy and long-term operation stability are under continuous research to support their further development. Some external factors, as temperature, cause errors that are difficult to overcome in a practical OCS [2].
This paper presents identification of uncertainty contributions to OCSs measurement uncertainty, using the methodology defined by ISO GUM [3]. Some of these contributions are also evaluated, but some are only identified due to the early stage of this research. Depending on the OCS structure, an uncertainty contribution can be relevant or not.

2. Optical current sensor
OCSs are usually based on Faraday effect and Ampere loop theorem, that is, polarized light with different polarization states in a material with magnetic optic properties has different propagation speed under the action of magnetic field generated by current, resulting in polarization plane rotation of polarized light [4]. The schematic diagram is shown in Figure 1.

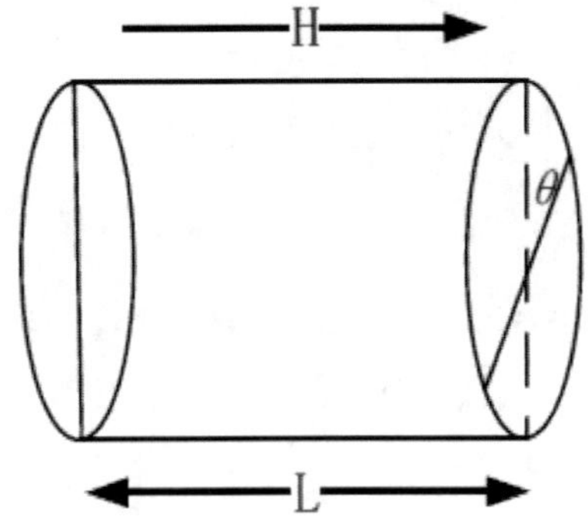

Figure 1. Schematic diagram of Faraday effect [4].

The rotation angle θ is directly proportional to the magnetic field strength H, L is the optic material length in the magnetic field and V is the Verdet constant of the optic material, as it can be seen in (1) [4]. From Ampere's Law, H is proportional to the electric current I, and R is the distance from the conductor carrying that current, as it can be seen in (2). The Verdet constant of the optic material is function of the light wavelength and the temperature.

$$\theta = V.H.L \tag{1}$$

$$H = {}^{1}/_{2\pi.R} \tag{2}$$

In an OCS, a light wave is launched from an optical source and passes through a linear polarizer and turns into an LP (linearly polarized) light. This LP light propagates through a magneto-optical material along the external magnetic field produced by the electric current. Due to Faraday effect, the plane of polarization (plane of the electric field in the light wave) rotates. The amount of rotation θ is proportional to the line integral of the magnetic field along the light path in the optic material [4]. The detector part of an OCS converts the measurement of the rotation angle to an electric signal and digitalize it through and analog-to-digital (A/D) card. A software then processes the digital signals and presents the measured current. Figure 2 shows a basic scheme of an OCS. If an OCS surrounds entirely the conductor of the electric current, it is closed loop type. Otherwise, if it surrounds the conductor partially, it is an open loop type OCS.

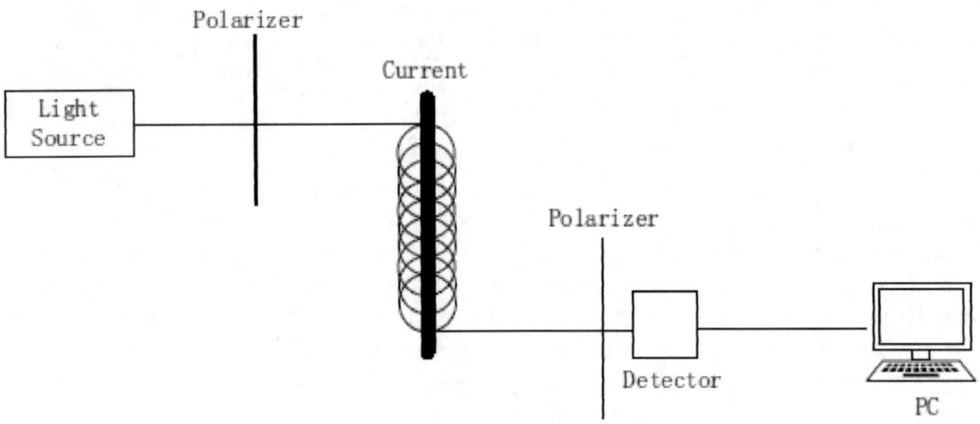

Figure 2. Basic scheme of an OCS [4].

Journal of Physics: Conference Series **2606** (2023) 012033 doi:10.1088/1742-6596/2606/1/012033

3. Uncertainty contributions

This section presents the most important uncertainty contributions to OCS measurement uncertainty.

3.1. Temperature

Temperature contributes to OCS measurement uncertainty due to the following mechanisms: (a) temperature fluctuations change the wavelength of light generated by the light source, thus modifying the Verdet constant of the optic material and the θ measured by the OCS; (b) As V depends on the temperature, it will fluctuate with temperature variations; (c) temperature fluctuations change metrological properties of optoelectronics components used in the OCS, as photodiodes and data acquisition units; (d) birefringence of the optic material varies with temperature variations. Figure 3 shows Verdet constant dependence on the wavelength for some materials (glasses), where this dependence can be seen in (3), where λ is the wavelength and a, b and λ_{0v} are the fitting parameters [5].

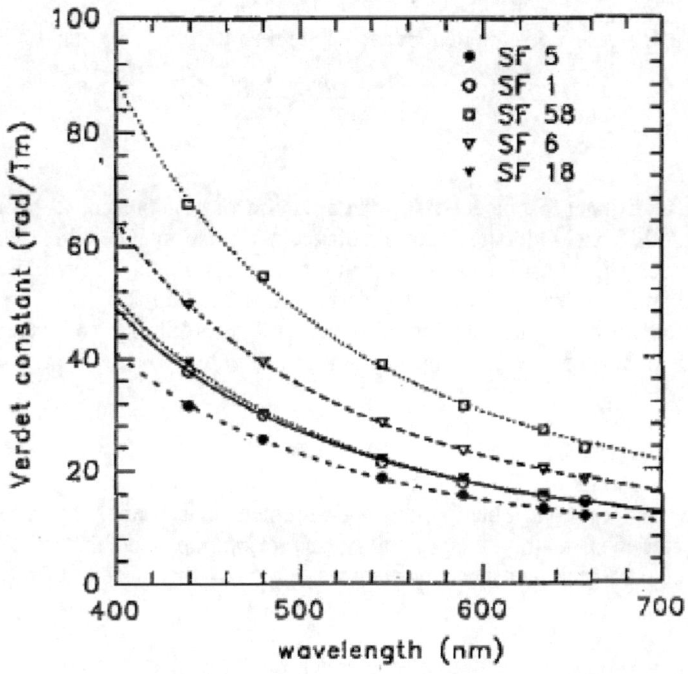

Figure 3. Verdet constant variation as function of wavelength [5]

$$V = \frac{\pi}{\lambda}\left\{a + \frac{b}{\lambda^2 + \lambda_{0v}^2}\right\} \quad (3)$$

Figure 4 shows Verdet constant change with temperature for SiO_2 and SF 57 glass. The slope of the data in that figure, assuming a uniform magnetic field and no birefringence, can be defined by (4), where V is the Verdet constant, L is the length of the sensor, V_0 and $(VL)_0$ are the values at 20°C and $\alpha = (dL/dT)/L$ is the thermal expansion coefficient of the glass for the range of temperatures tested [6]. It can be seen that SF 57 glass Verdet constant can change almost 2% with approximately 100 °C change.

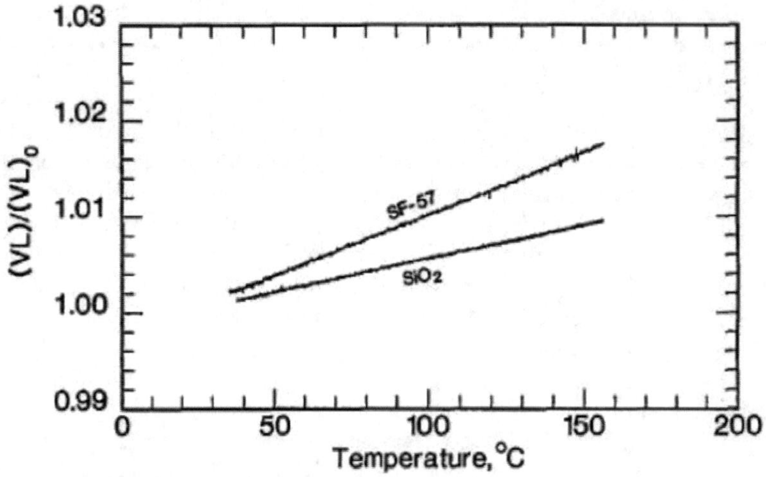

Figure 4. Verdet constant dependance on temperature for two materials [6].

$$\frac{1}{(VL)_0}\frac{d(VL)}{dT} = \frac{1}{V_0}\frac{dV}{dT} + \alpha \quad (4)$$

3.2. Birefringence

In a material with birefringence characteristic, polarization orthogonal components of the incident light wave travel at different velocities, due to different refraction index in each direction of the material. If the incident light is LP, then polarization state can change from linear to elliptical. If birefringence of the optic material changes the polarization state of a LP light, errors and uncertainties arise in the measurement of the rotation angle of an OCS. The amount of uncertainty due to the birefringence can be evaluated using special techniques in the rotation angle measurement algorithm or software.

3.3. Optoelectronics stability and resolution

The detector module of an OCS employs optoelectronics devices, as photodiodes, photodetectors and A/D cards. These devices contribute to the measurement uncertainty of an OCS due to the metrological characteristics, mainly their stability and resolution. The metrological data of these devices usually come from their datasheets or manuals.

3.4. Vibration and stray magnetic fields

When using OCSs in power systems, they are under the influence of vibration of the conductor carrying the electric current to be measured and stray magnetic fields. As open loop OCSs do not surround the conductor entirely, these factors cause variations on the distribution of the magnetic field on the OCS, causing uncertainties on the measurement of the magnetic field, and consequently on the measurement of the current. Influence of vibration and stray magnetic filed on closed loop OCSs is not relevant.

4. Conclusion

This paper presented the identification and evaluation of uncertainty contributions to optical current sensors measurement uncertainty. As they change the behaviour of the optic material employed by the sensors, temperature fluctuations and birefringence influence are very critic. Characteristics of optoelectronic devices used in the sensor, as stability and resolution, also contribute to the measurement uncertainty.

As further work, there is the quantification of each measurement uncertainty contribution, their combination, and the evaluation of the measurement uncertainty of the optical current sensor, using ISO GUM or other methods, as Monte Carlo simulation.

References

[1] Li Y, Zhang W, Liu X and Liu J 2019 *Applied Sciences* **9** 333.
[2] Samimi M, Akmal A and Mohseni H 2015 *IEEE Sensors Journal* **15** 4271
[3] Bureau International des Poids et Mesures (BIPM) 2008 *Evaluation of measurement data — Guide to the expression of uncertainty in measurement* (BIPM) Available from: https://www.bipm.org/documents/20126/2071204/JCGM_100_2008_E.pdf
[4] Wu J and Zhang X 2020 *7th International Forum on Electrical Engineering and Automation (IFEEA)* p 134
[5] Hoffmann H 1992 *J. Physique IV* **Colloque C2-21-C2-30**
[6] Williams P, Rose A, Day G, Milner T and Deeter M 1991 *Appl. Opt.* **30** 1176

Quality infrastructure for energy efficiency of household appliances in Latin America and the Caribbean

G A Kyriazis[1] **and K-C Göthner**[2]

[1] Instituto Nacional de Metrologia, Qualidade e Tecnologia (Inmetro), Duque de Caxias, Brazil
[2] Consultant, Physikalisch-Technische Bundesanstalt (PTB), Brauschweig, Germany

E-mail. gakyriazis@inmetro.gov.br, christian.goethner@gmx.net

Abstract. The main activities and results of the energy efficiency component of the regional project 'Quality Infrastructure for Renewable Energy and Energy Efficiency in Latin America and the Caribbean' are presented. This project was implemented by the Organization of American States (OAS), the Inter-American System of Metrology (SIM), the Pan American Standards Commission (COPANT), the Inter-American Accreditation Cooperation (IAAC), and the Physikalisch-Technische Bundesantalt (PTB). The main objective of its energy efficiency component consisted in improving the regional quality infrastructure for energy efficiency of household appliances.

Keywords. Energy efficiency, Technical regulation, Labeling, Quality infrastructure

1. Introduction

The evolution from tariff to non-tariff barriers brought about the need for emerging and developing countries to complete and upgrade national quality infrastructures (QI) to enable them to participate in global markets. Bilateral and multilateral free trade agreements make more and more reference to recognized technical competence through equivalent national QIs [1]. However, most emerging and developing countries still lack adequate QI to meet the needs of the fast-growing global economy [2].

In the Americas, the following regional QI organizations deal respectively with metrology, standardization, and accreditation: the Inter-American System of Metrology (SIM) [3], the Pan American Standards Commission (COPANT) [4], and the Inter-American Accreditation Cooperation (IAAC) [5]. We report here on recent efforts aimed at improving the QI in Latin America and the Caribbean (LAC), specifically with regard to the energy efficiency (EE) of household appliances. Such efforts have been implemented in the realm of the regional project 'Quality Infrastructure for Renewable Energy and Energy Efficiency in Latin America and the Caribbean' aimed at strengthening the capabilities of the abovementioned organizations, along with their national members, to provide services in selected fields of renewable energy sources and EE, and to promote the coordination among the different QI institutions in supporting the implementation of national energy policies. The project was funded by Germany´s Federal Ministry of Economic Cooperation and Development (BMZ) and executed in two phases between 2011 and 2020 by the Organization of American States (OAS), the three regional QI organizations, and the Physikalisch-Technische Bundesanstalt (PTB), Germany´s national metrology institute (NMI). The project was supported by strategic partners with regard to energy policy issues, namely the International Renewable Energy Agency (IRENA) and the Latin American Energy Organization (OLADE).

Journal of Physics: Conference Series **2606** (2023) 012034 doi:10.1088/1742-6596/2606/1/012034

Section 2 presents and discusses the objectives and actions of the aforementioned regional project with regard to the EE of household appliances. Proficiency tests (PT) and effective border control and market surveillance are reconsidered as important conditions for successfully implemented EE labeling programs. With this content as background, the main results and conclusions are presented in section 3.

2. Regional project

The regional project covered the following topics: quality assurance for photovoltaic and solar thermal energy, capacity building for smart grids, strengthening of reliable determination and certification (e.g. labeling) of the EE of electric household appliances, and raising awareness of the significant role of QI services in these fields. Here, we only mention those project activities regarding the EE of electric household appliances. Such appliances save a lot of domestic labor by consuming a large amount of electricity [6]. One of the project goals in this regard was to establish a solid metrological base in LAC for measuring the EE of household appliances. The gaps, challenges and proposed actions were identified during the kick-off workshop held in Fortaleza, Brazil, in May 2012, in combination with COPANT's Annual General Assembly. They constituted the framework for all the subsequent activities which were adjusted and implemented step by step during the execution of the project.

The workshop held in Itaipava, Brazil, in July 2013, co-organized and co-financed by the project and by the Instituto Nacional de Metrologia, Qualidade e Tecnologia (Inmetro), the Brazilian NMI, was an important step forward. It was possible to extend the attendance to regulatory agencies, companies, industrial associations and testing laboratories of different LAC countries, and to gather experts from different institutions: European Commission, Deutsches Institut für Normung (DIN), VDE Institut, Germany, United Nations Environment Programme (UNEP), Economic Commission for Latin America and the Caribbean (ECLAC), Organización Latinoamericana de Energía (OLADE), Agência para a Energia (ADENE), and Collaborative Labeling and Appliances Standards Program (CLASP). Most of them accompanied the project until its end. The main project goals related to EE of electric household appliances were defined in this workshop:

(a) EE labeling programs for electric household appliances were considered as the main instrument to promote QI services for EE programs.

(b) Harmonization of written standards adopted in the region and their correct application by the conformity assessment bodies (CAB) should be promoted with support of the national standardization bodies (NSB), the national accreditation bodies (NAB), and the national metrology institutes (NMI).

(c) Development and application of EE accreditation schemes should be supported.

(d) Technical competences of first- and third-party testing laboratories should be improved by PT's.

Impact studies should be elaborated and awareness events be realized.

The main project activities designed to achieve the above goals are related to labeling programs, standards, accreditation schemes, PTs, awareness and impact studies, and regional forums on labeling programs.

A first project activity was to identify the international standards relevant to performance, safety, and testing methods of refrigerators and domestic LED lamps. The activities were concentrated on the harmonization of written standards and correct application of the standards requirements related to testing methods (within the second activity discussed below). COPANT and SIM played a major role in this activity. The task forces of these institutions were particularly interested in the quantities to be measured, the measuring instruments, and the accuracy required for those instruments. Several written standards provide this information. As refrigerators make use of compressors, written standards relating to the latter were also consulted. In the case of refrigerators and associated compressors, the quantities temperature, pressure, electric voltage, electric current, electric power, refrigerant flow, water flow, speed, mass, time, length, torque, humidity, linear dimensions, volume, and electric energy, are of interest. Therefore, electrical, mechanical, thermal and fluid dynamics metrology are involved. In the case of LED lamps, a broad spectrum of quantities are of interest: length, electric voltage, harmonic distortion, electric current, electric power, displacement factor, luminous intensity distribution, luminous efficacy, chromaticity tolerance, correlated colored temperature, color rendering index and luminous flux. Thus, optical (photometry and radiometry), electrical and mechanical metrology are included.

A second activity was a regional survey on the competences of the laboratories for testing refrigerators and domestic LED lamps in accordance with the written standards already identified. As is well known, it is necessary that the quality management system (QMS) of those laboratories be in accordance with ISO/IEC 17025:2017 [7], traceability to the International System of Units (SI) being a requirement of this standard, so that IAAC and SIM played a leading role in this activity. An initial problem noted was the limited availability of accredited, competent, and experienced testing laboratories in the region ensuring reliable and comparable measurement results. Thus, it was decided to carry out two regional rounds of PTs with in-house (first party) and independent (third party) participating laboratories; the two rounds linked with initial and closing workshops, one for refrigerators and one for LED lamps for domestic use (see more details in subsection 2.1).

A third activity undertaken by the project was to ensure the information exchange among the QI institutions and the improvement of technical competences by trainings. An important outcome was an increased knowledge of the existing measurement capabilities and its dissemination throughout the region. To this end, resources were invested in training to measure the quantities of interest for the relevant appliances, involving especially SIM in this activity. The trainings were carried out at different levels according to the needs of NMIs, calibration laboratories, and testing laboratories. Measurements conducted by testing laboratories require calibrated standards and meters, and certified reference materials. Calibration laboratories, in turn, need reference standards traceable to NMI primary standards. And the NMIs need to participate in international comparisons of the quantities of interest mentioned (in addition to other quantities).

At the same time, training on laboratory management and accreditation as well as training for NABs were carried out with important support by IAAC and COPANT. In accordance with the demand expressed by IAAC and its members, the project supported workshops and training related not only to written standards with a direct link to electric household appliances but also to those of a more general content (for instance the new version of ISO/IEC 17025:2017 [7]). By doing so, it was possible to improve and upgrade the technical competences of the NABs, which also have an influence on the improvement of EE in the region.

Concerning standardization, the project supported COPANT members to: (a) strengthen their participation in the technical committees of ISO and IEC, in particular the ISO Committee on Conformity Assessment (ISO/CASCO); (b) strengthen their participation in COPANT´s ISO/CASCO Focus Group and COPANT TC 152 (Energy Efficiency); and (c) promote their participation in the International Electrotechnical Commission (IEC) work and, in particular, in the IEC Affiliate Conformity Assessment Status (IEC ACAS) program for smaller countries, especially for Central America and the Caribbean. Important for the standardization activities was a kick-off workshop held in Bogota (Colombia) in May 2014, and organized by PTB, COPANT, and the Colombian Institute of Technical Standards and Certification (ICONTEC) with the participation of 18 NSBs from LAC and three ISO/IEC experts. As a result, a Support Program for COPANT members was fostered by the project with virtual trainings, meetings and workshops. The last activity was realized in 2019 in coordination with the ISO/CASCO Plenary Session in Nairobi, Kenya.

A fourth project activity involved awareness events and impact studies. Governments and private sector need concrete data on the impact of QI services on energy savings and the economic and social consequences [8]. Impact studies on the energy saving effect of QI services should be carried out as evidence for convincing authorities on the need for financing the EE programs to be enacted by the governments in the region. Thus, by the end of 2013, nine impact studies were initiated in coordination with ECLAC. The impact studies were confronted with some serious methodological problems (among them missing data base, lack of common language between economists and QI experts). At least, it was possible that two studies (Argentina and Uruguay) were published by their

Journal of Physics: Conference Series **2606** (2023) 012034 doi:10.1088/1742-6596/2606/1/012034

own institutions. The main results of four studies (Argentina, Costa Rica, Panama, and Uruguay) were summarized on one page and published on PTB website.

The awareness activities carried out by the EE component were very successful. They can be divided in two parts: (a) national awareness seminars for the stakeholders of the labeling programs and (b) publications for different target groups. Whereas the awareness seminars were oriented to inform the stakeholders of the host country on the content and the main outcomes of the workshops and to strengthen the relationship between the QI institutions and their customers, the publications were directed to (a) EE experts, in the case of a study published by the European Council for an Energy Efficient Economy (eceee) and presented on the annual meeting of this organization [9], (b) energy ministries and energy agencies, in the case of a presentation during the 3rd OLADE Energy Week in December 2018, in Montevideo, and the elaboration of a Policy Position Paper together with OLADE [10], as well as to (c) the Brazilian public, in the case of [11].

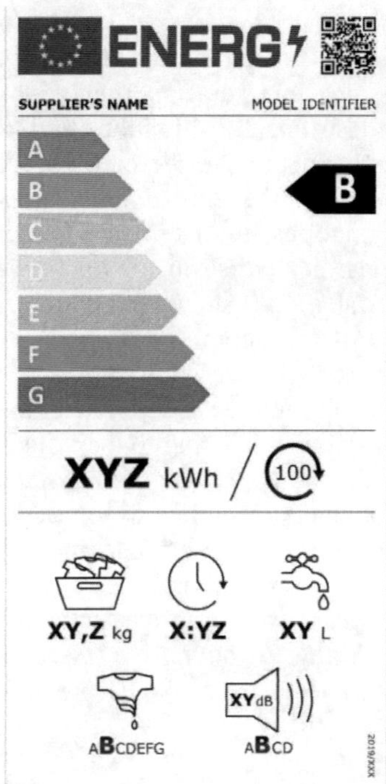

Fig. 1. The new European EE label (Ecolabel).

A fifth activity of the project was the monitoring of household appliance labeling programs [10] in force in the region aiming at: (a) improving cooperation among national and regional QI institutions and (b) strengthening the relationship and interaction of these institutions with regulatory agencies and government energy policies in the region. In total, four workshops were held with the attendance of representatives from NMIs, NSBs, NABs, CABs (testing laboratories, certification bodies and inspection bodies), regulatory agencies, market surveillance and control institutions, companies, and energy and industry ministries in LAC. Experts from European countries, the European Commission, and USA also participated. The events were carried out with the important support of representatives from the regional QI organizations and host institutions in the different countries.

The workshops offered an important forum for exchanging experiences, lessons learnt and discussing challenges, and helped to understand better such important technical issues like standards harmonization, label design (the new European label is shown in Fig. 1), the required testing laboratory competences, or the training of government officers that are designing and controlling the implementation of programs. The important topic of consumer education was also addressed [12]. In LAC, consumer organizations, organized consumer action, and consumer policy are social phenomena of recent date (for instance, the Brazilian consumer protection code was promulgated only in 1990.). The workshops were the starting point for many activities such as PTs or training in standardization and accreditation. They were also the platform where information on other project activities was shared. Surely, there were some insufficient results, particularly the impossibility to make large advances in the harmonization of written standards, conformity assessment procedures and labels. But the discussions clarified the procedural differences among countries and demonstrated their best practices. The differences have a lot to do with their main commercial relations.

More detailed information about the proficiency tests (PT) and the border control and market surveillance will be presented in the following subsections 2.1 and 2.2.

2.1 Proficiency tests

The objectives of the PTs were: (a) evaluate the technical capability of testing laboratories to report reliable measurement results within the stated measurement uncertainty; (b) review the correct application of the written standards, providing an opportunity to identify weaknesses (for instance, related to personnel performance, instrument calibration, and testing methods) and initiate the corresponding corrective actions; and (c) improve the capability of testing laboratories to measure electricity consumption and EE, as set out in the regulations in force in the region.

Nine Latin American testing laboratories, the German VDE Institut, who served as a reference laboratory, and a Mexican PT provider accredited by ISO/IEC 17043:2010 [13], participated in the refrigerator PT [14]. A typical test installation is shown in Figure 2. In order to perform the test, a total of 15 two-door refrigerators with 134A refrigerating gas cooling system and automatic temperature control were selected as items under test. The test items were previously characterized with respect to homogeneity and stability, and distributed in three groups according to the testing method selected by each participant. The groups were classified according to the different mains frequencies (50 Hz or 60 Hz), supply voltages (220 V or 115 V), and applicable written standards (ISO/IEC standards or North American standards) in LAC. The following quantities were measured during the PT: temperature (in degree Celsius), electric voltage (in volt), electric current (in ampere) and electric power (in watt). These quantities were measured as functions of time. The room temperature and the temperature of each item required by the written standard were also measured as functions of time. The measurand was the electric energy consumed during 24 hours, expressed in kilowatt hours. The measurand reference value was determined for each test item by the PT provider.

Fig. 2. Test chambers for refrigerators (CEPEL, Rio de Janeiro, Brazil). Photo: Karl-Christian Göthner.

Briefly, the PT results were as follows: (a) Group 1 (220 V, 50 Hz - ISO 15502:2005 [15]): all three laboratories presented satisfactory normalized error; (b) Group 2 (220 V, 60 Hz - ISO 15502:2005 [15]): three laboratories presented satisfactory normalized error; and (c) Group 3 (115 V, 60 Hz - AHAM HFR-1:2008 [16]): two laboratories presented satisfactory normalized error.

Evidently, due to the use of different types of refrigerators and testing methods, the measurement results could not be compared among all participants. The important thing is that each refrigerator has been characterized by the PT provider; a bilateral comparison being made between each participant and the test provider. A highly complicating issue was logistics, in particular the transport and customs procedures within the region, as export/import permissions were very often required. Only a few countries in the region have agreed mutual exemptions for PTs and tests, for instance Argentina and Uruguay.

The following reasons for the poor performance of three participating laboratories were identified: (a) the laboratories did not know the relevant standards in detail; (b) they did not know how to evaluate the uncertainty in measurement; (c) they did not use the measuring instruments allowed by the written standard (thermocouples, electric energy meters, etc.), and (d) consequently certain external influences (temperature, applied voltage, air speed) were not sufficiently controlled.

In the following PT, of LED lamps for domestic use, an attempt was made to prevent the difficulties encountered in the previous one from occurring again.

A total of 16 laboratories from Latin America and once more the German VDE Institut participated in the PT of LED lamps for domestic use [17]. The reference laboratory was the Instituto Nacional de Tecnología Industrial (INTI), the Argentinian NMI, who also acted as PT provider accredited by ISO/IEC 17043:2010 [13]. The following participation criteria were defined: (a) accredited laboratories, laboratories with accreditation in progress, or with proven experience in carrying out tests on LED lamps for domestic use; (b) capability for carrying out the tests; and (c) experienced personnel. A typical test installation is shown in Figure 3. In order to perform the test, a set of LED lamps with standard E27 fitting was selected and each participating laboratory received three of these lamps. The test items were previously characterized with respect to homogeneity and stability according to EN 50285:1999 [18]. The measurand was the electric energy consumed during one month, expressed in kilowatt hours. The following quantities were measured during the PT: electric power consumed (in watt), luminous flux (in lumen), efficacy (in lumen per watt), and EE (in percentage). In this case, logistics did not play an aggravating role.

Fig. 3. Testing bench for LED lamps (LABELO, Porto Alegre, Brazil). Photo: Karl-Christian Göthner.

Briefly, the PT results were as follows: Nine laboratories finished the PT with satisfactory results in the measurements made of the four quantities, five laboratories presented unsatisfactory (or questionable) results in some of these quantities and two laboratories had only unsatisfactory (or questionable) results. In general, the reported results were better than those in the refrigerator PT.

Supported by SIM, PTB and the VDE Institut, a closing workshop for each PT was realized together with the participants detailing the abovementioned results, the difficulties encountered, and helping them to overcome existing weaknesses in theory and practice. The following issues were discussed in the workshop on domestic LED lamps: (a) calibration of the measuring instruments; (b) knowledge of and experience with the testing methods by technical personnel; (c) consideration of the formulas used for corrections and conversion factors contemplated in the test methods – in some cases there were difficulties in the evaluation of uncertainty in measurement; (d) proper use of the reference standards employed by measuring instruments; and (e) the importance of having a laboratory QMS, especially with regard to the control of environmental conditions.

In summary, it was possible to verify with these PTs that: (a) there were problems (in some cases serious problems) with the application of the relevant written standards; (b) insufficient attention was paid to the environmental conditions of the laboratories in the PT of refrigerators; (c) there were problems related to the evaluation of uncertainty in measurement; and (d) in some cases, the traceability of measurements performed by laboratories was not sufficiently warranted. The recommendation is that similar PTs should be carried out in the region in the future. The repetition of those presented here and the organization of PTs for other electrical appliances will provide greater comparability between the measurements made in the region.

2.2 Border control and market surveillance

Currently, almost all countries in the region have labeling programs that regulate the supply of household appliances on the market according to EE. The devices on the marketplace are either domestically produced or imported. It is possible to differentiate between three groups of countries in terms of QI services: (a) countries with a large-scale market who produce, export and import electric products (Argentina, Brazil, Colombia, Mexico); (b) countries with limited markets, with small production and exports, but with a relevant import of electric products (Chile, Cuba, Dominican Republic, Ecuador, Guatemala, Peru, Uruguay); and (c) countries with limited markets that only import electric products (Bolivia, Caribbean countries, Central American countries, Paraguay).

The documentation required for a manufacturer´s model to enter a market can be of two types: (a) manufacturer´s self-declaration (requirement adopted in the European Union); and (b) certification of the product by a certification body accredited by ISO/IEC 17021:2015 [19] based on a test carried out at a testing laboratory accredited by ISO/IEC 17025:2017 [7] that evidences the application of relevant written standards and the adoption of good practices in carrying out the necessary tests (requirement adopted by most of LAC countries).

The issues of border control and market surveillance will be here addressed as a way of ensuring that only those household appliances remain on the market that comply with technical regulations (including labeling programs) for EE. In case of a foreign product´s entry in the national market, there are three possibilities:

a) Control of the imported product documentation by relevant authorities, particularly customs. For this to occur, government officers must be especially trained and competent to interpret the documents that accompany the imported batch. The technical competence of customs, at the time of writing this article, is relatively developed only in some LAC countries.
b) In some countries (for instance, in Argentina), a test carried out by a national testing laboratory is also required to verify that the product complies with the regulations. This is also usually the case if the border authorities have questions about the documentation. However, the topic of sample collection has not yet been resolved at the time of writing this article.
c) The application of one of the seven ISO/CASCO types of certification schemes listed in ISO/IEC 17067:2013 [20], which are based on the mutual recognition of product conformity assessment certificates at different scales. At the time of writing this article, the type 5 (five) is being applied to a limited extent in Argentina, Brazil and Chile.

A problem that is difficult to solve is the multiplicity of written standards adopted in the region. The requirements of all these written standards differ substantially, hampering comparability between test results. Mexico, as part of the United States-Mexico-Canada Agreement (USMCA), performs the international trade in large part with North America. The same occurs with the Central American and Caribbean countries. Such countries adopt North American standards (ANSI, ASTM, CSA, etc.), very often in the form of the Mexican obligatory standard (NOM), whereas South American countries adopt primarily the ISO/IEC standards. It is important that the imported product meets the requirements of the regulations of the recipient country. If this does not occur, additional testing by an accredited testing laboratory may be necessary, which means an additional importation cost.

Once on the marketplace, regulations provide control and surveillance measures that are indispensable to ensure that only compliant products remain on the market. The main methods are: (a) inspections directly on the marketplace; (b) verification of products by collecting and testing samples on the marketplace; and (c) denunciations.

Since there are millions of products on the market, it is recommended to establish priorities based on risk analysis to define those products which have the greatest impact on energy consumption. Market surveillance is not highly developed in LAC due to the cost involved and the lack of trained personnel. Inspections are predominant in the region and they focus on: (a) checking that the label is in the correct place and (b) checking that the information on label, documentation and product are consistent.

There is almost no standardized sampling method on the LAC marketplace to verify through tests that the product performance is in accordance with the documentation and label. It is a very costly activity that requires the purchase of minimum five replicas of each model and the testing in a competent and experienced laboratory (accredited by ISO/IEC 17025:2017 [7]). In the European Union, two rounds for verification of household appliances were carried out, one with refrigerators (Atlete I) and other with washing machines (Atlete II). The results were interesting: in Athlete I, with reference to EE, approximately half of the models did not comply with what was indicated on EE in the documentation and label. In Atlete II, two years later, in EE all models were in compliance with what was indicated in the documentation and label [21]. This indicates the need for more rigid market surveillance. Sometimes, private consumer protection organizations also collect samples for verification, as is the case with Stiftung Warentest (in Germany).

Denunciation is another important tool. It is practically the only measure adopted in USA. Competitors or consumer associations purchase products on the marketplace, test the products, and when there is something which does not comply with written standards, proceed to denunciation. In some Latin American countries, consumers can contact public consumer protection agencies (SEC and SERNAC in Chile, INDECOPI in Peru, and SIC in Colombia).

3. Outcomes and conclusions

The PTB-OAS-SIM-COPANT-IAAC-project addressed three main issues related to EE in electric household appliances: (1) the technical aspects of a sufficiently developed national QI which represent an essential foundation for the successful implementation of EE policy and related programs, (2) the awareness of policymakers and governmental institutions on the importance of sufficient technical competences within the national QI and (3) the necessity of cooperation and coordination among energy policy, regulation and QI institutions not only on a national but also on the regional level. The experiences collected by the project activities demonstrate that the EE policy needs to consider carefully the inherent technical aspects and in particular the actual status of the QI services available in the country for implementing such policy. Some of the most important project outcomes are summarized in the sequel.

The project period of 8 (eight) years and the focus on implementation and improvement of services for EE of electric household appliances provided a good opportunity to influence changes in the partner organizations and their members, to promote the relationship with their customers in the public and private sector, and to strengthen the awareness of the importance of a QI according to national needs. In a general way, the project supported a better understanding on the need of QI services when implementing EE programs and on the importance of including all stakeholders (NMIs, NSBs, national CABs, testing laboratories, certification bodies, surveillance and control institutions, inspection bodies, customs and industry) in the design, implementation and evaluation of such programs, for instance in Brazil, Chile, Colombia, Costa Rica, Mexico, and Uruguay. The following subsections 3.1-3.4 describe the most important outcomes of the EE component at three levels: (a) the regional and national QI organizations, (b) the CABs, and (c) the government institutions.

3.1 Regional and national QI organizations

Concerning metrology, the project allowed clarification of the need for an uninterrupted measurement traceability chain, EE was included in the relevant SIM working groups, and gaps between the demand and supply of calibration and measurement capabilities in the region were defined and proposals developed to overcome them, in particular by more regional cooperation.

Concerning standardization, the project allowed promotion of the COPANT-Focus Groups (in particular COPANT/CASCO), support to change the orientation of standardization work in COPANT towards information and dissemination of international standards instead of elaboration of regional standards, and improvement of the relationship between the NSBs and the IEC (for instance, IEC ACAS).

Concerning accreditation, the project allowed training of 18 NABs in ISO/IEC 17021-1:2015 [19] and IAF Mandatory Documents applicable, focused on the implementation of an Accreditation Program for Energy Management Systems (ISO 50001:2018 [22]).

3.2 Conformity assessment procedures

The project allowed improvement of the technical competences of first- and third-party laboratories for testing refrigerators and domestic LEDs during regional PTs: new accreditations, one laboratory in Uruguay, and 6 (six) in Colombia. It also allowed the improvement of technical competences of 19 laboratories in Argentina, Brazil, Ecuador, Guatemala, and Mexico.

3.3 Government policies

The project allowed support for a better understanding of the QI services needed for elaboration, implementation and evaluation of EE programs. NMIs, NSBs, NABs and CABs had increased participation in the elaboration and implementation of labeling programs (examples are Brazil, Colombia, Costa Rica, Peru, Uruguay and the Reglamento Técnico Centroamericano [RTCA]). The role of QI services in border control and market surveillance was clarified as well as the strengths and weaknesses of the different instruments. Government officers gained special knowledge of QI and its services for working with labeling programs (positive example: training program for officers in Colombia, including customs, during the implementation of the Reglamento Técnico de Etiquetado (RETIQ)).

The project also invited consumer protection associations and governmental consumer protection agencies, specially sharing the European experiences with the German DIN Consumer Council. The project also produced some indirect results (see next subsection 3.4).

3.4 Spin-off effects of the EE component

VDE Institut experts were hired by other PTB projects and by United Nations Development Programme (UNDP). Consumer Rights Protection Centre (Latvia) experts helped in other PTB projects to implement market surveillance measures.

INM–ONAC–ANDI / Cámara de Electro-domésticos program (Colombia) upgraded first- and third-party testing laboratories funded by UN Development Programme (UNDP) and supported by PTB. For the first time, 6 (six) testing laboratories for electric household appliances could be accredited in accordance with the new ISO/IEC 17025:2017 [7].

A Colombian electric household appliance manufacturer participated in the IEC meeting 2017 in Vladivostok (Russia).

Networks have been established between the professionals in the LAC region and experts from outside the region.

4. References

[1] Sanetra C and Marbán R M 2007 *The Answer to the Global Quality Challenge: A National Quality Infrastructure* Physikalish-Technische Bundesanstalt (PTB) Organization of American States (OAS) and Inter-American System of Metrology (SIM)

[2] APEC (2017) *Guide to Support Quality Infrastructure Incorporation in MSMEs*, Asia-Pacific Economic Cooperation - Sub-Committee on Standards and Conformance, APEC Project: CTI 13 2015A (report produced by Instituto Nacional de Calidad (INACAL) on the final output of the Workshop Supporting Micro, Small and Medium Enterprises (MSMEs) Trade Facilitation through Standardization Activities, August 2016, Lima, Peru).

[3] SIM available at https://sim-metrologia.org/ accessed on 2022-09-20

[4] COPANT available at https://www.copant.org/index.php/en/ accessed on 2022-09-20

[5] IAAC available at https://www.iaac.org.mx/index.php/en/ accessed on 2022-09-20

[6] Bertoldi P Ricci A and de Almeida A 2001 *Energy Efficiency in Household Appliances and Lighting* Springer-Verlag Berlin Heidelberg GmbH

[7] ISO/IEC 17025:2017 *General requirements for the competence of testing and calibration laboratories* available at https://www.iso.org/standard/66912.html accessed on 2022-09-20

[8] Gonçalves J and Peuckert J 2011 *Measuring the Impacts of Quality Infrastructure – Impact Theory, Empirics, and Study Design* Guide No. 07/2011 Physikalisch-Technische Bundesanstalt (PTB)

[9] Braungardt S and Göthner K-C 2017 Harmonization of MEPS and Energy Labelling in Latin America and the Caribbean - Opportunities and Challenges *Proceedings of the 2017 eceee Summer Study*

[10] PTB-OLADE 2019 *Infraestructura de la Calidad para Programas de Eficiencia Energética en América Latina y el Caribe* available at http://biblioteca.olade.org/ accessed on 2022-09-20

[11] Kyriazis G A and Göthner K-C 2020 Infraestrutura da qualidade e comparabilidade de desempenho de produtos no mercado *Cadernos de Metrologia* vol 1 no 2 Apr/Jun Instituto Nacional de Metrologia Qualidade e Tecnologia (Inmetro) available at https://asmetro.org.br/portalsn/2020/06/05/metrologia-as-medicoes-para-o-comercio-global/

[12] Jensen H R 1991 Consumer education as a parameter of consumer action in Latin America and the Caribbean *Journal of Consumer Policy* **14** 207

[13] ISO/IEC 17043:2010 *Conformity assessment — General requirements for proficiency testing*, available at https://www.iso.org/standard/29366.html accessed on 2022-09-20

[14] NYCE 2015 *Informe de Resultados de Ensayos de Aptitud – Comparación Interlaboratorios Eficiencia Energética en Refrigeradores* Informe No. INF00215M

[15] ISO 15502:2005 *Household refrigerating appliances — Characteristics and test methods* available at https://www.iso.org/standard/27428.html accessed on 2022-09-20

[16] AHAM HFR-1-2008 *Energy and internal volume of refrigerating appliances* available at https://www.aham.org/AHAM accessed on 2022-09-20

[17] INTI 2019 *Ensayo de Aptitud en Lámparas LED (Light Emitting Diode) de Uso Doméstico* Informe Final EAEE-02 30-05-2019

[18] EN 50285-1999 *Energy efficiency of electric lamps for household use. Measurement methods* available at https://www.en-standard.eu , accessed on 2020-07-13.

[19] ISO/IEC 17021:2015 *Conformity assessment — Requirements for bodies providing audit and certification of management systems* available at https://www.iso.org/standard/61651.html accessed on 2022-09-20

[20] ISO/IEC 17067:2013 *Conformity assessment — Fundamentals of product certification and guidelines for product certification schemes* available at www.iso.org/standard/55087.html/ accessed on 2022-09-20

[21] ADEME 2016 *Market surveillance of Energy Labelling and Ecodesign product requirements* Agence de l´Environnement et de la Maîtrise de l´Energie (ADEME) Valbonne available at https://www.iea-4e.org accessed on 2022-09-20

[22] ISO 50001:2018 *Energy Management Systems* available at https://www.iso.org/iso-50001-energy-management.html accessed on 2022-09-20

Acknowledgments

The outcomes of the EE component of the regional project could not be obtained without the support of many professionals in different countries. In particular, we thank the contributions of the following individuals: Juraj Krivošik (SEVen, Czech Republic), Christoph Tuerk (VDE Institut, Germany), Marcos Borges and Gustavo Kuster (Inmetro, Brazil), Ronnie Hernández, Cristián Baeza and Alejandro Onofri (SEC, Chile), Alfonso Herrera (Costa Rica), Luis Fernando López, Omar Baez and

Carlos Chica (all from Colombia), Hugo Guerrero (Bolivia) and Carolina Mena and her team (MIEM, Uruguay). The objectives of the project could not be achieved without the support of Héctor Laiz (SIM), Kory Eguino (COPANT), Victor Gandy (IAAC), and Amaury Santos (IEC-LACR), and the help of several national QI institutions like Inmetro (Brazil), LCM and ICE (Costa Rica), INEN (Ecuador), INM and ICONTEC (Colombia), and UNIT and LATU (Uruguay).

Uncertainty assessment of a resistive voltage divider for binary step-up method implementation in HVDC applications

P C de O Vitorio[1], L A A de Souza[1], R P Landim[1] and C B do Prado[1]

[1] Instituto Nacional de Metrologia Qualidade e Tecnologia - Inmetro

E-mail: pcoliveira@inmetro.gov.br

Abstract. The most used methods for determining the ratio of High Voltage Direct Current (HVDC) standard resistive dividers are described in this paper. The binary step-up method was chosen to metrologically characterize a HVDC resistive divider, in order to achieve local traceability to the SI from Josephson Voltage Standards. A Monte Carlo-based simulation was performed to estimate the uncertainty levels of each step from 1.25 kV up to 10 kV, considering that the main components are the uncertainty of the digital multimeter and the voltage source. It was found a variation up to 64-fold in the standard deviation of the voltage output, for a uniform distribution, when the voltage source uncertainty was increased up to 10-fold.

1. Introduction

The High Voltage Metrology Laboratory (Lamat) of the National Institute of Metrology, Quality and Technology (Inmetro) is legally responsible in Brazil for providing and maintaining the traceability to the International System of Units (SI) volt at high voltage levels (HVDC, mostly between 100 kV and 800 kV). Lamat has an HVDC system composed of: (a) a standard resistive voltage divider (RVD), manufactured to operate in the range from 1 kV to 150 kV (at both positive and negative polarities), which is periodically calibrated by the National Measurement Institute (National Institute of Metrology of Australia) and; (b) an 8 ½ digital multimeter (DMM), which is periodically calibrated at Inmetro using its national low voltage standards (Zener-based DC voltage reference standard) which, by their turn, are calibrated four times a year using Inmetro's Josephson primary voltage standards [1]. The RVD has a module with a rated voltage of 150 kV, which has 150 resistors of 1.5 MΩ each, with a total resistance of 225 MΩ per module. The RVD ratio is 15000:1. In order to achieve local traceability of the RVD to the SI volt using Inmetro's Josephson primary voltage standards (JVS), it is needed to metrologically characterize such RVD. That will be done based on the national low-voltage DC standards of the Electrical Metrology Division (Diele), at Inmetro, establishing an uninterrupted chain of traceability from Inmetro's JVS to approximately 10 kV (at both polarities). This is the first step to provide Inmetro with an HVDC Reference System, whose traceability chain is totally independent of other NMIs.

The RVD Scale Factor (SF) can be calculated from the ratio between the resistance of the high voltage unit and the resistance of the low voltage unit. In order to do that, it is necessary to use a method to determine the variation of the HVDC ratio of the dividers. Some methods will be analysed in this paper, in order to verify the best performance in the laboratory.

Before implementing physically, the chosen method (which is beyond the scope of this paper), a mathematical simulation was done using Matlab R2018a software to obtain the high voltage and low voltage estimated values, in addition to the contributions of the uncertainties involved, for 5 voltage levels up to 10 kV. Additionally, a Monte Carlo-based simulation was performed to estimate the uncertainty levels of each step. The uncertainty of the DMMs and of the voltage source were the main components considered, since a preliminary analysis showed they as the biggest ones. The simulation results may be a reference for future analysis of uncertainty components as the stability of the voltage source and of the DMM are smaller than the manufacturer specifications. This will allow to minimize their impact on the RVD measurement uncertainty.

2. Methods to determining the divider ratio in HVDC

Characterization of a RVD includes the measurements of the divider ratio and the ratio change with the applied voltage. The divider ratio is the ratio between the low voltage (LV) section resistance and the total resistance of the divider. The ratio error is the deviation of the divider ratio with the applied voltage. Lee et al. [2] reported that the ratio change is mainly affected by the following factors: temperature coefficient and voltage coefficient of the resistors; warm-up period of the resistors; aging and parameter drift of the resistors; leakage current; temperature and humidity; corona effect; overvoltage during operation etc. The interaction of these factors makes it difficult to achieve a perfect measurement of the ratio change. Methods of such characterization include the following [3]:

- Method 1: Qian Li at al. (National Center for High Voltage Measurement, State Grid Electric Power Research Institute) proposed a method used for the characterization of high voltage dividers in HVDC, to distinguish the effect of thermal heating from other influencing factors when analyzing the change in the voltage ratio of high voltage direct current dividers. Two dividers A and B are required. [3];
- Method 2: Nadežda Dragounová (Nuclear Physics Institute, Academy of Sciences of the Czech Republic, Rež, Czech Republic) put forward a method called "on-off", where a calibrated resistor is in parallel with a switch and is placed between the high voltage section and the low voltage section of the calibrator divider. The ratio error is obtained by two measurements when the switch is turned on and off, respectively [4];
- Method 3: Kim at al. (Korea Research Institute of Standards and Science - KRISS) introduced a method which determines the ratio change of the divider by measuring the voltage coefficients of two HV section resistors individually and their series combination at the same testing voltage. The power supply in this operation needs to be highly stable because the uncertainty of the measurement is affected by its stability. With the increase of the voltage level, it will be difficult to obtain such power supplies, and thus, the suitability of the method may be reduced [5];
- Method 4: Qian Li proposed a method called "sum of DC voltage", used to determine the change of ratio of DC high voltage dividers, when applying a desired voltage. The DC voltage sum method requires four dividers [3];
- Method 5: Yi Li (National Measurement Institute, Australia) introduced the method called binary step-up [2;6], where the voltage ratio of the divider is determined by measuring the voltage of two HV section resistors individually, and their combination in series, at the same test voltage. The power supply in this operation needs to be stable because the measurement uncertainty is affected by its stability [7].

After analyzing the methods 1, 2, 3, 4 and 5, it can be concluded that method 5, the Binary Step-Up, seems to be the most suitable for implementation at Lamat because the needed equipment is already available and only scientific investigation regarding the challenges imposed by the methodology will be used. In order to determine the variation of the voltage ratio of high voltage resistors in direct current, it will be necessary to turn the HVDC source off during each measurement resistor section transition, for safety purposes. This will bring more importance to the analysis of the stability of HVDC source.

3. Binary step-up method

The Binary Step-Up method allows, from the second "step", to obtain the voltage of the high voltage module of the RVD, keeping it traceable to the volt without the need to measure it directly [5;7]. The Scale Factor (SF) for each partition is determined from the relationship between the high voltage value obtained for the previous step and the low voltage values. Repeating the procedure and always maintaining the premise of using twice the value of the previous voltage, it is possible to determine at high voltages the scale factors of the two dividers of high voltage unit R_A and R_B, to obtain a curve of the SF behavior of the RVD.

The metrological characterization of a RVD was made, consisting of a high voltage module connected directly to a HVDC and a low voltage module, connected to low voltage measuring instruments. Voltage dividers are used to divide the applied voltage, in such a way that the output voltage is proportional to the input voltage. The standard resistive voltage divider (RVD) consists of two basic modules: the high voltage unit and the low voltage unit. While the high voltage module has high ohmic resistance and it is connected directly to the HVDC circuit, in which it is intended to measure the voltage, the low voltage module is connected to the instrumentation (multimeter) by means of coaxial cables, to minimize interference problems electromagnetic transmission of the signal to be measured [2;6]. Figure 1 shows the schematic diagram of the RVD.

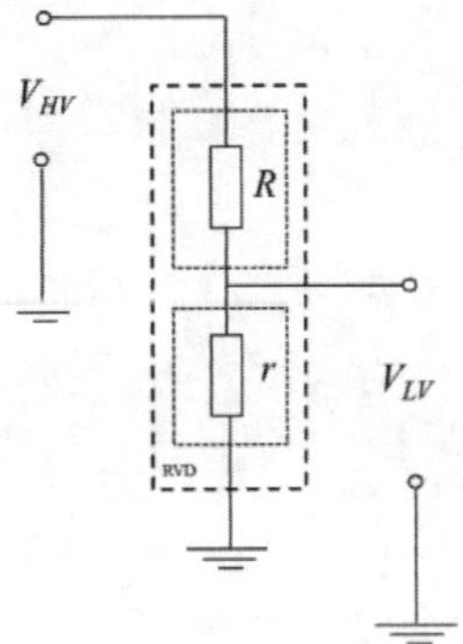

Figure 1. Schematic diagram of the RVD.

The conversion from high voltage (HV) to low voltage (LV) is carried out by the RVD, and the ratio between such voltages (*SF*) is the main characteristic of a voltage divider [2;6]. The *SF* value can be obtained using equation (1).

$$\boldsymbol{SF = \frac{V_{HV}}{V_{LV}} = \left(\frac{R}{r}\right) + 1} \quad \textbf{(1)}$$

Where *SF* is a scale factor of the resistive divider; V_{HV} is a high voltage applied to the divider (in V); V_{LV} is an output voltage of the low voltage unit (in V); *R* is resistance in the high voltage unit (in Ω) and *r* is a resistance in the low voltage unit (in Ω).

The measurement of LV values is performed using calibrated instruments traced to the SI through primary Josephson Voltage Standards, using a low voltage standard, such as a multimeter, a voltmeter, or other equipment capable of measuring voltages and that has the compatible nominal range. This method makes it possible to trace from an HVDC standard to LVDC standards. The range of the applied voltage and the measurement accuracy are limited only by the stability of the high voltage source but also by the divider under test. This limiting factor corroborates the need to study the linearity of the

HVDC source. Figure 2 show the Binary Step-Up method adopted to achieve traceability of High Voltage DC measurements.

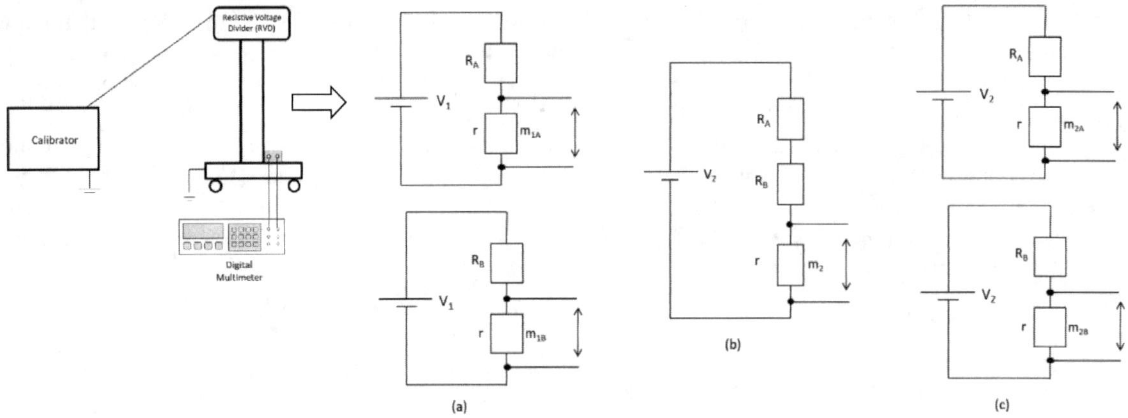

Figure 2. Schematic diagram for the RVD and the Binary Step-Up method.

Thus, the method consists of dividing the high voltage unit of the RVD into two sets of resistors, R_A and R_B, both with same nominal resistance value. The two sets, associated in series with the same low voltage unit *r*, act as two independent resistive voltage dividers [6], shown in figure 2.

$$SF_A(V_1) = \frac{V_1}{m_{1A}} \quad (2)$$

$$SF_B(V_1) = \frac{V_1}{m_{1B}} = SF_A(V_1)\frac{m_{1A}}{m_{1B}} \quad (3)$$

where, $SF_A(V_1)$ is a scaling factor of resistive divider A, at voltage V_1 (in V); $SF_B(V_1)$ is a scaling factor of resistive divider B, at voltage V_1 (in V); m_{1A} is an output voltage of the low voltage unit (in V), of resistance *r*, at voltage V_1 (in V) and m_{1B} is an output voltage of the low voltage unit (in V), of resistance *r*, at voltage V_1 (in V).

The Binary Step-Up method can then be applied to determine $SF_A(V_2)$ and $SF_B(V_2)$ for $V_2 \cong 2\,V_1$, and $SF_A(V_2) = 2\,V_1/m_{2A}$ and $SF_B(V_2) = 2\,V_1/m_{2B}$.

$$V_2 = m_2[SF_A(V_1) + SF_B(V_1) - 1] \quad (4)$$

$$V_2 = m_2\left[SF_A(V_1) + SF_A(V_1)\frac{m_{1A}}{m_{1B}} - 1\right] \quad (5)$$

where m_2 is an output voltage of the low voltage unit (in V), of resistance *r*, at voltage V_2 (in V).

The Binary Step-Up method is used by several national metrology institutes around the world, and it is effective for determining the voltage coefficient of the RVD, providing the use of it as a reference standard in HVDC. According to Lee et al. [2], the advantage of this method is the traceability to the Josephson Effect, in contrast with other existing methods.

4. Monte Carlo based simulation

Aiming to estimate an order of magnitude for the uncertainty levels resulting from the use of the binary step-up method, a Monte Carlo based simulation was implemented to estimate the uncertainties from the main component values. For the simulated measurements, the random input variables were considered to have a Gaussian (or Normal) distribution. Such adoption is quite reasonable considering that the measurements represented will be performed by high resolution and stable DMMs.

Journal of Physics: Conference Series **2606** (2023) 012035 doi:10.1088/1742-6596/2606/1/012035

For the case of the random variable that represents the uncertainty component of the voltage source, two approaches were implemented. In the first one, it was considered that the random variable has a normal distribution, and in the second one, it was considered a uniform distribution. We opted for the double simulation because the study of the stability of the voltage source is specific to the source used, and it is still unknown what behavior will be presented when the experiment is carried out in practice.

Figures 2 shows that the input voltage V_1 is generated by a calibrated calibrator. The output voltages m_{1A} and m_{1B} are measured with a calibrated precision digital multimeter (DMM). The voltages V_2, m_2, m_{2A} and m_{2B} are calculated by mathematical modeling using the Monte Carlo method to perform a numerical simulation from equation (4).

This simulation was performed in Matlab R2018a software, using four input voltage levels (V_2, V_3, V_4 and V_5):1250 V, 2500 V, 5000 V and 10000 V, respectively, considering the contributions of the predefined uncertainties of the source instability and the measurement standards used. The input voltage V_1 was 625 V and it was generated by a calibrated calibrator. The DVR ratio is 15000:1.
The value of output voltages m_{1A} and m_{1B} was V_1 divided by half the ratio, i.e., 625 divided by 7500. So, output voltages m_{nA} and m_{nB} are V_n divided by half the ratio. Figure 4 shows a schematic diagram of a Monte Carlo based simulation. This modeling was validated by a NIST web-based software [8].

Given that the uncertainty of the DMMs is defined in this simulation as 5 parts in 10^6 (typical value for the DMMs used at Inmetro), it is expected that the most significant contribution comes from the stability of the voltage source. The simulation was carried out to evaluate the impact of different values of uncertainty in this contribution, considering that the study to define this component has not yet been carried out.

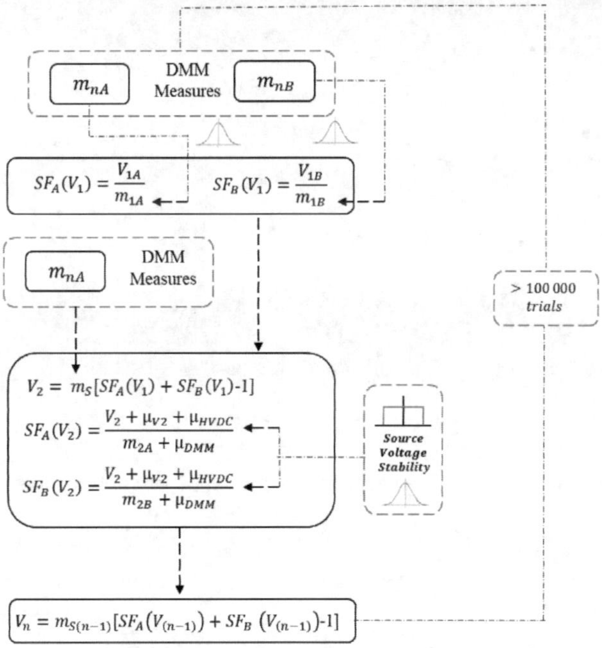

Figure 3. Schematic diagram of a Monte Carlo based simulation.

5. Results

Nominal input voltage value of 10 kV was simulated in Matlab R2018a software. Considering voltage source uncertainties of 0.0010 %; 0.0025 %; 0.0050 % and 0.0100 %; as well as normal and uniform distribution. Table 1 shows the standard deviation in the output voltage and the scale factor obtained for the random variable that represents the last simulated voltage step, using normal and uniform distributions. The last nominal input voltage was 10000V.

Table 1. Values of uncertainty, the last input voltage simulated and scale factor using normal and uniform distribution.

uncertainty (%)	standard deviation Vn+1 simulate Norm Dist	Vn+1 (V) simulate - Dist Norm Vn (V)	standard deviation Vn+1 simulate Unif Dist	Vn+1 (V) simulate - Dist Unif Vn (V)	mn (V)	Scale Factor (SF) Norm Dist	Scale Factor (SF) Unif Dist
0.0010	**0.0707**	**9997.30**	**0.040**	**9997.40**	**0.6667**	**14995.95**	**14996.10**
0.0025	**0.2338**	**9997.30**	**0.182**	**9997.50**	**0.6667**	**14995.95**	**14996.25**
0.0050	**0.4710**	**9997.30**	**0.343**	**9997.70**	**0.6667**	**14995.95**	**14996.55**
0.0100	**0.9436**	**9997.30**	**2.561**	**9998.10**	**0.6667**	**14995.95**	**14997.15**

From Table 1, it can be observed that there is no significative voltage variation using a normal distribution. Also, the standard deviation and scale factor did not increase at the same rate of the voltage source uncertainties. On the other hand, it was found a variation up to 64-fold in the standard deviation of the voltage output and scale factor, for a uniform distribution, when the voltage source uncertainty was increased up to 10-fold. Figure 4 is based on Table 1 and shows the scale factor simulated and uncertainty using a uniform and normal distribution.

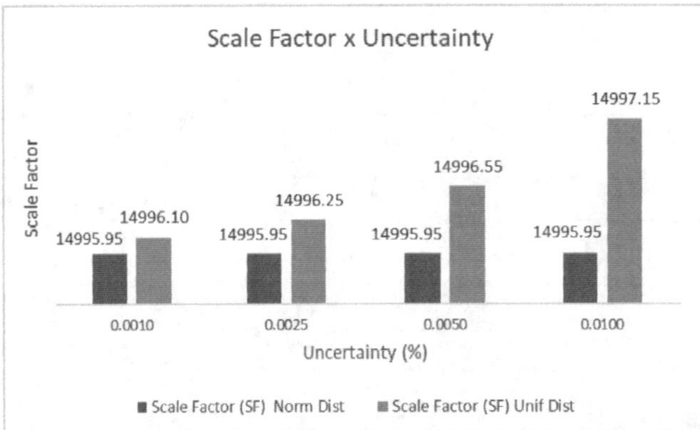

Figure 4. Scale factor simulated and uncertainty using a uniform and normal distribution.

From the Table 1 and Figure 4 it can be concluded that, in the uniform distribution case, with the increase in voltage level there is an increase in the standard deviation and scale factor, and consequently an increase in uncertainty due to the instability of the HVDC source. Also, the probability distribution brought strong influence in the standard deviation of the simulated measured voltage. This analysis reaffirms the need to study the linearity and stability of the HVDC source and the impact on the measurement.

6. Conclusion

The binary step-up method was chosen to metrologically characterize a HVDC resistive divider, in order to achieve local traceability of the RVD to the SI volt through Inmetro's Josephson primary voltage standards [1]. This is the first step to provide Inmetro with an HVDC Reference System, whose traceability chain is totally independent of other NMIs. A Monte Carlo-based simulation was performed to estimate the uncertainty levels of each step from 1250 V up to 10000V, considering that the main components are the uncertainty of the digital multimeter and the voltage source. It was found a variation up to 64-fold in the standard deviation of the voltage output, for a uniform distribution, when the voltage source uncertainty was increased up to 10-fold. It can be concluded that the probability distribution may bring strong influence in the standard deviation of the real measured voltage. This analysis reaffirms the need to study the linearity and stability of the HVDC source and the impact on the measurement.

7. References

[1] Landim R, Tang Y, Afonso E and Ferreira V 2011 A 10 V Josephson Voltage Standard Comparison Between NIST and INMETRO as a Link to BIPM *IEEE Transactions on Instrumentation and Measurement* **60** 7, pp. 2353-58, doi: 10.1109/TIM.2010.2099370

[2] Lee S-H, Yu K M, Kang J H, Kim M S and Kim K T 2012 A Josephson voltage-traceable DC high-voltage divider evaluation using the Binary Step-Up Method *Measurement*, **45** 3 pp. 305-340, doi: https://doi.org/10.1016/j.measurement.2011.10.024

[3] Li Q, Wang L, Zhang S, Tang Y. and Xu, Y 2012 Method to Determine the Ratio Error of DC High-Voltage Dividers *IEEE Transactions on Instrumentation and Measurement*, **61** 4, pp.1072-78, doi 10.1109/TIM.2011.2178672

[4] Dragounová N 2005 Precision High-Voltage DC Dividers and Their Calibration *IEEE Transactions on Instrumentation and Measurement*, **54** 5, pp.16911-15, doi 10.1109/TIM.2005.853222

[5] Kim K-T, Lee S-H, Jung J K and Song Y S 2003 Method to determine the voltage coefficient of a DC high-voltage divider *IEEE Transactions on Instrumentation and Measurement* **52** 2, pp. 469-473, doi: 10.1109/TIM.2003.810038

[6] Lee S-H, Yu K M, Kang J H and Jang S M 2012 DC High-Voltage Divider Evaluation Using The Binary Step-Up Method *Proc. XX IMEKO World Congress* (Busan: Republic of Korea)

[7] Li Y, Ediriweera M K, Emms F S and Lohrasby A 2011 Development of Precision DC High-Voltage Dividers *IEEE Transactions on Instrumentation and Measurement* vol. **60** 7, pp. 2211-16, doi: 10.1109/TIM.2010.2100655

[8] NIST 2020 The Nist Uncertainty Machine. https://uncertainty.nist.gov. Accessed in August 11th, 2021

AUTHOR INDEX